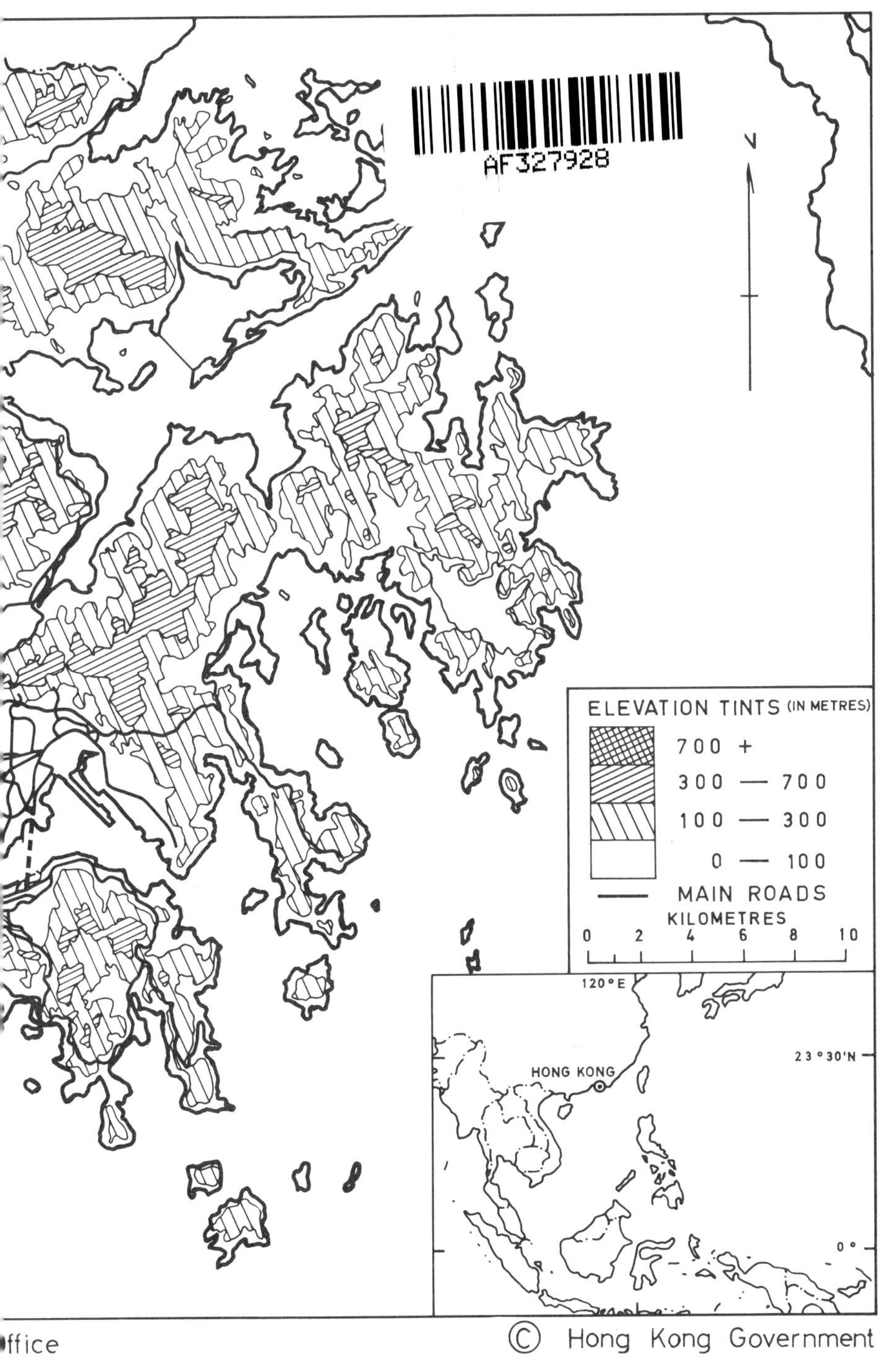

AF327928
N
ELEVATION TINTS (IN METRES)
700 +
300 — 700
100 — 300
0 — 100
MAIN ROADS
KILOMETRES
0 2 4 6 8 10
120°E
HONG KONG
23°30'N
0°
ffice
© Hong Kong Government

INSECTS OF HONG KONG

INSECTS OF HONG KONG

by

Dennis S. Hill

Phyllis M. Hore

Ian W. B. Thornton

HONG KONG UNIVERSITY PRESS

Printed in Hong Kong by LIBRA PRESS LTD
56 Wong Chuk Hang Road 5D, Aberdeen, Hong Kong

CONTENTS

CHART

PREFACE

The need for a book on the insects of Hong Kong is three-fold. First, entomology is taught as part of the general degree course in biology in both the local universities, but at present there is no local text on this subject. Second, there is an increasing interest and awareness in ecology, both at the universities and in the secondary schools. Because of the limited natural environment of Hong Kong, most studies in animal ecology here would of necessity lean heavily on the insects which is the most suitable animal group available, particularly in the fields of terrestial and freshwater ecology. Finally, an increasing number of the local populace are becoming aware of and interested in their biological surroundings, and it is hoped that this book will enable them to identify most of the common insects they are likely to encounter here. Throughout the book emphasis is placed upon the most common insects (i.e. those most likely to be encountered) and those of the most economic significance. However, some species are also mentioned because of their biological interest.

The systematic classification and arrangement of Orders and Families follows that in A.D. Imms' *A general textbook of entomology* (1964), and for use as an entomology text this book is to be regarded as complementary to that by Imms (1964).

Entomology has been a rather neglected science in Hong Kong, for although several entomologists have worked here in recent years, they have tended to restrict their activities to the few groups of which they had specialized knowledge. For example, Thornton (1959, 1961, 1962, etc.) studied the order Psocoptera, Hill (1967) the fig-wasps found in association with local wild fig species, and Marshall (1964, 1970, 1973) used various local plant bugs in his physiological and histochemical studies. The few local government entomologists have been fully employed with problems concerning household and agricultural pests, while the Pest Control Officer of the Urban Services Department and his staff are mainly responsible for advice and control measures concerning household, medical and local stored product pests. The field of agricultural entomology and crop protection is the province of the agricultural officers of the Department of Agriculture and Fisheries, and the former entomologist there had published *A preliminary list of the insects of agricultural importance in Hong Kong* before his retirement; this has now been considerably expanded in the publication by Lee & Winney (1982). However, several excellent butterfly collections have been made by amateur Lepidopterists. These are personal collections and only remained in the Colony while their collectors resided here, but J. C. S. Marsh (1968) did complete his book on butterflies before his departure from Hong Kong.

There is, regrettably, nothing entomologically unique about Hong Kong, and to appreciate this one must think of Hong Kong *per se* and not solely in comparison to the neighbouring countries, for the insects here intrinsically represent an impoverished South-east Asia fauna.

The major problem in the study of entomology is that of actually identifying the insects concerned. In Great Britain, Canada, U.S.A., and most European countries, entomology has been an established science for many years; the local insect fauna has been studied by several generations of entomologists, and published keys for identification are available for most groups of insects. However, because of the amount of work involved in the preparation of these illustrated keys, many, for example the British set of keys, are still not yet completed! The Hong Kong insect fauna is unfortunately not at all well known, with of course the exception of the butterflies which were admirably described by J. C. Kershaw (1901) and later by J. C. S. Marsh (1968). Only some of the occasional small groups here, such as termites, have been studied, in this case by W. V. Harris (1963); but such groups as these are few indeed. But one feature is certain: that in Hong Kong there are many species of insects, certainly several hundreds at least, new to science, making it a paradise for a taxonomic entomologist!

There has been no real attempt at the compilation of a formal collection for Hong Kong, partly because there is no local museum in which such a collection could be permanently housed. The Department of Agriculture and Fisheries is however, making a collection of crop pest species at its research farm at Tai Lung in the New Territories; and the Pest Control Advisory Unit of the Urban Services Department has a small collection of household, medical and stored products pests in their offices in the Causeway Bay Magistracy Building, including particularly a complete collection of local mosquitoes.

In general, if anyone wants to identify an insect here, the usual procedure is to send specimens to the Keeper of Entomology, British Museum (Natural History), South Kensington, London, or alternatively and preferably to the Director of the Commonwealth Institute of

Entomology at the same address, which provides an insect pest identification service for countries of the British Commonwealth. A collection of insects which have been named by the experts in London is gradually being amassed in the Department of Zoology at the University of Hong Kong; and it is to be hoped that not before too long most of the more common local insects can be identified with the aid of the collections in the Department of Agriculture and Fisheries and the University of Hong Kong. It is of course to be hoped that eventually there will be a local museum where a definitive collection of Hong Kong insects will be maintained.

ACKNOWLEDGEMENTS

The photographs in this book have all been taken in Hong Kong during the period 1973–1978 by the author D. S. Hill, and the line drawings of certain insect and mite pest species were drawn by Mrs Hilary R. Broad for a previous publication (Hill, 1975). Thanks are due to the staff of the Commonwealth Institute of Entomology, London, and the British Museum (Natural History), London, for identification of many of the local insects. Some of the information regarding local insect pests of agricultural plants is derived from the publication by So (1967), and acknowledgement is made to the Department of Agriculture and Fisheries, Hong Kong.

The stored products pest illustrations are copyright of Degesch Company who kindly permitted their inclusion in this book. We are grateful to Mr R. Wong of Bayer China Company Ltd., who supplied the illustrations and obtained permission for their use.

The key to the major groups of terrestrial invertebrates was arrived at by modification of the key published in 1967 by Kitty Paviour-Smith and Dr J. B. Whittaker and their kind permission to use their work is gratefully acknowledged.

We are particularly indebted to Jeremy D. Hill for his assistance in collecting and rearing of many local insects. Thanks are also due to the following: Dr S. Asahina, Mr. Y. K. Chan, Dr A. D. M. Dudgeon, Dr. I. J. Hodgkiss, Mrs G. Johnston, Mr W. Kightley, Dr V. Lance, Dr H. Y. Lee, Mr T. C. Leung, Dr B. S. Morton, Mrs Janice Morton, Miss Sylvia Ng, Dr W. L. Peters, Mr J. D. Romer, Dr. E. F. Riek, Dr E. S. Ross, Dr H. H. Ross, Dr A. Sommerville and Mr. R. Winney, who all assisted in various ways, and to a number of colleagues and neighbours who collected insects occasionally on our behalf. Finally we wish to thank Prof. B. Lofts in whose Department of Zoology this work was carried out, and for his support of the project.

CLASSIFICATION OF THE CLASS INSECTA§

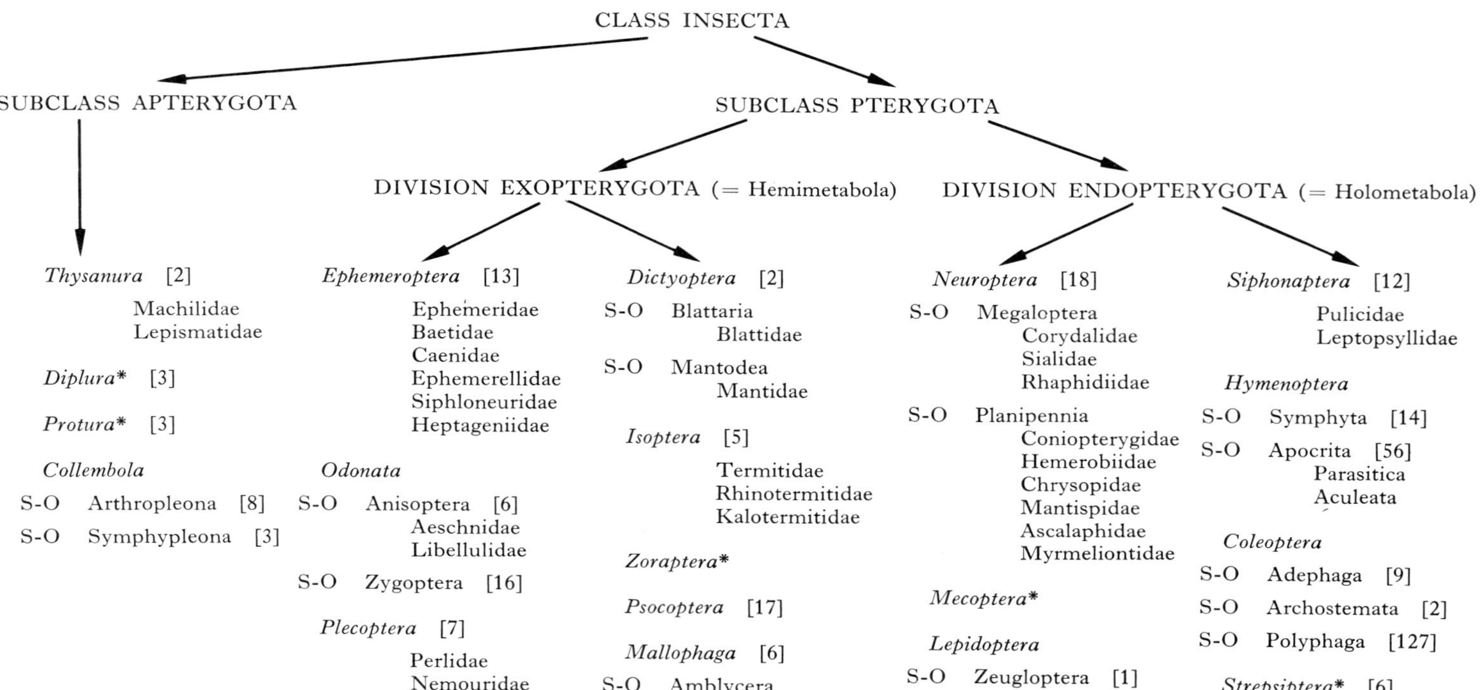

*Grylloblattodea** [1]

Orthoptera [15]

S-O Ensifera
       Tettigoniidae
       Gryllidae
       Gryllotalpidae

S-O Caelifera
       Acrididae
       Tetrigidae

Phasmida [3]

       Phylliidae
       Phasmidae

Dermaptera [8]

S-O Forficulina
       Labiduridae
       Forficulidae

S-O Arixeniina
       Arixeniidae

S-O Hemimerina
       Hemimeridae

Embioptera [7]

       Oligotomidae

S-O Ischnocera
       Philopteridae
       Trichodectidae

S-O Rhynchophthirina
       Haematomyzidae

Siphunculata [6]

       Pediculidae
       Haematopinidae

Hemiptera

S-O Homoptera [43]
S-O Heteroptera [45]

Thysanoptera

S-O Terebrantia [5]
       Thripidae
S-O Tubulifera [7]
       Phlaeothripidae

S-O Ditrysia [68]

Trichoptera [22]

       Hydropsychidae
       Limnephilidae
       Sericostomatidae
       Molannidae
       Odontoceridae
       Leptoceridae

Diptera

S-O Nematocera [18]
S-O Brachycera [14]
S-O Cyclorrhapha [49]
       Aschiza
       Schizophora
       Pupipara

§ The number in brackets after each group indicates the number of families in that group. Sub-orders (S-O) and families listed under each order are the main ones which appear in Hong Kong and are reasonably distinguishable.

* Do not occur in Hong Kong or not yet recorded.

CHAPTER ONE

INTRODUCTION TO THE STUDY OF INSECTS

MORPHOLOGY AND ANATOMY

The internal structure of an animal is referred to as its anatomy, whereas the study of the animal's shape and body form externally is known as morphology. Some basic knowledge of insect morphology is rewarding for then it is possible to appreciate the full extent of the adaption and evolution shown by many members of the group, which accounts partly for the tremendous success of the insects.

The body of an insect is composed of three distinct regions, the head, thorax and abdomen (Fig. 1). The head is generally a smooth round capsule showing little trace of its origin from six embryonic segments which completely fuse during development. So far as is known the Insecta have evolved from primitive worm-like animals which probably inhabited litter and top-soil, under rocks and fallen tree trunks. These ancestral animals showed metameric segmentation in that the body was divided into a number of similar segments, each segment bearing a pair of crudely articulated appendages (limbs), a pair of excretory organs opening at the base of the appendages, and various paired or segmented internal structures. The mouth was probably terminal in position (like a worm), and some of the appendages at the extreme anterior end of the

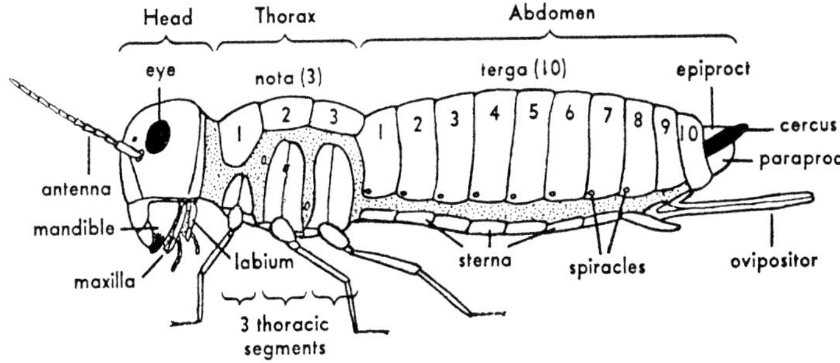

Fig. 1.   A generalized primitive insect showing the main parts of the body.

body eventually became either sensory in function or used in feeding, or both, rather than being just ambulatory. This specialization would of course lead the animal to becoming more efficient and hence giving it definite survival advantages. This trend continued in the insects as they evolved from the basic arthropod stock and its worm-like ancestor. The fusion of body segments together for greater functional efficiency is termed tagmosis. Thus in present day insects, the head tagma represents the highest level of functional modification. The head bears the major sense organs, eyes and ocelli for light perception (sight), and of the original articulated segmental appendages there remains the sensory antennae (for touch and smell/taste), and the mandibles, maxillae and labium, all for food manipulation and ingestion. If an early insect embryo is examined microscopically then some trace of the original head segmentation can be seen, but during development (ontogeny) the original segments fuse so completely that in the final head tagma the only evidence of its segmental nature are the paired appendages. At the same time, during development, the terminal mouth moves posteriorly until it occupies a ventral position under the head. The head is thus a specialized structure bearing the major sense organs (eyes, antennae) and the complex mouth-parts used for feeding. Most insects have their compound eyes situated on the sides of the head in order to give them the maximum field of vision, for detection of approaching predators.

The thorax is always composed of three segments, called the prothorax, mesothorax and metathorax. Usually the three segments are quite distinct, but occasionally they are fused together and appear as a single structure. Sometimes one segment, usually the prothorax (Fig. 2), may

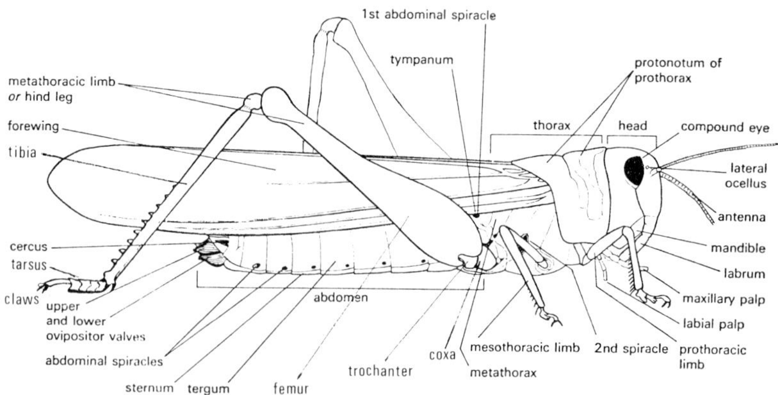

Fig. 2.   A common unspecialized insect in side view: female grasshopper (Orthoptera; Acrididae).

be disproportionately larger, and it may extend posteriorly as a definite projection. The function of the thorax is to house the muscles for leg articulation and locomotion, and the alary muscles for flight. Each thoracic segment bears ventrally a pair of long jointed legs, and the two posterior segments bear the wings dorso-laterally.

Each leg consists of five major segments which articulate upon each other to make the leg flexible. The basal segment is the coxa, usually firmly fixed or fused on to the thoracic sternum for anchorage. Next is the small trochanter for flexibility, followed by the large elongate femur. The lower leg is composed of the thin elongate tibia which often bears characteristic rows of stout spines, or else a long spur. The 'foot' is formed of the tarsus, typically as four small squarish segments, followed by the elongate pretarsus bearing terminally a pair of claws with a central adhesive areolum. The function of the 'foot' is clearly for anchorage and gripping. The number of tarsal segments may vary from leg to leg, and also from species to species, and can be a useful taxonomic character.

The wings are borne on the meso- and metathorax of most insects. The very primitive Apterygota, so far as is known, never evolved wings. In the more primitive Pterygota the two pairs of wings are equal in size but do not touch and function as two separate pairs of wings. To have one wing operating behind the other is aerodynamically very inefficient as the second wing is operating in turbulent air and consequently produces little lifting force. Thus the more advanced insects have evolved different flight systems for greater efficiency. The primitive termites (Isoptera), some Neuroptera, web-spinners (Embioptera), stoneflies (Plecoptera) and some Lepidoptera all have retained the primitive condition of two pairs of equal-sized wings and they fly in a slow laborious manner when they take to the air. Mayflies (Ephemeroptera) and Psocoptera fly rather poorly and most of them have the hind pair of wings reduced and small (or absent). Among the more advanced groups the various systems adopted for improved flight include the joining of the larger anterior with the smaller posterior wing so that together they function as a single aerofoil; in the Lepidoptera and Hemiptera this union is effected by basal overlap and sometimes a crude form of locking device; in the Hymenoptera the wings are coupled by a series of small hooks and the equivalent of 'eyes', in a most effective manner. The flies (Diptera) have not only completely reduced the hind wings so that only the fore-wings remain, but have modified the hind wings into special balancing organs called halteres which generally make their flying more efficient. Some groups of insects have changed the fore-wings into protective covers by a process of thickening and hardening; at rest they completely cover the enlarged and folded hind wings, and in flight the fore-wings are held out at an angle away from the body and have no motion. In the Orthoptera and Dictyoptera the relatively less thickened fore-wings are termed tegmina, whereas in the Coleoptera the much thickened fore-wings are called elytra. The very mobile and fast-flying predatory dragonflies (Odonata) have retained both pairs of wings but have achieved remarkable flight efficiency by having the wings beat out of phase, with the hind wings beating slightly before the fore-wings. In most groups of insects, however, will be found species which have in the course of evolution lost their wings, either in one sex alone, or in both sexes. The commonest reason for this secondarily wingless (apterous) condition is the adoption of a parasitic mode of life (e.g. Mallophaga, Siphunculata, Siphonaptera, etc.).

The wing is primitively formed with a regular series of longitudinal veins, which are hollows down which run tracheae and also blood, and they give considerable support to the thin wing membrane which is composed entirely of two thin layers of cuticle stuck together by the remains of the epidermal cells (Fig. 3). Each vein is named, and it either protrudes dorsally (marked +) or projects ventrally (marked −), and this corrugation presumably adds to the rigidity of the wing surface. In some groups of insects (Ephemeroptera, Odonata, Neuroptera), there are many cross-veins to be found, and sometimes even additional longitudinal veins; but sometimes a group is characterized by having

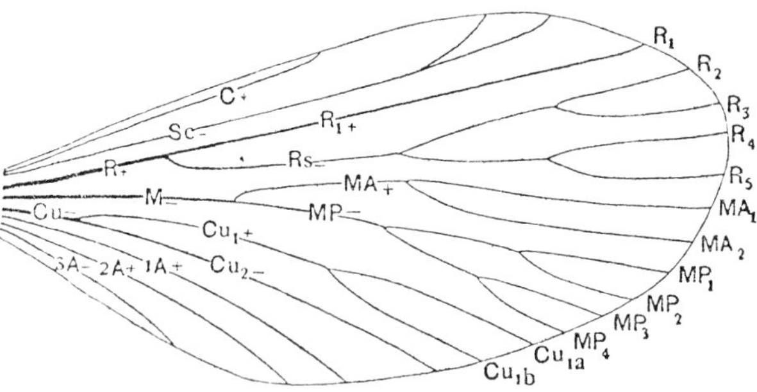

Fig. 3. A theoretical primitive insect wing showing the basic venation; veins marked + are protruding upwards, and those marked—project ventrally.
C = Costa, M = Median (Anterior & Posterior), Sc = Subcosta, Cu = Cubital series, R = Radial series, A = Anal series.

a reduction in the number of wing veins, as shown in some flies (Diptera) and chalcid wasps (Hymenoptera). In general wing venation is regarded by many taxonomists as being an important character for classification of families, super-families, sub-families as well as at generic and specific level.

The insect abdomen is thought to consist primitively of eleven segments, but the last segment (11th) is always small, and is lost during development in most higher insects. It is represented by the median epiproct (11th tergum) in some primitive species (Fig. 4), and, for example, this forms the median 'tail' in the three-tailed species of mayflies (Ephemeroptera). In all but a very few primitive insects (e.g. *Lepisma* in the Thysanura) there is no trace of abdominal appendages (c.f. shrimps and most Crustacea), except for the external genitalia and the cerci found in grasshoppers (Orthoptera), cockroaches (Dictyoptera), earwigs (Dermaptera) and the like. *Lepisma* does however bear a series of small styles along the abdomen which are thought to represent vestiges of abdominal appendages. The cerci arise from the last (10th) abdominal segment, and may be short and stubby as in grasshoppers, rather longer and more slender in cockroaches and mantids, pincer-shaped and movable in earwigs, or long and slender as in both adults and nymphs of mayflies and stoneflies.

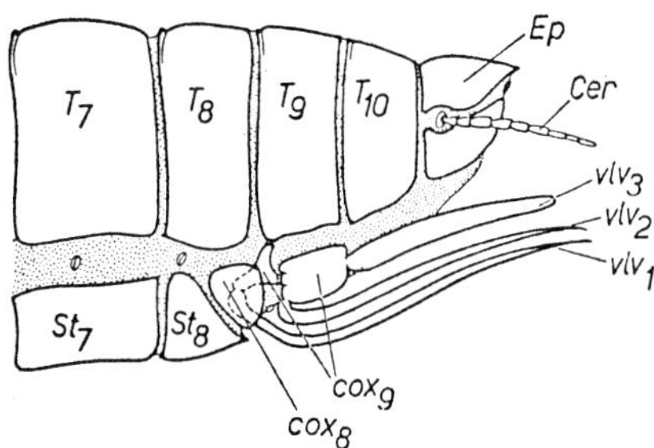

Fig. 4.   Posterior region of abdomen of a generalized primitive female insect showing the basic structure of the ovipositor.
T = Terga, Ep = Epiproct, Cox = Coxite, St = Sterna, Cer = Cercus, Vlv = Valves.

The ovipositor in some female insects can be very long, and sometimes stout or else quite slender and delicate. It is constructed essentially of two pairs of valves which fit together to form a long tube down which the eggs will be laid; usually a third pair of valves fuse together to form an ovipositor sheath which encloses the actual ovipositor for protection. The tip of the ovipositor valves often bears sharp teeth or spines, and usually sensory sensilla, and the two halves can operate with a piercing or sawing action, which enables penetration of plant stems, leaf tissues, animal bodies, etc., so that the eggs can be laid deep in plant tissues (wood even), or inside the bodies of other insects, or in deep soil. In the aculeate (stinging) Hymenoptera the ovipositor has lost its primary function, and is connected with internal poison glands and has become a sting, used in defence or for immobilizing prey. Bees (Apidae) generally only sting in defence, and if they sting a human the elastic nature of human skin prevents the bee from withdrawing the sting, with the result that when the startled victim swats the bee the whole stinging apparatus is torn out from the bee's body, which of course causes its death. Wasps however, and some ants, are carnivorous in habits and use the sting to kill or immobilize their prey on capture. One of the authors, Dr Hill, was once attacked by two *Polistes* wasps, each of which stung him approximately 18 times in about 20 seconds, and the final stings were just as painful as the first, indicating the efficiency of this system.

The insect skeleton consists of a series of hollow tubes joined together, or else a series of hard plates (sclerites) joined together by membranous regions for flexibility. The dorsal plates are called terga, the ventral ones, sterna, and the smaller lateral plates on the side of the thorax, pleurites (see Fig. 1). This type of skeleton combines very successfully hardness and rigidity with a measure of flexibility, and is physically strong and very effective as a means of protection.

The surface of the cuticle is often covered with small setae, large bristles, and sometimes spines, and frequently bears distinctive sculpturing of many different types. Insect coloration is basically of two different types; chemical colours due to specific pigments present in or under the cuticle, or physical colours formed as optical effects by special cuticular structures; sometimes an insect is coloured by combined physico-chemical colours. Many of the irridescent colours to be seen on some tropical beetles and butterflies are optical in origin, and can be temporarily destroyed by wetting the insect body with water or alcohol.

Insect mouth-parts are primitively of a biting and chewing nature, and those of cockroaches are usually regarded as being typical for teaching purposes (Figs. 5 and 6). The mouth-parts represent the original head appendages and as such are arranged in a definite sequence, with the inclusion of two unpaired median structures with no appendage affinity. The labrum is median and single and represents the functional upper lip being broadly hinged at the base and somewhat flexible. Still

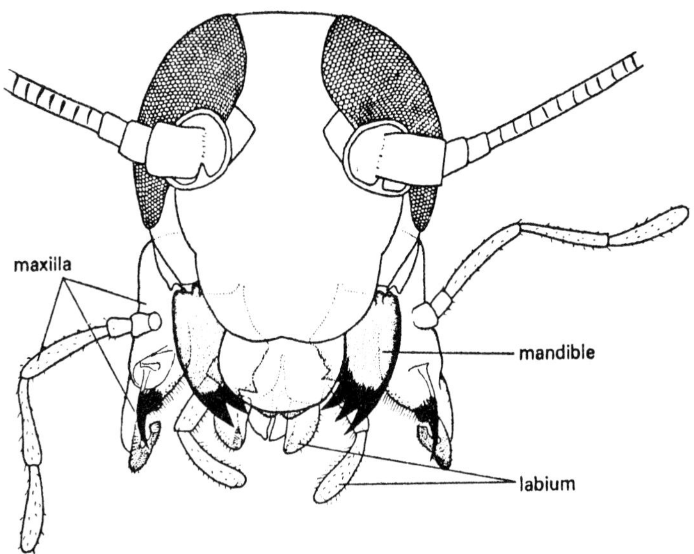

Fig. 5   Face view of a Cockroach (Dictyoptera; Blattidae) showing the relative positions of the mouth-parts.

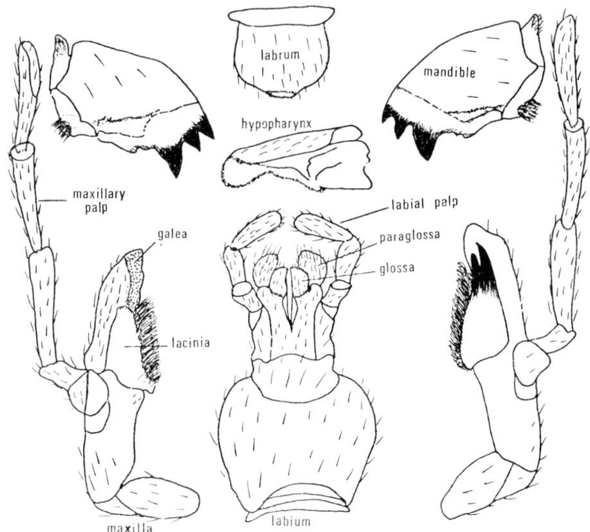

Fig. 6.   Cockroach mouth-parts, representing a primitive unspecialized insect with biting and chewing mouth-parts. The galeae of the maxillae form the proboscis of the Lepidoptera. The glossae and paraglossae of the labium form the sucking 'tongue' of bees, and the outer proboscis sheath of mosquitoes.

in front of the mouth articulate the broad, hard, sharp-edged (toothed) mandibles responsible for cutting up the food into pieces small enough to be swallowed. In most insects the mandibles show really no trace of their appendage origin, as distinct from most mandibles in the Crustacea. Lateral to the mouth hang the paired maxillae, on either side. The basal part of the maxilla is the lacinia which is ventrally hard and pointed and is used for breaking up the food and also for physically pushing the food particles into the mouth. The distal galea is covered with pressure receptors (mechanoreceptors) and taste organs (chemoreceptors) and is responsible for testing the anticipated food for hardness, texture and taste. The long delicate maxillary palps are distally covered with chemoreceptors and other sensory structures, and, with their internal musculature, are quite flexible. On the floor of the mouth lies the single median hypopharynx in which the median salivary duct opens; the saliva being used for lubrication of the food pieces, and for starting the digestive process. Finally, behind the mouth, functioning as a lower lip lies the median labium, hinged at the base and capable of movement back and forth. The labium shows clearly its origin as a pair of segmented appendages fused basally through evolution for greater efficiency. Its function appears to be a combination of sensory perception for tasting and testing food substances, and for pushing the food particles into the mouth.

Many insects have retained this type of biting and chewing mouth-parts; some are undoubtably primitive species but many are not. The major mouth-part modifications that have occurred in the evolution of the insects are connected with the adoption of a liquid diet, of several different types. The bugs (Hemiptera) have the mouth-parts drawn out into a long proboscis or beak, which is composed of the elongate labium wrapped around the paired needle-like mandibles (stylets) with the paired maxillae (also terminally piercing) in the centre, and between the maxillae lie two canals—the salivary canal to take saliva down to the feeding site, and the food canal up which the liquid food (plant sap, blood, digested body contents, etc.) is sucked by the pharynx. Mosquitoes have a rather similar type of proboscis but here the labrum forms the roof of the food canal and the hypopharynx the floor, and the maxillae function like a second pair of piercing mandibles and are also referred to as stylets. The salivary canal lies in the centre of the elongate hypopharynx. The bloodsucking insects require two canals through their proboscis, the first to take saliva down to the puncture site where its enzymes with anticoagulant properties prevent the blood from clotting, and then the food canal up which the blood and other fluids are sucked by the action of the muscular pharynx. Somewhat different constructions are shown in the case of the fleas (Siphonaptera), Stable Fly (*Stomoxys* sp.), and other biting flies.

For the insects that imbibe liquid food but only have sucking mouth-parts (no piercing required) the construction is often simpler. In the Lepidoptera, where the vast majority feed on nectar or juices from plants, the proboscis is formed entirely from the paired galea of the maxillae which are greatly elongated and they have internal concavities which when pressed together form the food canal. Since they feed mostly on nectar there is no salivary gland required. In the larger Hawk Moths (Sphingidae) the fully extended proboscis can be as long as 25 cm. A few of the most primitive Lepidoptera show vestiges of mandibles. At the distal end of the labium lie the paired glossae and paraglossae, together collectively termed the ligula. Bees feed almost entirely on nectar from flowers, and honey, and so again no salivary canal is required in the 'tongue'. As the mandibles have not been incorporated into the 'tongue', they have been free to remain so long as the bees have a use for them. Honey Bees use their small mandibles for eating pollen grains and for chewing the wax in the construction of the combs. The local Carpenter Bees have retained large mandibles, which are used to excavate the nest holes, either in a stem of bamboo or a dead tree trunk. The final type of sucking mouth-parts typical of a large group of insects is the suctorial type to be seen in House Flies and Bluebottles (Diptera—Muscidae; Calliphoridae). These flies feed by placing their sponge-like proboscis end onto some food material, then pouring out saliva with its contained digestive enzymes, and after some of the food material has been digested into a liquid state, this liquid is sucked back into the labellum initially by capillary action and then by suction from the pharynx. The short stout proboscis is composed mostly of the elongate labium, but the dorsal food canal is roofed by the elongate labrum and floored by the hypopharynx, through which runs the narrow salivary canal. The distal end of the proboscis bears a swollen two-lobed sponge-like structure termed the labellum, formed from the tips of the labium. The labellum is ramified with fine channels and canals up which liquids flow by capillary action. If the fly dips its proboscis into or onto a liquid surface, obviously external digestion is not required and the liquid is imbibed directly.

It can be seen that a study of insect mouth-parts in relation to their feeding habits is an excellent example of evolutionary adaptation, or as it is sometimes called—form and function.

Internally, insects show several peculiar features which are characteristic of the group. The blood system is of the 'open-type', which means that there are very few actual vessels and the blood is found in a series of large inter-connecting cavities (haemocoel) bathing the organs of the body. The heart is a long narrow tube lying dorsally in the abdomen and it extends anteriorly as the aorta (main blood vessel). Blood percolates into the heart via a series of small paired ostia (holes with one-way valves) along its length, and the blood is then pumped forward to the head/thorax region. This causes a gradual general circulation of the blood in the haemocoel from anterior to posterior. In some insects as much as 75% of the body weight may be blood. In the majority of cases the blood plays no role in respiration at all, which again makes the insects quite unique. A few aquatic insect larvae such as Chironomids ('bloodworms') have respiratory pigments (haemoglobin —red; haemocyanin—green) but these are exceptional.

Respiration is by a unique system of delicate tubes (tracheae) which ramify throughout the insect body terminating in a fine web of minute tubes (tracheoles) covering the main organs of the body and actually penetrating the cells. The system opens to the exterior through spiracles, which are small paired holes on each body segment. Basically there are ten pairs of spiracles—two pairs on the thorax (meso and metathorax only) and eight pairs on the abdomen (Fig. 7). In more advanced insects this number of spiracles is generally reduced. A tracheal system that relies solely on diffusion is inevitably not very efficient if the insect species is an active one. Thus during evolution there has developed a system of interconnecting dorsal, ventral and lateral trunks which link together all the segmental spiracles and tracheae. Also a series of air sacs were developed and linked into the system. The spiracles have a mechanism whereby they can be automatically closed off if the relative humidity is too low and there is danger of desiccation, and they can also be closed by direct control from the brain. Insects such as grasshoppers,

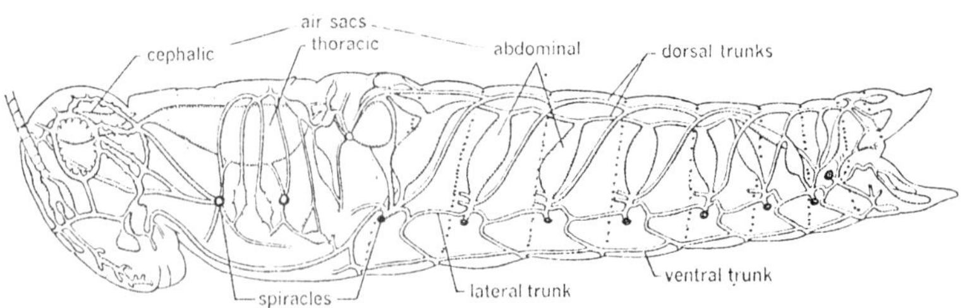

Fig. 7.   Grasshopper respiratory system, showing tracheal trunks, air sacs, two thoracic and eight abdominal spiracles.

bees and wasps, which are quite large and active have a system whereby they close some spiracles and by muscular contraction they force air to flow along the system of tracheal trunks and through the air sacs. This forced ventilation increases the efficiency of respiration many-fold and permits the high level of activity found in some insect species. The muscular action in grasshoppers generally results in the abdomen being depressed (i.e. dorsal-ventral flattening) but the bees and wasps have a longitudinal in-and-out movement for the same purpose. As would be expected, some of the more advanced insects have reduced their number of spiracles, some only have them on the thorax (propneustic), some only at the end of the abdomen (metapneustic), and some at both ends (amphipneustic), and some aquatic larvae have lost all their spiracles (apneustic). The normal condition of having a pair of spiracles on almost all segments is termed peripneustic. Mosquito larvae and some water bugs are examples of metapneustic insects.

Virtually all adult insects are air-breathers, and this includes the aquatic ones, but the aquatic larvae and nymphs acquire their oxygen by diverse means. Aquatic larvae (and nymphs) all have air-filled tracheal systems. Some have a respiratory siphon or tube used to connect the tracheal system to the atmosphere during periodical visits to the water surface (e.g. mosquito larvae, water-scorpions, Rat-tailed Maggot, etc.), and the others have gills. The spiracles are closed, and the gill is usually a thin-walled flap or an extension of the body surface ramified with tracheae; the cuticle over the gill is so thin that oxygen can diffuse directly across into the air in the tracheae. But as mentioned, a few benthic insect larvae have blood gills (Chironomidae—'bloodworms') as they live in the mud on lake bottoms and polluted rivers where oxygen is scarce. The Chiromomids which inhabit the most anaerobic conditions (deep lakes and polluted river bottoms) have haemoglobin as their respiratory pigment, but those living under better conditions have the less efficient but more camouflaging green pigment haemocyanin. A few small aquatic insect larvae living in conditions of reasonable aeration have no special gills and they obtain their oxygen by direct diffusion across their thin body cuticle into their air-filled tracheal system.

Most insects are active and mobile and are equipped with well-developed sense organs. The brain lies in the head capsule with direct connections to the eyes and antennae. The eyes are compound and composed of hundreds or thousands (as in dragonflies) of independent facets, each recording only the presence of light or dark, but together constructing a picture of light and dark spots in the same way a newspaper or T.V. picture is formed. Most adult insects are therefore quite sensitive to shape, and particularly to movement, which of course makes them difficult to catch. The larvae, however, have only simple lateral ocelli with which they seem to be able to detect some movement.

The antennae in the Insecta occur only as a single pair (two pairs in Crustacea) situated on the front of the head usually between the eyes. But during the course of evolution many changes in form and structure have taken place. Primitively, they are many-segmented, long and tapering (setaceous), but have in some cases been reduced to such an extent as to be scarcely visible. Their functions are for touch and smell and the distal ends are covered with tiny sensilla. The pressure receptors are called mechanoreceptors and those for smell and taste chemoreceptors. Many insects whilst walking or looking for food can be seen to be constantly waving their antennae and touching objects with them. The highest Diptera (Muscidae, etc.) have tiny reduced, three-segmented antennae in sockets on their face and these flies have taste sensilla on their feet—presumably to compensate for the loss of this function by their antennae. This also applies to some beetles. The Scavenging Water Beetle (*Hydrophilous* sp.) uses its antennae as breathing tubes when it rises to the water surface, and the functions of the antennae have been taken over by the elongated and flexible maxillary palps.

LIFE CYCLES AND DEVELOPMENT

One of the major characteristics of Insecta is that their hard rigid exoskeleton does not permit continuous growth—they can only physically increase in size at the time of moulting when the old skeleton is shed and the new one underneath is still soft and flexible. This is termed growth by ecdysis (moulting).

The body integument consists of a single cellular layer, the epidermis, overlying which is the non-cellular cuticle, the greater part of which is secreted by epidermal cells. The cuticle is actually composed of three distinct layers, a thin hard outermost epicuticle ($c.$ 5μ) covered by wax which makes the cuticle waterproof and a very thin outer skin of cement to protect the wax from abrasion. Underneath is a thicker layer of exocuticle ($c.$ 20μ) which is dark coloured and quite hard, and the innermost layer is the endocuticle which is quite thick ($c.$ 60μ) and is soft, colourless and flexible, and overlies the epidermis. At moulting the soft endocuticle is dissolved by epidermal enzymes and the material resorbed into the body. In the space left under the hard exocuticle and epicuticle the new cuticle then develops. At ecdysis the old exocuticle splits along a line of weakness down the back of the thorax and the insect in its coat of new soft flexible cuticle crawls out of the old skin which is discarded. The old cuticle skin is called an exuvium, and the exuviae of many species can be found, for example, those of cicadas usually attach to tree trunks and vegetation in the spring, and exuviae of dragonflies, stoneflies, mayflies often attach to rocks and vegetation in freshwater streams. On exposure to air the new pale, soft cuticle gradually darkens and hardens as a chemical tanning process takes place. However, the insect body physically swells and enlarges while the cuticle is still soft. After a few hours the cuticle is fully hardened and rigid once more.

One advantage of growth by ecdysis is that it permits metamorphosis, which can be of great significance in the life of the insects. Metamorphosis is a change of body form during development. The most primitive insects, the subclass Apterygota (see pp. 17 & 18) show virtually no metamorphosis during their development and their development is direct. But for the winged insects (subclass Pterygota) this change in body form can be either gradual, in the more primitive Exopterygota, or else can be most dramatic and sudden in the more highly evolved Endopterygota. In the Exopterygota, sometimes called Hemimetabola, metamorphosis is slight; the young is referred to as a nymph and it resembles the adult except it does not have wings nor external genitalia. At each moult the nymph gradually increases in size and during the final stages the wing-buds become progressively larger (Fig. 8). This is sometimes

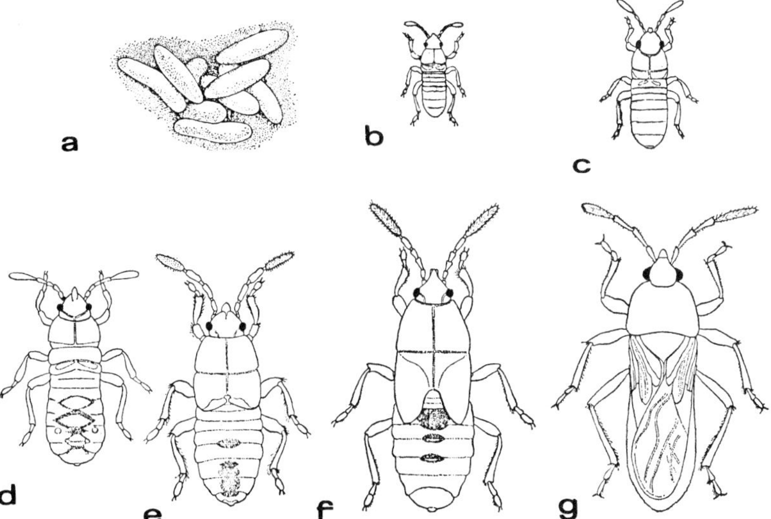

Fig. 8. Gradual development and metamorphosis of a typical Exopterygote insect (Homoptera; Heteroptera). (a) Eggs, (b)–(f) Nymphal instars, (g) Adult bug.

referred to as direct development. The period between successive moults is known as a stadium, and the body form of the insect during each stadium (stage) is referred to as an instar. The wing buds are quite evident during the final nymphal instar but of course there is disproportionate growth at the final moult when the buds become wings.

In the more advanced insects termed the Endopterygota, or the Holometabola, the young are called larvae and they are quite different from the adults in body form, and often in habits. At the final moult the last-stage instar changes into a pupa, which is a quiescent developmental stage, usually protected inside a coccoon, chrysalis, or puparium, either fixed firmly to some substrate or lying in vegetable debris or buried in the soil. From the pupa emerges the adult insect, or imago.

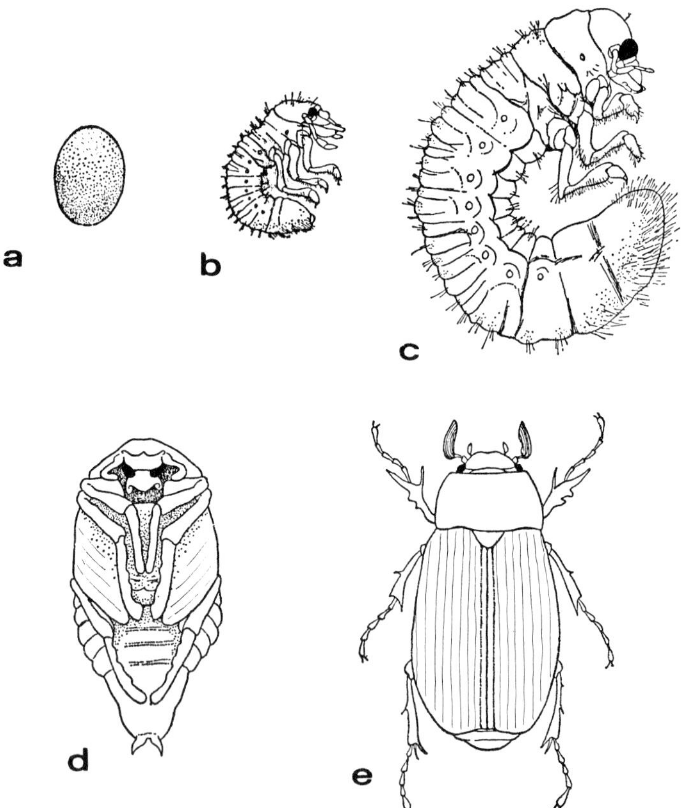

Fig. 9. Development of Endopterygote insect (Coleoptera-beetle) showing complete metamorphosis. (a) Egg, (b) Young larva, (c) Mature larvae, (d) Pupa, (e) Adult beetle.

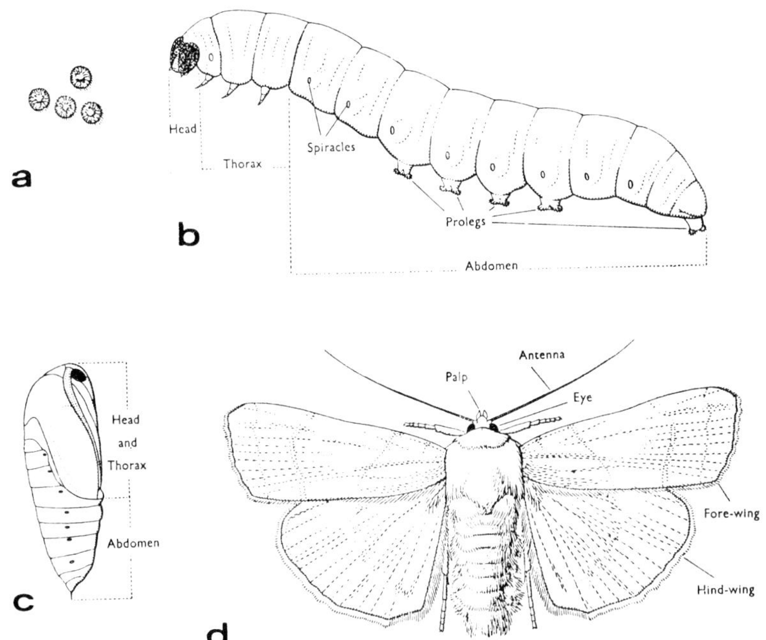

Fig. 10. Development of a moth (Lepidoptera), showing complete metamorphosis. (a) Eggs, (b) Mature caterpillar, (c) Pupa, (d) Adult moth.

The main orders of insects in the Endopterygota are Lepidoptera, Diptera, Hymenoptera, and Coleoptera. The larvae that are frequently encountered are given different common names. Thus the larvae of Lepidoptera (moths) are caterpillars, of Diptera (flies) are maggots, and of Coleoptera (beetles) are grubs. Larvae of Hymenoptera are seldom encountered as they live in nests, often underground, or are parasitic, but they are also referred to as grubs. Figs. 9 and 10 show examples of indirect development involving definite pupal stages and complete metamorphosis.

Metamorphosis offers tremendous evolutionary and ecological advantages in that the larvae of Holometabolous insects can inhabit totally different habitats and have completely different lives from the adults, thereby occupying a wide range of ecological niches. Thus caterpillars are mostly phytophagous and live on plant foliage, inside stems or fruits, bore in trees, or live in the soil, while the adults are free-flying and feed on nectar. The flies (Diptera) show perhaps the most divergence, with the larvae being aquatic, parasitic, soil-inhabiting and either predacious, phytophagous or scavening, while the adults are predacious, ectoparasitic and blood-suckers, or nectar-feeders, and of course aerial. In the Hymenoptera the insects are mostly social with the young being raised in nests and feed more or less on what the adults eat, although a number are parasitic as larvae. The beetles (Coleoptera) show some divergences, but frequently adults and larvae are found together and they often have the same type of diet. Some beetles are aquatic both as adults and larvae, and usually are predacious, while others are phytophagous and can be found together on the same host plant leaves. But the longhorn larvae are timber borers; the weevils have legless grubs in soil or inside plant bodies and the adults are often defoliators. Similarly the scarabs have large fleshy grubs in soil or in rotting vegetation and the adult chafers eat leaves and damage fruit.

However, some of the Hemimetabola have nymphs that are aquatic and lead quite different lives from the adults, and they possess a number of special structures required for their particular mode of life, such as gills, the dragonflies 'mask', burrowing appendages, and so on.

The economic consequences of metamorphosis are quite considerable in that most of the important medical and agricultural pests are only pests at one stage in their life-cycle. Thus the Lepidoptera are pests as caterpillars, the flies can be pests either as adults (mosquitoes) or as larvae (fruit fly maggots), and the beetles can in addition both be pests as adults and larvae, either together on the same foliage (leaf beetles) or else the adult on the foliage and the larvae in the soil (scarabs and chafer grubs). Sometimes for control purposes it may be easier or more effective to destroy the pest species at the stage when it is actually not the pest.

CLASSIFICATION

The Insects represent one class within the phylum Arthropoda collectively characterized by their long jointed appendages and hard exoskeleton. The other classes are as follows:

Classes   (1)   Trilobita   — fossil forms only
          (2)   Onychophora — primitive, worm-like
          (3)   Crustacea   — shrimps, crabs, etc.
          (4)   Myriapoda   — centipedes & millipedes
          (5)   Insecta   — insects
          (6)   Arachnida   — spiders, scorpions, mites, etc.

Appendix to the Arthropoda
          (7)   Pycnogonida — sca 'spiders'
          (8)   Pentastomida — tongue-worms
          (9)   Tardigrada   — water 'bears'

The Insecta are distinct in having the body divided clearly into three separate regions, the head, thorax, and abdomen. The head bears one pair of antennae, and one pair of compound eyes. The thorax bears three pairs of legs and usually two pairs of wings, although the Apterygota are wingless and a few other groups have lost their wings secondarily. There are usually no appendages on the abdomen, except for sometimes a pair of terminal cerci and the ovipositor in some females. The great majority of insect species are terrestrial, although many are aquatic and are found living in or near freshwater bodies. A few species live on the seashore under saline conditions but generally they are not really marine in distribution. The marine habitat is the province of the Crustacea.

The classification of insects has been subject to several changes over recent years. With increasing knowledge there has been a tendency to increase the number of orders and to group various orders together in higher taxa. Here we are following the classification used by O. W. Richards and R.G. Davies in the 9th edition (1964) of *A general textbook of entomology* by the late A. D. Imms. In the 10th edition (1977) a revised classification is used.

Sub-class I Apterygota

There is general agreement that they are all primitive and have never developed wings. The Thysanura are clearly primitive insects but systematists have argued that the Protura, Diplura and Collembola are possibly not really insects. However, here they are regarded as primitive insects. As a group there is little or no metamorphosis (Ametabolous). They are small or minute in size and usually found in soil or leaf-litter. Consequent to a subterranean life some species have lost their eyes and have only very small antennae.

Sub-class II Pterygota

By definition these are winged insects, although some orders and some species are secondarily apterous. Metamorphosis is varied, but never slight. They vary in size considerably and many are large. According to the nature of their metamorphosis they are subdivided into two groups: Exopterygota and Endopterygota.

Division I **Exopterygota (= Hemimetabola)**

It is generally agreed that this group represents the more primitive winged insects, sometimes referred to as the lower orders in that they are less highly evolved than the next group. Metamorphosis is simple and most of the young nymphs resemble the adults quite closely. The aquatic nymphs however have gills and other structures developed for their life in freshwater—these include Ephemeroptera, Odonata, and Plecoptera. Wings develop externally as wing-buds, and there is seldom a pupal stage.

Class Insecta

Sub-class I Apterygota

				Total species in world
Orders	1.	Thysanura	(Bristle-tails)	700
	2.	Diplura*†		400
	3.	Protura*†		90
	4.	Collembola*	(Spring-tails)	2,000

Sub-class II Pterygota

Division (1) Exopterygota (= Hemimetabola)

Palaeopteran Orders	5.	Ephemeroptera	(Mayflies)	1,500
	6.	Odonata	(Dragonflies & Damsel-flies)	4,870
Orthopteroid Orders	7.	Plecoptera	(Stoneflies)	1,490
	8.	Grylloblattodea†		6
	9.	Orthoptera	(Grasshoppers, Crickets, etc.)	10,000+
	10.	Phasmida	(Stick- & Leaf-insects)	2,000
	11.	Dermaptera	(Earwigs)	1,100
	12.	Embioptera	(Web-spinners)	150
	13.	Dictyoptera	(Cockroaches & Mantids)	5,300
	14.	Isoptera	(Termites)	1,700
	15.	Zoraptera†		19
Hemipteroid Orders	16.	Psocoptera	(Book-lice & Bark-lice)	1,100
	17.	Mallophaga	(Biting Lice)	2,675
	18.	Siphunculata	(Sucking Lice)	250
	19.	Hemiptera	(Bugs)	56,000+
	20.	Thysanoptera	(Thrips)	3,170

Division (2) Endopterygota (= Holometabola)

Panorpoid Complex	21.	Neuroptera	(Lacewings, Ant-lions, Alder-flies)	4,670
	22.	Mecoptera†	(Scorpion-flies)	350
	23.	Lepidoptera	(Moths & Butterflies)	112,000
	24.	Trichoptera	(Caddis-flies)	4,450
	25.	Diptera	(Flies)	85,000
	26.	Siphonaptera	(Fleas)	1,000
	27.	Hymenoptera	(Wasps, Bees, Ants, etc.)	103,000
	28.	Coleoptera	(Beetles)	276,000
	29.	Strepsiptera†	(Stylops)	300

---

* According to some authorities these are possibly not truly insects.

† Not yet recorded in Hong Kong and not included in this book.

Division II **Endopterygota (= Holometabola)**

These are the higher orders of insects, and are thought to be the most highly evolved. They are characterized by having a complex metamorphosis including a definite pupal stage. As the name suggests, the wings develop internally and are first seen at the time of pupation. The immature stages are larvae which differ extensively from the adults in structure and often in habits, and there is often great ecological separation between larval and adult stages.

In some classifications the orders of insects are grouped into further subdivisions as indicated below; the Ephemeroptera and Odonata are separated off from the remainder as belonging to the Palaeoptera—their wing venation is regarded as being particularly primitive, and dragonflies are well-represented in fossil beds as old as Carboniferous Period. The remainder of insect orders are regarded as Neoptera, but within this grouping the orders from Plecoptera to Zoraptera inclusive are termed Orthopteroid, and the Psocoptera to Thysanura are Hemipteroid. Within the Endopterygota the orders Neuroptera to Siphonaptera are termed the Panorpoid complex after the group of insects in the Mecoptera.

NOMENCLATURE AND NAMES

The scientific system of naming animals and plants is termed the Binomial System and it was first fully propounded by Linnaeus in Sweden in 1758 in the 10th edition of his *Systema Naturae*. Under this system each animal (and plant) is given a generic name in Latin or Greek followed by a specific name in the same gender. The genus is always written with an initial capital letter and the specific name now is always in lower case. In typescript or writing both names are underlined, but in print they are italicized. To be punctillious these two names should be followed by the name of the authority who named that species and the date of publication (year only) of that new name together with the species description. In some insect groups it is common to find the development of geographical races, and then a third name can be added to the binomial—the subspecific name. The important point to note is that each generic name can only be used once throughout the entire animal kingdom; it then has nomenclatorial priority and should the same name be accidentally used again the latter name will be suppressed. Any specific name may only be used once with each genus, so the result is that the scientific name of each animal species is unique. Thus *Apis mellifera* is the name of the ubiquitous Honey Bee, and *A. mellifera africanus* is the African subspecies of notoriety.

The terms taxonomy and systematics cause problems in definition; taxonomy has a Greek origin and was originally proposed in 1813 for the theory of plant classification, whereas systematics is from Latinized-Greek and was used to describe the systems of classification used by early naturalists, notably Linnaeus (*Systema Naturae,* 1735). Some biologists now use both terms interchangeably in both animal and plant classification. But in some institutions there is a tendency to use taxonomy for the identification of the species (and genus), and systematics for the study of the higher classification of the group. Anyway, the different systematic categories are referred to collectively as taxa!

Animals are classified as belonging to the Kingdom Animalia, as distinct from plants, and the insects then form part of the phylum Arthropoda as the class Insecta. From the level of order in descending rank each taxon has a typical suffix which is characteristic of that taxon. For example:

Order Hymenopter*a*	
Sub-order Apocrit*a*	
Superfamily Ap*oidea*	
Family Ap*idae*	typical suffixes
Subfamily Ap*inae*	
Tribe Ap*ini*	

The insect being classified here would be:

Apis mellifera Linnaeus, 1756

For animals other than insects the suffixes for taxa higher than family may not be consistent. The whole system of scientific nomenclature and classification of animals is now governed by the 'International code of zoological nomenclature' adopted by the XV International Congress of Zoology, held in London, July 1958.

The family unit in zoology (and botany) is particularly important, for all members of any one family generally resemble each other in several major aspects of morphology and anatomy, as well as in their life-cycles and ecology. Ecologically many studies treat each family of insects collectively. Clearly families will vary, and some will show striking ecological similarities whereas in others there may be considerable divergence. But for most zoologists and many ecologists the family-group is the one readily recognizable in the field. No matter which part of the world one might be in, it takes little effort to recognize a cockroach (family Blattidae), or a mantid (Mantidae), or a long-horned grasshopper

(Tettigoniidae), an ant (Formicidae), and so on. However, it is not always so easy, and it is quite another matter for the general entomologist to place to family many dragonflies (Odonata), mayflies (Ephemeroptera), caddis-flies (Trichoptera), biting lice (Mallophaga), and a very large number of small flies, beetles and moths.

The genus is often a fairly distinctive category and with practice it is possible to recognize many insect genera in the field. But some groups of genera which have been in the hands of taxonomic 'splitters' have been established on very esoteric grounds and these do present a problem to field biologists. The descriptive definition of the genus is ultimately the sum of the closely related species allocated to that genus. But the problem with insect taxonomy is that there are thousands of new species being described annually and so we are clearly quite a long way from having all the insect genera definitively described. As new species are recorded so often, the limits of present-day genera have to be changed and some new genera erected. By comparison the nomenclature for mammals and birds, and some other vertebrates, is very stable for no new species have been described for several decades, or more.

The species is the basic biological unit of animal communities; it is a group of animals morphologically similar (but not identical), occurring in the same general area, and which naturally interbreed and have fertile young. It may or may not be separated physically from other species in the same genus, but is invariably separated ecologically in some definite way. Within the distribution range of a species there may be populations physically separated (islands, etc.) which have evolved slight but consistent morphological differences from the main population, but if returned to the main population they will interbreed and have fertile young— these isolated or separated populations are termed sub-species. They are usually formed as a result of geographical separation. Identification of the species is very difficult and the initial determinations are best left to the taxonomic experts at the British Museum (Natural History), or the Commonwealth Institute of Entomology, London, or the Smithsonian Institution in Washington and other places. However, the initial determinations do enable the establishment of a regional collection of insects such as is being developed in the Department of Zoology of the University of Hong Kong. Also there is a collection of named crop pests at the Department of Agriculture and Fisheries research station at Tai Lung Farm, New Territories. Using these regional collections for comparison of specimens it is generally possible to identify many of the local insect species. But these reference collections are still not large and so many local insects cannot be identified yet.

Common names are given to insects when they are well-known and of popular interest enough to warrant the creation of a name in the local language. This creation has generally been done because the scientific name, being in Latin or Greek, is often unfamiliar or regarded as awkward to use by amateur entomologists and agriculturalists. Obviously the first animals to be given common names were mammals, birds, some fish and other vertebrates, partly because they are larger and more conspicuous and also they are fewer in number and thus more easily given different names. In the Insecta the first groups to be collected and studied by amateurs were Lepidoptera and Coleoptera. In Europe and North America almost all the Lepidoptera and Coleoptera have local names, but of the Hemiptera, Orthoptera, Diptera, Hymenoptera, and other groups, they tend to have common names only if they are of importance as pest species or for other reasons such as mayflies which are used for trout fishing.

Common names also apply to families and here they are usually appropriate, consistent and useful. For example Coccidae are Scale Insects; Pseudococcidae—Mealybugs; Chrysomelidae—Leaf Beetles; Cicadellidae—Leafhoppers; Cicadidae—Cicadas; and most of these names are consistent throughout the English-speaking parts of the world.

Scientific names are unique and of worldwide application, whereas common names for species are sometimes only of use in the country of origin since they differ from country to country. For example *Delia platura* (Diptera; Anthomyiidae) is the Bean Seed Fly in U.K., but is the Corn Seed Maggot in the U.S.A.; and is the Onion Fly in Australia; whereas *Delia antiqua* is the Onion Fly in U.K. and the Onion Maggot in the U.S.A. Most large countries have published a list of approved common names for insects in that country, but most of the species listed are pest species. In Hong Kong entomology is very much in a state of infancy and as yet we have not developed our own system of local common names. For butterflies we are using Malayan (English) and Indian common names, with a few alterations to suit our situation. Similarly some of our most common species are actually major crop pests in other contexts and there is some international agreement about their common names. For other abundant local species we are proposing to start the establishment of our own common names.

CHAPTER TWO

INSECT ECOLOGY IN HONG KONG

The study of animal distributions is called zoögeography, and in 1858 Sclater divided the world up into a number of zoögeographical regions, based in part upon faunal lists showing the animal species found there, and with the geological history of the areas (see map). The regions are Nearctic (North America); Palaearctic (Europe, North Africa, North and East Asia); Ethiopian (Africa South of the Sahara Desert, and Madagascar); Oriental (India, and South-east Asia); Neotropical (Central and South America); and Australasian (Australia, New Zealand, and the Pacific Islands). Because of the strong similarities between the Palaearctic and Nearctic regions they are sometimes grouped together as the Holarctic.

In virtually all parts of the world the classic zoögeographical boundaries are sharp and well-defined, and are represented by seas, deserts or mountain chains. The boundary between the eastern Oriental region and the eastern Palaearctic, which is traditionally drawn in South China, is an exception and is very ill-defined. Here there are no insurmountable natural barriers to animal dispersal, and so in the broad area from South China northwards to Korea and South Japan there is a continuous graduation from Oriental species in the south to Palaearctic species in the north. The classical dividing line between Oriental and Palaearctic faunas might more suitably be drawn above the southern islands of Japan, for Taiwan is quite tropical and there is a considerable tropical element in both flora and fauna in Kyushu. Thus it can be seen that Hong Kong occupies a unique position zoögeographically (together with South and East China of course) which is clearly reflected in the composition of the local fauna. Thus in the local insect fauna, most species are Oriental, but there is a distinct Palaearctic element.

There are two main problems in studying insect distribution. First most insects can fly and so the usual barriers (rivers, seas, lakes, deserts, mountains etc.) which restrict the dispersal of many animals are far less effective against flying insects. Second, most insects are phytophagous (i.e. they feed on plants) and a large proportion feed on plants grown for

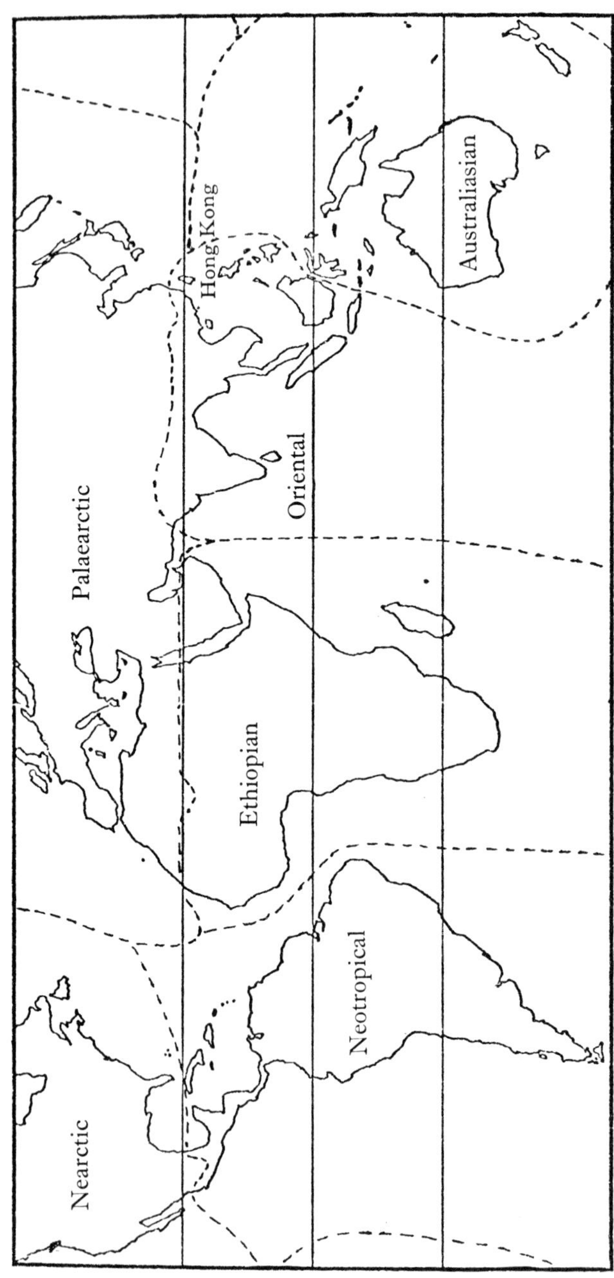

The Major Zoogeographical Regions of the World.

agricultural or horticultural purposes. Thus when a crop plant is introduced into a new country for local cultivation usually a whole series of insects follow it eventually, or are accidently introduced at the same time. For example, peach and *Citrus* are two fruit crops which originated in the area of South China and are now widely cultivated throughout the warmer parts of the world, and most of the Citrus and peach-eating insects are now just as widely distributed, although originally they were native to South China. Similarly tobacco was originated in South America (probably in the Argentine) and maize in southern U.S.A. and now both are widely cultivated in the tropics and subtropics.

The insects that cause damage to plants which are grown as crops or as ornamentals are known as insect pests, although in its widest sense an insect pest is a species which does harm to man, his crops, domestic animals, or possessions. So insect pests may be species important as medical, veterinary, agricultural, or household pests.

Insect pests are very widely distributed throughout the world since their distribution is a reflection of the distribution of their host plants, or host animals. For example, where man dwells will be found the Bedbug, Human Flea, Head and Body Lice, Crab Louse, and various domestic mosquitoes, for all these feed upon human blood; and in addition there will always be several species of cockroach, ants, house flies, clothes moths, and flour and grain beetles, all feeding upon man's foodstuffs and possessions. Not all insect pests are so widely distributed because some only thrive within a certain set of prescribed conditions, for example Tsetse Flies only occur in tropical Africa, and many other biting flies are restricted to certain parts of Africa, Asia or on the Pacific Islands.

With phytophagous insects one point of great significance is the extent to which they are dependent upon a particular host plant, or group of plants, for food (i.e. their host specificity). For example, Banana Stem Weevil feeds only on banana pseudostems, and Bamboo Weevil only in the shoot tip of the larger species of bamboo; these insects are termed monophagus ('monos'—alone; 'phagein'—to eat). Insects that feed upon a closely related group of plants, for example the family Malvaceae which includes cotton, okra (ladies' fingers), mallow and *Hibiscus* as their food, are called oliogophagous ('oligos'—few). Other insects such as locusts, some aphids, some beetles, and some caterpillars go to the other extreme and will feed on a vast range of different plants; these are termed polyphagous ('poly'—many). In general most insect species come into the oligophagous category in that they normally feed on a restricted group of (usually) closely related hosts. This is of particular interest locally because, for example, although no cotton is cultivated in Hong Kong it is a major crop in South China, and a number of well-known cotton pests are found here feeding on the very abundant *Hibiscus*. Similarly some important polyphagous crop pests in other parts of the world are found locally feeding on wild relatives of major crops or else on wild alternative hosts. For example many of the major cereal pests will feed quite readily on various species of wild grasses. The result is that many of the most common local species of insects are in fact major crop pests in South-east Asia and other parts of the world.

The insect fauna of Hong Kong is predominantly that of tropical South-east Asia, but somewhat more sparse in view of the cooler conditions and less diverse and luxuriant vegetation here. There is, however, a distinct Palaearctic element in the fauna, and furthermore there are many species of insect here that are either polyphagous in their feeding habits and so very adaptable, or else they feed on common agricultural crop plants and are now either cosmopolitan in distribution or else pantropical, according in part to the type of host plant on which they feed.

The two most important Palaearctic (in this case actually Holarctic) species which occur here are the Gypsy Moth *(Lymantria dispar)* whose hairy caterpillars defoliate forest trees, and the Small White Butterfly *(Pieris rapae)* which is a pest of *Brassica* crops and cruciferous plants. The common European Large White Butterfly apparently reaches Yunnan but so far no further into China. The Painted Lady which is the most widely distributed butterfly in the world and is completely cosmopolitan does occur here but is not common. The cosmopolitan insect species found here which are agricultural pests include the Peach-Potato Aphid, Turnip Aphid, Groundnut Aphid, Corn Leaf Aphid, Cotton-Melon Aphid, Onion Thrips, Black Tea (Greenhouse) Thrips, Tobacco White-fly, Coconut Scale, Cottony Cushion Scale, Green Stink Bug, Pea Leaf-miner Fly, Diamond-back Moth, American Bollworm, Pink (Cotton) Bollworm, and Black (Greasy) Cutworm.

There are a few groups of insects which are conspicous by their absence; three of the most notable are the Holarctic bumble bees, the leaf insects which are so well represented in South-east Asia and Australia, and the large mound-building termites so characteristic of tropical regions. Actually, there are two local species of these termites (family Termitidae) in Hong Kong, but their colonies are small and the characteristic mound non-existent, the entire nest being completely underground.

For ecological studies in general the most important systematic group is without doubt the family. Most species within a family unit bear considerable resemblance to each other, and have similar life-cycles and occupy similar ecological niches, the same trophic level, and the same position in local food webs. Provided that the fundamental taxonomy has been done well, using significant characters and not just superficial resemblances, then most families of insects can be identified by a general entomologist or by a competent ecologist. Clearly, some families are difficult to distinguish and really can only be separated by a taxonomic expert, for example some families in the Lepidoptera, especially within the microlepidoptera, but the number of such families is not big.

What causes some confusion is that there are a few well-established families which have distinctive sub-families with important ecological and in some cases also taxonomic differences. Some of these sub-families are very distinctive morphologically, but others are difficult to recognize at this level by non-specialists. For example within the Coleoptera there is the family Chrysomelidae (Leaf Beetles) with its very distinctive sub-families Cassidinae (Tortoise Beetles), Sagrinae, Eumolpinae (Leaf Beetles), Halticinae (Flea Beetle), Hispinae (Leaf Miners) etc. There is also the Scarabaeidae with its sub-families Cetoniinae (Rose Chafers), Rutelinae (Flower Beetles), Melolonthinae (Cockchafers), Coprinae (Dung Beetles), and Dynastinae (Rhinoceros and Unicorn Beetles), with their equally striking morphological resemblances. In some of these cases it is really necessary to consider the sub-families separately from an ecological point of view.

In some groups it is not really necessary to descend below the level of order or sub-order for ecological studies locally; for example Psocoptera, Mallophaga, Siphunculata, Thysanoptera, and Siphonaptera where virtually all the local species occupy the same broad ecological niche. Naturally the important point is the level at which the study is being conducted, for in a general community study, for example, all Mallophaga can be regarded as bird feather-lice (except for a few mammalian ectoparasites), but in a detailed ornithological study then it is immediately apparent that different birds, and groups of birds, have different genera and species of feather-lice and that on any one bird the different species occupy different areas of the body.

In some families there is great ecological uniformity, for example the Pseudococcidae (Mealybugs) and Acrididae (Short-horned Grasshoppers). But in other families there may be considerable ecological differences, as in the Psyllidae (Jumping Plant Lice), Noctuidae (Owlet Moths), Pyralidae, Muscidae (House Flies), Curculionidae (Weevils), etc. If an ecological study is taken to ultimate detail then obviously every species of insect will occupy a separate ecological niche and is distinct from the others.

As this book is intended as a general insect text, it is not possible to delve into too much detail at the species level. The species mentioned are the commonest, for they are the most frequently encountered, the more interesting and the most spectacular ones.

In the various terrestrial and freshwater habitats the dominant element in the fauna is undoubtably the Insecta, either as adults or larvae, or both.

To be consistent with the first local ecology text already produced (Hill, Gott, Morton & Hodgkiss, 1978), the same series of habitat designations is followed here in the discussion of insect ecology in Hong Kong.

(1)   Soil insects (including litter)
(2)   Grassland insects
(3)   Scrubland insects
(4)   Forest (= woodland) insects
(5)   Freshwater insects
(6)   Marine insects
(7)   Urban insects (including stored products pests) (= Domestic insects)
(8)   Agricultural insects
(9)   Symbiotic insects

Each section has a systematic list of species under their appropriate families, with brief notes on their ecology, hosts, and their relative abundance. Not all of the insect species so far identified have been included; the lists generally contain the most abundant, conspicuous, important and ecologically interesting species found locally.

(1) SOIL INSECTS (INCLUDING LITTER)

All local soils are broadly considered together in this category. Clearly this will result in an over-simplification, because as the soil and overlying litter will vary from habitat to habitat, so will the insect fauna to some extent change. However, at present so little is known about Hong Kong soil fauna that it is pointless to attempt to be more specific. Hopefully as more ecology research is undertaken locally our knowledge of these cryptozoic animals will increase and then it may prove possible to consider the different types of soil fauna comparatively. Since these insects are generally of little economic importance they are invariably the last local species to be identified and studied in any area.

This habitat also contains a number of different life forms. Some insects live permanently quite deep in the soil, whereas others live on the soil surface or in surface crevices, or under stones and logs, and some build nests underground but feed on the surface, while some only inhabit the litter.

Faunistically soil is a peculiar habitat with regard to the Insecta as it characteristically contains a disproportionately large number of immature stages, which are present only for a period of development. Sometimes the period of larval development may be prolonged, and it is thought that some of the larger local cicada nymphs may remain underground for several years. Some local Scaraboid Beetle larvae certainly spend a year or more as subterranean grubs, but many Lepidoptera only use the soil substratum as a medium for pupation, as do all the thrips (Thysanoptera), most weevils (Curculionidae), and many flies (Diptera). Then there is the intermediate group where the whole larval and pupal stages are spent underground, while the adult is aerial or arboreal; these include many flies (Diptera; Tipulidae, Asilidae, Anthomyiidae, etc.), some Lepidoptera (Noctuidae), and some Coleoptera (Elateridae, Scarabaeidae, Curculionidae, etc.).

Ecologically, the soil fauna can be classified according to the size of the insects, the type of feeding, and the degree of dependence upon the soil medium as a habitat. On the basis of size, the larger elements are termed *Macrofauna* (2–20 mm) and *Megafauna* (above 20 mm in body length); the small insects from 200 μ to 2 mm are *Mesofauna* (= meiofauna), and the *Microfauna* seldom contains any insects, being mostly protozoans. From the feeding point of view the insects can be carnivores (subdivided into predators and parasites), phytophagous (on green plant material from above ground, plant roots, and underground woody material), saprophagous (on dead and decaying organic material), microphytic feeders (on fungal hyphae, algae, lichens, etc.), and miscellaneous polyphagus feeders. On the basis of time spent in the soil, the insects can be classified as permanent, periodic, temporary, or transient.

The aerial forms which hibernate, aestivate, or just spend the night in litter are not included in this category since most will not be encountered in soil and litter samples for laboratory extraction. Similarly, insects which just lay their eggs in or on the soil are not included, for example grasshopper egg-pods, and some Lepidoptera (Noctuidae);

neither are the bees, ants and wasps that only occasionally build underground nests included.

It must also be remembered that during the process of sampling soil and litter there may be captured some insects that just happened to be down on the ground for various reasons. The method of sampling will influence the quantity of such extraneous species included. If the extraction process is lengthy (e.g. Berlesé Funnel), and particularly if heat is applied, a number of insect pupae present in the soil and litter will hatch, yielding adult flies, moths, beetles etc.

(2) GRASSLAND INSECTS

For purposes of biological and ecological investigations the International Biological Programme (I.B.P.) made a series of definitions for the various types of botanical habitats, as published by Peterkin (1967). However, in 1968 the Hong Kong Government published a short treatise entitled 'Land Utilization in Hong Kong' where slightly different definitions of 'grassland', 'scrubland', and 'forest' were given, and these are the definitions used in this work.

The two definitions for grassland are as follows:

I.B.P.—vegetation principally made up of grasses and other graminoid plants.

H.K. Government—land covered with grass and low scrub generally under 1 ft (0.3 m) in height.

Grassland occurs where conditions are too dry, too exposed, or the soil too shallow to support woody plants. In Hong Kong regular burning combined with animal grazing and cutting of shrubs for firewood maintain a plagioclimax of grassland.

Two different types of grassland occur in Hong Kong:

1. Hillside grassland—on the exposed upper slopes of mountains, usually above 300 m, but descending to sea level on some islands and areas unfavourable for plant growth.

2. Old field grassland—paddy or other irrigated soils usually develop a hard-pan just beneath the surface resulting from continuous wet cultivation, and many of these previously cultivated areas in the New Territories have later been abandoned and allowed to revert by colonization of local plants. However, the soil that is effectively so shallow will support only a cover of grasses and woody plants will find it difficult to become established.

The insect fauna of grasslands is essentially confined to species which actually prefer Gramineae as host plants, or else species which are predacious and prefer to hunt in the open, highly illuminated, grassland vegetation. Some are species whose larvae live in grassland soil, such as the Scarab Beetles, and the adults found in grassland are either just emerging from pupation in the soil or else are ovipositing in the soil.

The groups of insects found in Hong Kong grassland are mostly Short-horned Grasshoppers (Acrididae), some Leafhoppers (Cicadellidae) and Planthoppers (Delphacidae), a few Shield Bugs (Pentatomidae), stem-boring moth caterpillars (Pyralidae, Noctuidae) and various species of Brown Butterflies (Satyrinae) and Skippers (Hesperiidae) which feed on grasses as larvae; a few flies are found boring in shoots and stems of grasses, or mining the leaves, and some predatory beetles (Carabidae and Staphylnidae) occur in addition to the various Scarabaeidae whose larvae feed on grass roots in the soil.

(3) SCRUBLAND INSECTS

Scrubland in Hong Kong is defined as follows:

I.B.P.—woody vegetation, predominantly of shrubs, ranging between 0.5 and 5.0 m in height.

H.K. Government—land with a fairly continuous cover of shrubs and bushes from 1 ft (0.3 m) to 8 ft (2.9 m) in height.

This is a difficult habitat to delimit as it is basically a transition between grassland and forest and as such is essentially of short duration. However there are some hillside areas where a natural shrub climax vegetation occurs—the shrubs concerned are False Tea *(Eurya sinensis),* and other members of the Theaceae (Tea, Camellia, etc.), and the Rhododendrons and Azaleas.

And it does have some characteristic woody shrub species, such as *Rhodomyrtus,* which find grassland too exposed and dry, and forest too shady. A few forest plant species are found in scrubland where they assume a dwarfed form, such as *Schefflera octophylla.*

The insect fauna of scrubland appears to consist almost entirely of phytophagous species whose host plants occur in this habitat. For example, in many coastal scrubland areas the Lemon Butterfly *(Papilio demoleus)* can be found in large numbers as one of its preferred host plants, *Atalantia buxifolia,* grows as a semi-prostrate shrub in these areas, and this butterfly is equally common in agricultural areas and on the edges of woodland where *Citrus* species are grown.

(4) Forest (=Woodland) Insects

Forest (woodland) as a biological habitat in Hong Kong.has two rather different definitions:

I.B.P.—Open or closed vegetation with the principle layer consisting of trees, these averaging over 5 m in height.

H.K. Government—Land with a continuous cover of shrubs and trees over 8 ft (2.4 m) in height, with interlocking canopies, not being within afforested areas.

By virtue of its geographical location, climate and geology, Hong Kong is naturally a tropical rain forest area and this type of vegetation must once have been almost completely dominant. However there is an ancient local tradition of deforestation in China (for a number of different reasons) and for many years most of Hong Kong and the New Territories area have been devoid of extensive woodland. Since the Second World War there has been a programme of reafforestation and *Pinus* and *Acacia* and a few other species have been both aerially sown and planted. Now in many parts of Hong Kong Island and the New Territories there are extensive areas of scrubland that are virtually forest by both definitions.

Woodland in most countries refers to relatively large areas of land extensively covered with large trees of various types, with a certain amount of undergrowth, the extent of which depends upon the height and density of the trees, whether deciduous or coniferous, etc. In these woods the fauna will vary considerably according to the floristic composition. For example a pine forest will have a quite different fauna from a beech or oak forest, although they will have certain common faunistic features. There will be similarities at the family level of fauna, but with great specific differences.

In Hong Kong, apart from some areas of planted *Pinus,* areas of established woodland are mostly very small and sparse. The exception is that of Tai Po Kau Forestry Reserve which was partly afforested many years ago, and the trees there are now large and the vegetation makes a good semi-natural mixed forest. Since Hong Kong is at the edge of the Palaearctic region there is a northern element to our forests which consist of oaks (family Fagaceae), such as *Quercus* and *Lithocarpus,* and members of the more tropical family *Lauraceae,* such as *Cinnamomum* and *Litsea.* As already mentioned, there are now quite extensive areas on the Hong Kong Island and in parts of the New Territories where the previous scrub has grown to such an extent as just about to constitute forest, and as the trees grow each year so the forest will become more apparent. In some parts of the New Territories the 'Fung Shui' woods behind the villages are small patches of seemingly native vegetation, with the addition of a few fruit trees and bamboo because of their utility. The undergrowth is usually cut for firewood, and this prevents natural regeneration of trees and alters the nature of the habitat.

A characteristic of tropical rain forest is that no one species of tree is dominant, as is usually the case in temperate forests, but that some 10–20 tree species are co-dominant together. The forest is distinctly layered, or storied, with the tree canopy at the top, and lianas, creepers and epiphytes climbing up the tree trunks to reach the sunlit canopy. Shade tolerant shrubs form an understory, and on the floor grow the herbs and grasses which are adapted for life in very low light intensities.

Thus the woodland insect fauna has several distinct components other than the obvious taxonomic groupings. First there are the species associated with the trees, including leaf-eaters, those coming to the flowers and fruits, sap-suckers, trunk and bark borers, etc.; then those associated with the creepers (vines), and with the undercover shrubs and the woodland herbaceous plants; and finally a number of insects specific to Gramineae on the forest shade grasses and on the forest bamboos. Also there are the adults of the species whose larvae and pupae live in the soil and litter, a large number of flies, beetles, and moths which come into the forest for shelter and for the more equable microclimate, and all the predators and parasitic insects that feed on all the preceding species.

Bearing in mind that most of the phytophagous insects show a high degree of host specificity in their choice of food plant, the monophagous ones are restricted to a single species or genus of host plant, and the oligophagus to a family (or two), the basic fauna of a woodland area will depend largely upon the species of trees present and the species of undershrubs and vines. In Hong Kong a number of trees are both widespread and common, so that in many areas of local woodland the basic floral composition does tend to be relatively constant qualitatively. One obvious exception is the distribution of most of the vines of the families Asclepiadaceae and Aristolochiaceae upon whose poisonous leaves various rather rare local butterfly caterpillars feed (*Danaus* spp., *Troides*, etc.). These vines are only found in a few restricted habitats in the more distant areas of the New Territories.

(5) Freshwater Insects

Freshwater habitats in Hong Kong are limited because local rivers are now so polluted that they are lifeless. There are a few major permanent stream systems in the New Territories, and there are no natural lakes or large ponds; but there are reservoirs, of which Plover Cove is both large and biologically quite productive. Also there are a number of small local ponds, many extensive fish ponds, and a considerable area of flooded paddy fields with their associated ditches.

The freshwater insects tend to be found in two biological groups— those present in the swiftly flowing streams (lotic environments) and the others from the quieter or stagnant waters (lentic environments). The same insect groups will tend to occur in both environments (e.g. dragonfly and damselfly nymphs) but the species composition will differ. In the streams there will be plenty of oxygen and few plants, but the danger of physical dislodgement and subsequent sweeping downstream will be a constant hazard. In the still water oxygen generally will not be so abundant, although there will be more aquatic vegetation and no danger of dislodgement.

The insect fauna of freshwater habitats consists of several basic types; in some cases both adults and larvae are aquatic, as in some Coleoptera and Heteroptera, but in many species the adults are aerial and only the larvae are aquatic (e.g. Odonata, Ephemeroptera, Trichoptera, Diptera). The main types are as follows:

(a) Surface dwellers, some are only as adults (Gyrinidae), but some are both as adults and larvae (Gerridae, Hydrometridae, Veliidae).

(b) Mid-water swimmers, including adults which have to visit the surface for oxygen (Dytiscidae), adults and nymphs which visit the surface (Notonectidae, Nepidae, Corixidae); larvae with metapneustic type respiration which visit the surface (Cuclicidae; Dytiscidae), and larvae with gills (Chaoborinae) which obtain their oxygen from that dissolved in the water.

(c) Bottom dwellers (benthic) on the rocks, submerged vegetation and the surface of the substrate, such as the mayfly nymphs, Odonata, caddis larvae, *Simulium, Chironomous* larvae, Alderfly larvae, Psephenidae, all of which use either tracheal gills or blood gills for respiration.

(d) Burrowing benthic forms in the substrate, which includes some chironomid larvae and some beetle larvae, and the genus *Ephemera* in the Ephemeroptera. Here of course repiration can be a problem as the substrate region is often virtually anaerobic, especially in lentic environments.

(e) Occasional visitors such as amphibious Cockroaches and Grouse-locusts, and certain species of adult dragonflies which take up territories over particular stretches of water. Into this category could be placed the adults of Ephemeroptera, Odonata, Trichoptera, Coleoptera and Diptera which emerge from aquatic larvae and remain in the vicinity of the water body for mating and eventual oviposition purposes.

Alternatively, aquatic insects can be classified according to the amount of time they spend in the water, as:

(a) Permanent
(b) Temporary
    (i) All stages
    (ii) Larvae only
    (iii) Larvae and pupae
    (iv) Adult only
(c) Occasional visitors

Some of the local freshwater habitats are rather hazardous for insect life, in that some of the streams are only temporary and dry up in the winter to a series of stagnant pools. The streams on Hong Kong Island and in the Kowloon area are regularly oiled by the U.S.D. anti-malaria teams, and some of the streams and ponds in the New Territories are heavily polluted with excrement from chickens, ducks and pig farms. Heavy organic pollution results invariably in depleted oxygen levels in both streams and ponds and only certain insect species can thrive under such conditions. In the polluted parts of the streams the stonefly (Plecoptera) nymphs and caddisfly (Trichoptera) larvae will invariably be absent as both are very sensitive to organic pollution and its associated oxygen depletion. However, this situation is rather complex in that these insects are characteristic of the torrent regions of streams, whereas in Hong Kong the major organic pollution usually occurs in the middle and lower reaches of the streams.

(6) MARINE INSECTS

The marine habitat can be regarded as being composed of three fairly distinct regions: the area of continuous water (sub-littoral and pelagic), the shore on the littoral zone, and the back-of-beach area of semi-halophytic vegetation. Each region can of course be meaningfully sub-divided, but for entomological purposes there is no advantage.

Generally, the marine habitat is the one major ecological area not colonized by the Insecta. However, a few species have made a slight encroachment. On the open sea in the Indo-Pacific can be found the Sea

Skater, a close relative of the freshwater Pond Skater (*Gerris* spp.), and this insect can be found on the sea in Deep Bay, congregating amongst the mangrove plants at high tide where the water is still and food plentiful. A few mosquitoes typically breed in saline rain pools amongst the rocks, and in other parts of the world there are chironomids to be found breeding in the seaweed beds. Seaweed (*Sargassum* spp., etc.) washed up as flotsam on the shore is typically inhabited by various flies, the most frequent being members of the family Ephydridae. A few predacious beetles and flies are to be found hunting on the sand at the beach above the high water mark, and in Hong Kong there is apparently a species of Tiger Beetle to be found exclusively on sandy shores.

Of course, the back-of-beach vegetation is composed of certain plants only to be found in this habitat (e.g. *Hibiscus tiliaceous, Scaevola,* etc.), as well as a number of vigorous highly adaptable species (such as *Ficus microcarpa, F. superba* var. *japonica* and *Asparagus cochinchinensis*), and a number of species best regarded as weeds (or ruderals), such as *Ampelopsis* spp., *Lactuca* spp. and *Oxalis* spp. Similarly the mangrove flora is a mixture of species, mostly plants especially adapted to this specialized habitat.

Entomologically, the only species mentioned in the marine environment are those that are apparently living there more or less exclusively and are adapted to the prevailing saline conditions, or else are phytophagous and are specific to typical back-of-beach or mangrove plants. Insects casually or accidently located on the shore, and those feeding on the ruderal plants and on the salt tolerant banyans, etc., are not included.

(7) URBAN (=DOMESTIC) INSECTS

This habitat includes all human dwelling places, stores, warehouses, godowns and shops. Thus domestic insects will include insects that feed upon man, his pets, his foodstuffs, books, clothing, leather, etc., and those that find a human dwelling a suitable microhabitat for their inhabitance.

In books on domestic insects the phytophagous insects feeding on house plants and ornamentals in pots would be included, but in this text these insects are included under 'agricultural insects'.

There is a recent trend, especially in the U.S.A., to establish a new field in 'urban entomology'. By definition this includes all aspects of entomology in relation to man's activities except agriculture. Thus in urban entomology domestic premises, gardens, parks, municipal gardens, roadsides and even nature reserves are considered. However, since all the major ecological habitats of Hong Kong are being considered in the present treatment, it is felt that this very broad 'urban' approach is, in this case, not profitable.

(8) Agricultural Insects

Agriculture in the broad sense refers to the cultivation of plant and animal species for food, for ornamental purposes, or for specific products.

The only insect actually reared locally is the ubiquitous Honey Bee *(Apis mellifera)*. Hives are kept in small numbers on Hong Kong Island and on some of the other islands, but most of the apiculture is practiced in Fanling and Sheung Shui in the New Territories. There is probably some local rearing of Silk Worms *(Bombyx mori)* as at intervals these caterpillars are on sale to school children in the streets, but this cultivation appears to be for educational purposes rather than for silk production.

In various fish ponds in the New Territories chironomid midges are encouraged to breed and at regular intervals the mature larvae are collected from the mud and sold as 'blood-worms' for feeding live aquarium fish and other aquatic animals. A few cases are known where the chironomids are actually cultivated in small concrete ponds and are fed daily with a special mixture of dried food.

In certain local agricultural situations care is taken to ensure that when spraying fruit tree with insecticides against fruit tree insects and mite pests, some trees are not sprayed, and on these trees the predators and parasites of the pests are allowed to remain unharmed. Natural control of crop pests by insects that are predatory or parasitic is vitally important in any agricultural situation. It is however, much more important in the tropics and especially on long-term orchard crops such as fruit trees. Also the Cottony Cushion Scale *(Icerya purchasi)* in Hong Kong is being kept under a measure of control by imported Ladybirds (Coccinellidae) from India. So the 'natural control' elements of the Insecta fauna can be regarded as agriculturally desirable species to be encouraged in their population development whenever possible.

However, the usual concern over agricultural insects stems from the fact that an enormous number of species are agricultural pests. They are pests of agriculture (in the narrow sense, that is of field crops, such as cereals, sugarcane and crop roots), pests of horticulture (of fruit trees, fruits and vegetables generally, flowers, shrubs and ornamental trees), pests of forestry in that they attack the trees being grown for timber or providing cover for reservoir catchment areas, and pests of animals and birds reared for food or skins, etc.

Agriculture is not an extensive industry in Hong Kong because primarily of the shortage of suitable land. There is in total only about 32,000 acres under cultivation, but a sizable population of pigs, ducks and chickens is maintained under conditions of intensive production. But if the pests of ornamentals (flowers, shrubs, trees) and shade trees as well as forest trees are included, then the study is an extensive one.

It should however be pointed out that many of the common local insects are in fact major crop pests in other parts of the world whereas here they are usually on alternative (sometimes wild) hosts and often not causing appreciable damage. For example several of our commonest moths are major rice pests (*Cnaphalocrocis, Mythimna, Spodoptera, Chilo*, etc.) but on the Hong Kong Island they are feeding on wild grasses. So the Hong Kong list of international crop pests is formidable in length but deceptive, as relatively little damage is done locally.

(9) SYMBIOTIC INSECTS

In an ecological study of the habitats of Hong Kong it is interesting to consider the different associations that occur between insects and insects, insects and man, and insects and plants, where either another insect, man (or animal) or plant is the host organism and as such constitutes a microhabitat for the parasitic (or symbiotic) insect. A wide range of associations of insects with another organism is evident in a series of different types of relationship.

Recently there has been a tendancy to redefine these different associations as *symbiosis :* 'a condition in which two organisms live together, sometimes in beneficial partnership'. This is now the all-embracive term which includes within its scope the following relationships (it used to be a synonym for mutualism):

(a) Commensalism—an organism *living* with another and often sharing the food; the one species thus benefiting from the association, but the other is not harmed in any way.

(b) Inquilinism—usually refers to an animal living in the *nest* or home of another, and obtaining a share of its food; it can be regarded as a form of commensalism.

(c) Mutualism—A relationship between two organisms in which *both* parties derive a definite *advantage ;* it can in fact be complete physiological interdependence.

(d) Parasitism—refers to an organism living on or within another, to its *own advantage* in food or shelter. Previously it is used to be stated that the parasite actually injured the host in some way, but recent studies have revealed that many well-established parasites in fact do not cause detectable harm to their hosts. Of course the less well adapted parasites do still cause their host damage and in some cases even death.

The above definitions are in some cases very subjective in application and some situations are difficult to classify.

So far as parasitism is concerned, this relationship can be sub-divided into a number of different categories, such as, ectoparasite and endoparasite, obligatory or facultative, permanent or temporary, and so forth. In recent years there has been a tendency to classify some predacious insects such as mosquitoes and assassin bugs as micro-predators rather than temporary ectoparasites; this is a moot distinction!

Because of their inclusion in the previous section, gall-making insects are only briefly mentioned by group. Most insects are phytophagous and feed on plants, and they show varying levels of host specificity, from complete monophagy through oligophagy to extensive polyphagy. All these plant-eating insects could be regarded as being plant parasites of one form or another, but clearly such a treatment would not be either desirable or feasible in such a book as this.

Finally, without delving into detail unduly, the pollination of entomophilous flowers by insects can be regarded as a form of mutualism since both parties benefit from the activity.

CHAPTER THREE

COMMON INSECTS IN HONG KONG LISTED BY HABITATS

SPECIES	COMMON NAME	FAMILY	REMARKS	ABUN-DANCE*

(1) SOIL AND LITTER INSECTS

Order **Thysanura**

SPECIES	COMMON NAME	FAMILY	REMARKS	ABUNDANCE
Machilis spp.	Woodland Bristletails	Machilidae	Large eyes; litter	C
Several species	Bristletails	Lepismatidae	Small eyes; litter	C

Order **Collembola**

SPECIES	COMMON NAME	FAMILY	REMARKS	ABUNDANCE
Several species	Springtails	Sminthuridae	Round body; litter	C
Hypogastrura communis	—	Hypogastruridae	Bodies more elongate; more species burrow in the soil	C
Many species	—	Several families		VC

Order **Orthoptera**

SPECIES	COMMON NAME	FAMILY	REMARKS	ABUNDANCE
Brachytrupes portentosas	Big-head Cricket	Gryllidae	Agricultural soils, N.T.	U
Teloegryllus testaceus	Field Cricket	Gryllidae	Agricultural soils, N.T.	C
Gryllulus spp.	Field Crickets	Gryllidae	Burrow in soil	C
Tridactylus spp.	Litter crickets	Gryllidae	Small; in litter	C
Many species	Litter crickets	Gryllidae	Mostly in litter	VC
Gryllotalpa africana	Mole Cricket	Gryllotalpidae	Burrows in soil	C

Order **Dermaptera**

SPECIES	COMMON NAME	FAMILY	REMARKS	ABUNDANCE
Forficula spp.	Common Earwigs	Forficulidae	Mostly in litter	VC
Labidura spp.	Long-horned Earwigs	Labidurudae	Mostly in litter	C

---

* The abundance of the respective species is indicated by the following abbreviations:

    VC = very common   C = common   U = uncommon   R = rare

These designations are obviously subjective and somewhat arbitrary, but are based on many years of local experience and do serve to give an indication of the relative abundance of the different species locally.

SPECIES	COMMON NAME	FAMILY	REMARKS	ABUN-DANCE*
Opisthoplatia orientalis	Litter Cockroach	Blattidae	Wingless; litter	VC
Blattella germanica	German Cockroach	Blattidae	Small; brown; striped	U
Blattella lituricollis	Small Forest Cockroach	Blattidae	Forest litter species small; brown; active	VC
Rhabdoblatta humeralis	Field Cockroach	Blattidae	Large brownish	U
Pycnoscelis surinamensis	Litter Cockroach	Blattidae	Smallish; black	C
Nauphoeta cinerea	Speckled Cockroach	Blattidae	Smallish; brown; litter	C
Several species	Litter Cockroaches	Blattidae	Small black & brown spp.	C

Order **Phasmida**

SPECIES	COMMON NAME	FAMILY	REMARKS	ABUN-DANCE*
Datames sp.	Twig Insect	Phasmidae	Forest leaf litter	U

Order **Isoptera**

SPECIES	COMMON NAME	FAMILY	REMARKS	ABUN-DANCE*
Macrotermesbarneyi	Bark Termite	Termitidae	Largest species	VC
Capritermes fuscotibialis	Small Brown Termite	Termitidae	Small, brown, fungus feeder	U
Odontotermes formosanus	Large Brown Termite	Termitidae	Large species	U
Copotermes formosanus	Wet-wood Termite	Rhinotermitidae	Small, orange	VC
Reticulitermes affinis	—	Rhinotermitidae	Small, brown	U
R. labralis	—	Rhinotermitidae	Small, brown	U

Order **Hemiptera**

SPECIES	COMMON NAME	FAMILY	REMARKS	ABUN-DANCE*
Several species	Cicada nymphs	Cicadidae	Nymphs fossorial	VC
Prociphilus sp.	Bamboo (Root) Aphid	Pemphigidae	On roots of some trees	U
Planococcus citri	Root Mealybug	Psuedococcidae	On roots of several plants	U

Order **Thysanoptera**

SPECIES	COMMON NAME	FAMILY	REMARKS	ABUN-DANCE*
Many species	Thrips pupae	Thripidae etc.	Pupate in soil	C
Many species	Fungus Thrips	Thripidae etc.	Fungus feeders in litter	VC

Order **Neuroptera**

SPECIES	COMMON NAME	FAMILY	REMARKS	ABUN-DANCE*
Myrmeleon spp.	Ant-lions	Myrmeleonidae	Larvae in pits	VC

SPECIES	COMMON NAME	FAMILY	REMARKS	ABUN-DANCE*
		Order **Lepidoptera**		
Many species	Hawk Moths	Sphingidae	Pupate in soil	C
Several species	Loopers	Geometridae	Pupate in soil	C
Many species	Cutworms	Noctuidae	Cutworms in soil	C
Many species	Leafworms etc.	Noctuidae	Pupate in soil	VC
		Order **Diptera**		
Several species	Leatherjackets	Tipulidae	Larvae in soil	C
Several species	Sciarids	Sciaridae	Larvae in soil	C
Several species	—	Bibionidae	Larvae in soil	C
Several species	Fungus Anats	Mycetophilidae	Larvae in soil	C
Some species	Gall Midges	Cecidomyiidae	Some pupate in soil	U
Several species	Robber Flies	Asilidae	Larvae in soil	C
(Species uncertain)	Root maggots	Anthomyiidae	Larvae in soil	(?)
Several species	(Fly Maggots)	(Other families)	Larvae in soil	VC
Many species	House Flies	Muscidae	Pupate in soil	C
Many species	Bluebottles, etc.	Calliphoridae	Pupate in soil	C
		Order **Hymenoptera**		
Many species	Ants	Formicidae	Underground nests	VC
Sphex fulvohirtus	Solitary Wasp	Sphecidae	Underground nests	C
Pompilus spp.	Spider-hunting Wasps	Pompilidae	Underground nests	U
Species uncertain	Soil Bees	Halictidae	Underground nests	R
Several species	Bees	Apidae	Underground nests	U
Sedia azurea	—	Scoliidae	Parasitise scarab beetle larvae in soil	U
		Order **Coleoptera**		
Several species	Ground Beetles	Carabidae	Adults and larvae	VC
Cicindela spp.	Tiger Beetles	Cicindellidae	Larvae in soil tunnels	C
Several species	Rove Beetle	Staphylinidae	Adults and larvae	C
Protaetia spp. *Anomala* spp. *Phyllophaga* spp.	Chafer Grubs	Scarabaeidae	Larvae and pupae found in soil	C
Coprinus spp.	Dung Beetles	Scarabaeidae	Larvae coprophilous	U
Catharsius molossus	Giant Dung Beetle	Scarabaeidae	Larvae coprophilous	U
Compsosternus auratus	Large Green Click Beetle	Elateridae	Larvae are wireworms	U
Chenicera sp.	Wireworm	Elateridae	Larvae in soil	U
Megapenthes sp.	Wireworm	Elateridae	Larvae in soil	U

SPECIES	COMMON NAME	FAMILY	REMARKS	ABUN-DANCE*
Mylabris spp.	Banded Blister Beetles	Meloidae	Larvae in soil; feed on Acridid egg-pods or bee nests	U
Epicauta spp.	Striped Blister Beetles	Meloidae		
Several species	Beetles	(Other families)	Larvae in soil	C
Several species	Beetles	(Other families)	Pupate in soil	C
Several species	Weevils	Curculionidae	Larvae & pupae in soil	C
Several species	Weevils	Curculionidae	Pupate in soil	C

Order **Acarina**

SPECIES	COMMON NAME	FAMILY	REMARKS	ABUN-DANCE*
Several species	Beetle Mites	Oribatidae	All stages in soil	VC
Several species	Predatory Mites	Parasitidae	All stages in soil	C

(2) GRASSLAND INSECTS

Order **Orthoptera**

SPECIES	COMMON NAME	FAMILY	REMARKS	ABUN-DANCE*
Acrida turrita	Large Green Grasshopper	Acrididae	Widespread; large	C
Acrida spp.	—	Acrididae	Widespread; small	VC
Gonista sp.	—	Acrididae	Widespread; small	C
Attractomorpha spp.	—	Acrididae	Widespread; larger	VC
Patanga succincta	Bombay Locust	Acrididae	Hillsides mostly	C
Gastrimargus marmoratus	Common Green Grasshopper	Acrididae	Main birdshop species	VC
Geracris spp.	—	Acrididae	Widespread; small	C
Tristria spp.	—	Acrididae	Widespread; small	C
Oxya sinensis	Small Rice Grasshopper	Acrididae	On paddy rice mostly	VC
Many other species	Grasshoppers	Acrididae	Widespread; on grasses	VC

Order **Dictyoptera**

SPECIES	COMMON NAME	FAMILY	REMARKS	ABUN-DANCE*
Opisthoplatia orientalis	Field Cockroach	Blattidae	Mostly in litter	VC
Some spp.	Cockroaches	Blattidae	Small species	C
Tenodera sp.	Large Brown Mantid	Mantidae	Lowland areas usually	C

Order **Homoptera**

SPECIES	COMMON NAME	FAMILY	REMARKS	ABUN-DANCE*
Pauropsalta sp.	Grass Cicada	Cicadidae	On grasses	U
Mogannia spp.	Grass Cicadas	Cicadidae	On grasses	C
Cicadella spp.	Leafhoppers	Cicadellidae	On Gramineae	C
Bothrogonia sp.	Large Red Leafhopper	Cicadellidae	On *Miscanthus*, etc.	C

SPECIES	COMMON NAME	FAMILY	REMARKS	ABUN-DANCE*
Nephotettix spp.	Green Leaf-hoppers	Cicadellidae	On Gramineae	VC
Clovia spp.	Spittlebugs	Cercopidae	On grasses	C
Machaerota coronata	Tubicolous Spittlebug	Cercopidae	On *Helicteres angustifolia*	C
Phenacaspis dendrobii	White Grass Scale	Diaspididae	On ornametal grass	U
Several spp.	Armoured Scales	Diaspididae	Not widespread	U
Ricania sp.	Brown Planthopper	Ricaniidae	On *Miscanthus*	C
Sogatella furcifera	Brown Ricaniid Planthopper	Delphacidae	On rice & grasses	C
Inazuma dorsalis	Zig-zag Winged Planthopper	Delphacidae	On rice & grasses	C
Several spp.	Planthoppers	Delphacidae	On Gramineae	VC
Livia nigra	Juncus Psyllid	Psyllidae	On *Juncus prismatocarpus;* inflorescence galls	U
Longiuguis sacchari	Sugarcane Aphid	Pemphigidae	On Miscanthus	U
Macrosiphum spp.	Grass Aphids	Aphididae	On Gramineae	C
Rhopalosiphum maidis	Maize Aphid	Aphidae	On Gramineae	U

Order **Heteroptera**

SPECIES	COMMON NAME	FAMILY	REMARKS	ABUN-DANCE*
Nezara viridula	Green Shield Bug	Pentatomidae	Polyphagous	C
Scotinophara spp.	Black Paddy Bugs	Pentatomidae	Gramineae	U
Megarrhamphus hastatus	Miscanthus Shield Bug	Pentatomidae	Grasses only	U
Leptocorisa acuta	Paddy Bug	Coreidae	Rice & grasses	VC
Cletus spp.	Grass Bugs	Coreidae	Many grasses	C

Order **Neuroptera**

SPECIES	COMMON NAME	FAMILY	REMARKS	ABUN-DANCE*
Myrmeleon spp.	Ant-lions	Myrmeleontidae	Adults only	C

Order **Thysanoptera**

SPECIES	COMMON NAME	FAMILY	REMARKS	ABUN-DANCE*
Several spp.	Grass Thrips	Thripidae	Various grasses	U
Several spp.	Fungus Thrips	Thripidae	Grass tussocks	VC

Order **Lepidoptera**

SPECIES	COMMON NAME	FAMILY	REMARKS	ABUN-DANCE*
Several spp.	Grass Moths	Crambidae	Larvae eat leaves	U
Chilo spp.	Graminaceous Stem-borers	Pyralidae	Gramineae	C
Tryporyza spp.	Graminaceous Stem-borers	Pyralidae	Rice & grasses	C

SPECIES	COMMON NAME	FAMILY	REMARKS	ABUN-DANCE*
Sesamia spp.	Stem Borers	Noctuidae	Thicker grasses	U
Melanitis spp.	Brown Butterflies	Satyridae	Larvae eat grasses	C
Mycalesis spp.	Small Browns	Satyridae	Larvae eat grasses	C
Ypthima baldus	Common Five-ring	Satyridae	*Miscanthus* spp.	C
Parnara guttatus	Common Straight Swift	Hesperiidae	Larvae eat grasses	VC
Pelopidas spp.	Small Brown Skippers	Hesperiidae	Larvae eat grasses	VC
Ampittia dioscorides	Bush Hopper	Hesperiidae	Larvae eat grasses	VC
Suastus gremius	Indian Palm Bob	Hesperiidae	Larvae on *Phoenix hanceana*	U
Several spp.	Skippers	Hesperiidae	Larvae eat grasses & bamboo leaves	C
Zizeeria maha	Bush Blue	Lycaenidae	Flies in open grassland	C
Amictoides sp.	Grass Bagworm	Psychidae	Bags found on grasses in N.T.	U

Order **Hymenoptera**

SPECIES	COMMON NAME	FAMILY	REMARKS	ABUN-DANCE*
Polyrhachis dives	Black Tree Ant	Formicidae	Nests in tall grass	C
Several spp.	Ants	Formicidae	Some nests in grass	C
Sphex fulvohirtus	Solitary Wasp	Sphecidae	Nest burrows in grassland	C
Pompilus spp.	Spider-hunting Wasps	Pompilidae	Nest burrows in grassland	C

Order **Diptera**

SPECIES	COMMON NAME	FAMILY	REMARKS	ABUN-DANCE*
Ctenophra flavibasis	Crane Fly	Tipulidae	Adults only	C
Holorusia spp.	Common Crane Flies	Tipulidae	Adults only	C
Trypanoides indigenus	Robber Fly	Asilidae	Adults only	C
Cholorops sp.	Grass Stem Fly	Chloropidae	Swollen shoot galls	C
Hydrellia griseola	Rice Leaf Miner	Ephydridae	Mine grass leaves	C
Atherigona spp.	Grass Shoot Flies	Muscidae	Grass shoots	C

Order **Coleoptera**

SPECIES	COMMON NAME	FAMILY	REMARKS	ABUN-DANCE*
Several spp.	Ground Beetles	Carabidae	Adults only	C
Cicindela spp.	Tiger Beetles	Cicindelidae	Adults only	C
Several spp.	Rove Beetles	Staphylinidae	Adults only	C
Monolepta signata	Flea Beetle	Chrysomelidae	Adults only; on leaves	U
Colasposoma metallicum	Black Leaf Beetle	Chrysomelidae	Adults only; on leaves	C
Dicladispa spp.	Hispid Beetles	(Hispinae)	Larvae mine leaves	U

SPECIES	COMMON NAME	FAMILY	REMARKS	ABUNDANCE*
Protaetia orientalis	Green Rose Chafer	(Cetoniinae)	Adults only	C
Phyllogphaga spp.	Cockchafers	(Melolonthinae)	Adults only	C
Melolontha serrulata	Large Brown Cockchafers	(Melolonthinae)	Adults only	C
Holotricha geilenkenseri	Large Brown Cookchafers	(Melolonthinae)	Adults only	C
Anomala spp.	Flower Beetles	(Rutelininae)	Adults only	C
Catharsius molossus	Giant Dung Beetle	(Coprininae)	Coprophilous	U
Coprinus spp.	Dung Beetles	(Coprininae)	Adults only	U
Several spp.	Click Beetles	Elateridae	Adults only	C
Mylabris cinchorii	Small Yellow-banded Blister Beetles	Meloidae	Adults on herbaceous flowers	C

Order **Acarina**

Various grass-flower mites, and mammal parasitic ticks are to be expected to occur in local grassland but have not yet been searched for or noticed.

(3) SCRUBLAND INSECTS

Order **Orthoptera**

Calyptotrypus sp.	Bush Cricket	Gryllidae (Eneopterinae)	Widespread; arboreal	C
Some species	Long-horned Grasshoppers	Tettigoniidae	Arboreal in bushes	C

Order **Dictyoptera**

Tenodera sp.	Large Brown Mantid	Mantidae	Widespread	C
Onychostylus vilis	Arboreal Cockroach	Blattidae	Arboreal in bushes	U

Order **Hemiptera**

Mogannia spp.	Grass Cicadas	Cicadidae	On grasses & shrubs	C
Ceroplastes sp.	Pink Waxy Scale	Coccidae	On *Rhodomyrtus* twigs	C
Gascardia sp.	White Waxy Scale	Coccidae	On *Eurya* & *Rhodomyrtus*	C
Marsipococcus sp.	Soft Brown Scale	Coccidae	On *Rhodomyrtus* leaves	C
Machaerota coronata	Tubicolous Spittlebug	Cercopidae	On *Helicteres angustfolia*	C
Cosmocarta abdominalis	Melastoma Spittlebug	Cercopidae	Larvae on *Melastoma*	VC

SPECIES	COMMON NAME	FAMILY	REMARKS	ABUN-DANCE*
		Order **Heteroptera**		
Mictis tenebrosa	Large Brown Coreid Bug	Coreidae	On shrub shoots	C
		Order **Thysanoptera**		
Sp. indet.	Thrips	Thripidae	*Gardenia* flowers	C
		Order **Lepidoptera**		
Papilio demoleus	Lemon Butterfly	Papillionidae	On *Atalantia buxifolia*	VC
Chilades lajus	Lime Blue	Lycaenidae	On *Atalantia buxifolia*	C
Potanthus confucius	Common Dart	Hesperiidae	On *Bambusa multiplex*	C
Abisara echerius	—	Riodinidae	In scrub areas and near woods	C
Trabala vishnou	Rose Myrtle Lappet Moth	Lasiocampidae	On leaves of *Rhodomyrtus* and *Melastoma*	C
		Order **Coleoptera**		
Labidostomis sp.	Leaf Beetle	Chrysomelidae	On *Lespedeza*	C
Cryptocephalus trifasciatus	Leaf Beetle	Chrysomelidae	On *Lespedeza*	C
Hespera sp.	Flea Beetle	Chrysomelidae	*Melastoma* flowers	
Nodina chalcosoma	Flower Leaf Beetle	Chrysomelidae	*Rhodomyrtus* flowers	C
Popillia histeroides	Black Flower Beetle	Scarabaeidae	*Melastoma* flowers	C
		Order **Acarina**		
Eriophyes sp.	Schlefflera Gall Mite	Eriophyidae	Galls *Schlefflera* leaves	VC

(4) FOREST INSECTS

SPECIES	COMMON NAME	FAMILY	REMARKS	ABUN-DANCE*
		Order **Collembola**		
Many spp.	Arboreal Springtails	Many families	On foliage & dead wood	VC
		Order **Odonata**		
Many spp.	Dragonflies	Anisoptera	Fly in lee of tall trees	VC
Several spp.	Damsel-flies	Zygoptera	Hunt in woods	VC
		Order **Orthoptera**		
Conocephalus spp.	Point-headed Grasshoppers	Tettigoniidae	Arboreal; cryptic	VC

SPECIES	COMMON NAME	FAMILY	REMARKS	ABUN-DANCE*
Euconocephalus spp.	Point-headed Grasshoppers	Tettigoniidae	Arboreal & cryptic	VC
Mecopoda elongata	Large Leaf Grasshopper	Tettigoniidae	Very cryptic; often brown	C
Holochlora sp.	Smaller Leaf Grasshopper	Tettigoniidae	Cryptic; usually green	U
Sympaestria spp.	Long-horned Grasshopper	Tettigoniidae	Arboreal	U
Letana sp.	Long-horned Grasshopper	Tettigoniidae	Arboreal; small species	U
Elimaea punctifera	Long-horned Grasshopper	Tettigoniidae	Arboreal	C
Pseudorhynchus sp.	Long-horned Grasshopper	Tettigoniidae	Arboreal	C
Several spp.	Long-horned Grasshopper	Tettigoniidae	Arboreal	VC
Gryllus spp.	Crickets	Gryllidae	Terrestrial	C
Acheta spp.	Crickets	Gryllidae	Terrestrial	C
Oecanthus longicauda	Tree Cricket	Gryllidae (Oecanthinae)	Arboreal	U
Calyptotrypus sp.	Bush Cricket	Gryllidae (Eneopterinae)	Arboreal	C
Tetrix spp.	Grouse Locusts	Tetrigidae	On woodland paths	C

Order **Dictyoptera**

SPECIES	COMMON NAME	FAMILY	REMARKS	ABUN-DANCE*
Opisthoplatia orientalis	Litter Cockroach	Blattidae	Terrestrial mostly	VC
Onychostylus vilis	Arboreal Cockroach	Blattidae	All stages arboreal	C
Rhabdoblatta humeralis	Arboreal Cockroach	Blattidae	Arboreal	C
Blatella lituricollis lituricollis	Small Forest Cockroach	Blattidae	Litter species	VC
Hierodula patellifera	Large Green Mantid	Mantidae	Arboreal	C
Tenodera sp.	Large Brown Mantid	Mantidae	Arboreal	U
Mantis religiosa	Small Green Mantidae	Mantidae	Arboreal	C
Statilia maculata	Small Brown Mantid	Mantidae	Arboreal	C
Acromantis sp.	Small Brown Mantid	Mantidae	Arboreal	U

SPECIES	COMMON NAME	FAMILY	REMARKS	ABUN-DANCE*
Spilomantis occipitalis	Tiny Brown Mantid	Mantidae	More terrestrial	C
Leptomantella sp.	Leaf Mantid	Mantidae	Green; flat on leaves	U

Order **Phasmida**

SPECIES	COMMON NAME	FAMILY	REMARKS	ABUN-DANCE*
Entoria sp.	Stick Insect	Phasmidae	Arboreal in foliage	C
Datames sp.	Twig Insect	Phasmidae	Usually on ground	U

Order **Dermaptera**

SPECIES	COMMON NAME	FAMILY	REMARKS	ABUN-DANCE*
Forficula spp.	Earwigs	Forficulidae	Under bark; litter	VC

Order **Isoptera**

SPECIES	COMMON NAME	FAMILY	REMARKS	ABUN-DANCE*
Macrotermes barneyi	Bark Termite	Termitidae	Largest species; eat tree bark	VC
Capritermes fuscotibialis	—	Termitidae	Ground nester	U
Odontotermes formosanus	Large Brown Termite	Termitidae	Large species	U
Coptotermes formosanus	Wet-wood termite	Rhinotermitidae	Nest in wet wood	VC
Reticulotermes fukienensis	—	Rhinotermitidae	Nest in old tree stumps	U
Cryptotermes brevis	Dry-wood Termite	Kalotermitidae	Nest in dry wood	R

Order **Embioptera**

SPECIES	COMMON NAME	FAMILY	REMARKS	ABUN-DANCE*
Oligotoma spp.	Web-spinners	Oligotomidae	} On tree trunks and	VC
Aposthonia spp.	Web-spinners	Oligotomidae	rock surfaces	VC

Order **Psocoptera**

SPECIES	COMMON NAME	FAMILY	REMARKS	ABUN-DANCE*
Several spp.	Web-spinning Psocids	(Several families)	Webs on tree trunks and rocks	C
Several spp.	Bark-lice	(Several families)	Solitary in foliage	C

Order **Homoptera**

SPECIES	COMMON NAME	FAMILY	REMARKS	ABUN-DANCE*
Tacua sp.	Giant Brown Cicada	Cicadidae	Seldom seen	R
Gaeana maculata	Spotted Black Cicada	Cicadidae	Occurs March/April	VC
Cryptotympana mimica	Large Brown Cicada	Cicadidae	Occurs April/June	VC
Dundubia sp.	Green Clearwing Cicada	Cicadidae	Occurs April/July	C

SPECIES	COMMON NAME	FAMILY	REMARKS	ABUN-DANCE*
Platypleura hilpa	Brown Speckled Cicada	Cicadidae	Occurs April/July	C
Scieroptera sanguinea	Red-nosed Cicada	Cicadidae	Occurs April/July, & October	C
Typhlocyba spp.	White Leafhoppers	Cicadellidae	On leaves of trees	C
Coelidia sp.	Citrus Leafhopper	Cicadellidae	On leaves of trees	C
Bothrogonia sp.	Large Red Leafhopper	Cicadellidae	Under leaves of trees, etc.	C
Cosmocarta abdominalis	Red-banded Froghopper	Cercopidae	Adults feed on trees	C
Pyrops candelaria	Longan Lantern-fly	Fulgoridae	On Litchi & Longan	C
Pyrops lathburii	Eucalyptus Lantern-fly	Fulgoridae	On *Eucalyptus*	R
Paurohita fuscovenosa	Bamboo Planthopper	Delphacidae	Large bamboo species	U
Dictyophara sp.	—	Dictyopharidae	On trees	C
Tricentrus spp.	Tree-hoppers	Membracidae	On foliage twigs	C
Kallitaxila macaona	—	Tropiduchidae	On leaves of trees	U
Salurnis marginellus	Small Green Moth Bug	Flattidae	In foliage	U
Pulastya discolorata	Large Green Moth Bug	Flattidae	In foliage	C
Ricania speculum	B & W Ricaniid	Ricaniidae	In foliage	U
Coccus spp.	Soft Green Scales	Coccidae	On leaves & twigs	VC
Pulvinaria spp.	Soft Brown Scales	Coccidae	On leaves & twigs	VC
Saisettia coffeae	Hemispherical scale	Coccidae	On twigs mostly	VC
Saisettia oleae	Black (Olive) Scale	Coccidae	On twigs mostly	C
Ceroplastes rubens	Pink Waxy Scale	Coccidae	On twigs & leaves	VC
Ceroplastes sp.	White Waxy Scale	Coccidae	On twigs & leaves	C
Pseudococcus citriculus	Long-tailed Mealybug	Psuedococcidae	In shoots; on foliage	VC
Saccharicoccus sp.	Sugarcane Mealybug	Psuedococcidae	Bamboo leaf sheaths	C
Pseudococcus spp.	Mealybugs	Pseudococcidae	In shoots; on twigs	VC
Pseudaulacaspis spp.	Mussel Scales	Diaspididae	On leaves	C
Aspidiotus destructor	Transparent Scale	Diaspididae	On twigs & leaves	U
Lepidosaphes spp.	Mussel Scales	Diaspididae	On twigs mostly	VC
Several spp.	Armoured Scales	Diaspididae	On twigs & leaves	C

SPECIES	COMMON NAME	FAMILY	REMARKS	ABUN-DANCE*
Icerya purchasi	Cottony Cushion Scale	Margarodidae	Twigs usually	VC
Icerya aegyptiaca	Egyptian Fluted Scale	Margarodidae	On *Litsea rotundifolia*	C
Drosicha sp.	Giant Mealybug	Margarododae	On leaves & twigs	U
Aleyrodes lonicerae	Oxalis Whitefly	Aleyrodidae	On *Oxalis* leaves	C
Several spp.	White/Black-flies	Aleyrodidae	Leaves of trees & herbaceous plants	C
Macrohomotoma striata	Fig Shoot Psyllid	Psyllidae	*F. microcarpa* shoots	VC
Pauropsylla udei	Fig Gall Psyllid	Psyllidae	*F. variegata* leafgall	VC
Paurocephala sp. nov.	Fig Leaf Psyllid	Psyllidae	*F. hispida* leaves	C
Trioza camphorae	Camphor Psyllid	Psyllidae	Camphor leaf-pits	VC
Megatrioza vitiensis	Syzygium Psyllid	Psyllidae	*Szygium* leaf-pits	VC
Psylla fatsiae	Shefflera Psyllid	Psyllidae	Shoots of *Shefflera*	C
Tyara sp.	Sterculia Psyllid	Psyllidae	Shoots of *Sterculia*	C
Cinara spp.	Large Black Aphids	Aphididae	On *Pinus* species	C
Myzus persicae	Green Peach Aphid	Aphididae	Polyphagous	VC
Aphis craccivora	Groundnut Aphid	Aphididae	Polyphagous	VC
Aphis gossypii	Melon (Cotton) Aphid	Aphididae	Malvaceae mostly	VC
Aiceona litseae	Litsea Aphid	Aphididae	On *Litsea* spp.	C
Toxoptera aurantii	Black Citrus Aphid	Aphididae	Rutaceae, *Ficus,* etc.	VC
Greenidea ficicola	Fig Tree Aphid	Aphididae	*Ficus* species	C
Several spp.	Aphids	Aphididae	Various hosts	C
Eriosoma spp.	Woolly Aphids	Pemphigidae	Various hosts	U
Chaitoregma sp.	Woolly Aphid	Pemphigidae	Stem galls on privet	C
Pseudoregma bambusicola	Bamboo Woolly Aphid	Pemphigidae	On large Bamboos	C

Order **Heteroptera**

SPECIES	COMMON NAME	FAMILY	REMARKS	ABUN-DANCE*
Erthesina fullo	Tallow Stink Bug	Pentatomidae	Widespread	VC
Nezara viridula	Green Shield Bug	Pentatomidae	Polyphagous	VC
Several spp.	Shield Bugs	Pentatomidae	Various hosts	C
Leptoglossus spp.	Leaf-footed Bugs	Coreidae	Various hosts	C
Paradasynus spinosus	Green Coreid Bug	Coreidae	Various hosts	C
Acanthocoris scabrator	Brown Coreid Bug	Coreidae	Various hosts	C

SPECIES	COMMON NAME	FAMILY	REMARKS	ABUN-DANCE*
Notobitus meleagris	Bamboo Bug	Coreidae	On large Bamboos	VC
Leptocoris spp.	Red Coreid Bug	Coreidae	Bright red/black	C
Several spp.	—	Anthocoridae	Predacious bugs	C
Sycanus croceovittatus	Large Black Assassin Bug	Reduviidae	Predacious bug	C
Triatoma sp.	Assassin Bug	Reduviidae	Predacious bug	U
Oncocephalus sp.	Assassin Bug	Reduviidae	Predacious bug	U
Several spp.	Assassin Bugs	Reduviidae	Predacious bugs; small	U

Order **Thysanoptera**

SPECIES	COMMON NAME	FAMILY	REMARKS	ABUN-DANCE*
Gynaikothrips ficorum	Common Fig Thrips	Phlaeothripidae	Fold *F. microcarpa* leaves	VC
Gynaikothrips kuwanai	Aporusa Thrips	Phlaeothripidae	*Aporusa chinensis* leaves	VC
Gigantothrips elegans	Giant Fig Thrips	Phlaeothripidae	Under *F. microcarpa* leaves	C
Several spp.	Thrips	Thripidae	Various hosts	C
Many spp.	Fungus Thrips	Several Families	On dead branches etc.	C

Order **Neuroptera**

SPECIES	COMMON NAME	FAMILY	REMARKS	ABUN-DANCE*
Myrmeleon spp.	Ant-lions	Myrmeliontidae	Adults; larvae in dry sheltered soil	C
Chrysopa spp.	Green Lacewings	Chrysopidae	All stages	C
Hemerobius spp.	Brown Lacewings	Hemerobiidae	All stages	C
Hybris subjacens	—	Ascalaphidae	All stages (?)	U
Sp. indet.	—	Coniopterygidae	All stages on foliage	U

Order **Lepidoptera**

SPECIES	COMMON NAME	FAMILY	REMARKS	ABUN-DANCE*
Indarbela disciplaga	Wood Borer Moth	Metarbelidae	Larvae eat tree bark	VC
Conopis sp.	Camphor Clearwing	Sessidae	Larvae bore trunk of camphor	U
Zeuzera coffeae	Red Coffee Borer	Cossidae	Larvae bore branches of trees and shrubs	C
Several spp.	Goat Moths	Cossidae	Larvae bore branches	U
Phyllocnistis spp.	Leaf-miners	Gracillariidae	Larvae mine leaves	C
Cydia pulverula	Fig Budworm	Tortricidae	*F. microcarpa* buds	U
Several spp.	Budworms	Tortricidae	Larvae bore buds	C
Clania spp.	Bagworms	Psychidae	Hanging larval cases on trees	VC
Parasa lepida	Stinging Caterpillar	Limacodidae	Spiny larvae on foliage	C
Thosea sinensis	Slug Caterpillar	Limacodidae	Lumpy larvae on foliage	C

SPECIES	COMMON NAME	FAMILY	REMARKS	ABUN-DANCE*
Cyclosia papilionaris	Black-veined Moth	Zygaenidae	Larvae on foliage	U
Several spp.	Day-flying Moths	Zygaenidae	Various host plants	C
Sylepta derogata	Cotton Leaf-roller	Pyralidae	Malvaceae leaves	C
Several spp.	—	Pyralidae	Various hosts	VC
Many spp.	Leafworms, etc.	Noctuidae	Various hosts	C
Attacus atlas	Atlas Moth	Saturniidae	Larvae on camphor, *Sapium, Schefflera*, etc.	U
Samia cynthia	Lesser Atlas Moth	Saturniidae	Larvae on camphor, *Sapium,* etc.	C
Arctias selene	Moon Moth	Saturniidae	Larvae on trees, camphor, etc.	C
Eriogyna pyretorum	Indian Silk Moth	Saturniidae	Larvae on camphor trees	C
Several species	Emperor Moths	Saturniidae	Tree defoliators	C
Trichola ficicola	Fig Silk Moth	Bombycidae	Larvae on *Ficus microcarpa* leaves	C
Danais spp.	Milkweed Butterflies	Nymphalidae	Larvae eat leaves	C
Euploea spp.	Crow Butterflies	Nymphalidae	Larvae eat leaves	C
Charaxes polyxena	Tawny Rajah	Nymphalidae	Larvae eat camphor leaves	C
Neptis spp.	Sailor Butterflies	Nymphalidae	Larvae eat leaves	C
Hypolimnas spp.	Egg-flies	Nymphalidae	Larvae eat leaves	C
Precis spp.	Pansy Butterflies	Nymphalidae	Larvae eat leaves	C
Many spp.	Four-footed Butterflies	Nymphalidae	Larvae eat leaves	C
Melanitis spp.	Woodland Browns	Nymphalidae	Shade lovers (ads); larvae on *Setaria palmifolia*	C
Faunis eumeus	Common Faun	Nymphalidae	Shade lovers; larvae on *Smilax*	C
Mycalesis spp.	Small Eyed-Browns	Nymphalidae	Woodland species	C
Hebomoia glaucippe	Great Orange Tip	Pieridae	Larvae on trees	C
Delias parsithoe	Common Black Jezebel	Pieridae	On *Loranthus parasiticus* and *Dendrotrophe*	C
Pieris spp.	White Butterflies	Pieridae	On herbs mostly	C
Catopsila spp.	Migrant Butterflies	Pieridae	Larvae eat leaves of *Cassia* spp.	C
Eurema spp.	Small Yellows	Pieridae	*Cassia, Cratoxylum* and Leguminosae mostly	C
Many spp.	Small Blues	Lycaenidae	Various hosts	C
Chilasa clytia	Common Mime	Papilionidae	*Litsea glutinosa* leaves	VC
Papilio polytes	Common Mormon	Papilionidae	*Citrus,* etc.	C

SPECIES	COMMON NAME	FAMILY	REMARKS	ABUN-DANCE*
Papilio demoleus	Lemon Butterfly	Papilionidae	*Citrus, Atalantia, Fortunella* leaves	VC
Papilio memnon	Great Mormon	Papilionidae	*Citrus*, etc.	C
Papilio paris	Paris Peacock	Papilionidae	*Citrus*, etc.	U
Papilio helenus	Red Helen	Papilionidae	*Citrus*, etc.	C
Graphium sarpedon	Blue Triangle	Papilionidae	*Michelia alba*	U
Several spp.	Swallow-tails	Papilionidae	Various hosts	C
Telicota ohara	Dark Darter	Hesperiidae	*Setaria palmifolia*	C
Borbo cinnara	Formosan Swift	Hesperiidae	*Setaria palmifolia*	U
Several spp.	Skippers	Hesperiidae	Adults feeding; larvae on bamboo shoots & monocots	C
Abraxas sp.	Magpie Moth	Geometridae	Adults resting in foliage	C
Several spp.	Looper Caterpillars	Geometridae	Larvae eat leaves; adults rest on trees	C
Nyctalaemon sp.	Tailed Moth	Uranidae	Large; nocturnal	U
Macroglossa belis	Humming-bird Hawk	Sphingidae	Adults diurnal	U
Macroglossa saga	—	Sphingidae	Adults semi-diurnal	C
Hemeris tityus	Narrow-bordered Bee Hawk	Sphingidae	Adults diurnal	U
Hyloicus pinastri	Pine Hawk Moth	Sphingidae	Larvae on *Pinus* leaves	U
Hyles lineata	Silver-striped Hawk	Sphingidae	Larvae polyphagous	C
Agrius convolvuli	Convolvulus Hawk	Sphingidae	On Convolvulaceae	VC
Panacra mydon	Alocasia Hawk Moth	Sphingidae	Larvae on *Alocasia*	C
Several spp.	Hawk Moths	Sphingidae	Various hosts	C
Dysphania militaris	Common Tiger Moth	Arctiidae	Woolly caterpillars	C
Obeidia tigrata	Brown Tiger Moth	Arctiidae	Woolly caterpillars	U
Several spp.	Small Tiger Moths	Arctiidae	Various host plants	U
Syntomis sperbius	Yellow-banded Wasp Moth	Amatidae	Widespread	VC
Syntomis polymita	Brown Wasp Moth	Amatidae	Widespread	VC
Perina nuda	Banyan Tussock Moth	Lymantriidae	*F. microcarpa* leaves	VC
Several spp.	Tussock Moths	Lymantriidae	Various hosts	C

SPECIES	COMMON NAME	FAMILY	REMARKS	ABUN-DANCE*
		Order **Diptera**		
Aëdes albopictus	B & W House Mosquite	Culicidae	Breed in tree holes, tin cans, etc.	VC
Toxorhynchitis splendens	Predatory Mosquito	Culicidae	Predatory on former	C
Culex sp.	—	Culicidae	Adults only	VC
Anopheles spp.	Walarial Mosquitoes	Culicidae	Adults only	U
Asphondylia morindae	Gall Midge	Cecidomyiidae	Galls *Aporusa chinensis*	VC
Several spp.	Gall Midges	Cecidomyiidae	Plant galls	C
Bradysia sp.	Dark-winged Fungus Gnats	Sciaridae	Leaf galls	C
Several spp.	Fungus Gnats	Mycetophilidae	Adults only	C
Tabanus spp.	Horse Flies	Tabanidae	Adults bloodsucking	C
Stratiomya sp.	Soldier Flies	Stratiomyidae	Adults aerial	VC
Ligyra tantalus	Bee Fly	Bombylinidae	Adults at flowers	U
Anthrax sp.	Bee Fly	Bombylinidae	Adults at flowers	U
Trypanoides indigenus	Robber Fly	Asilidae	Adults hunt in woods	U
Several spp.	Robber Flies	Asilidae	Adults hunt in woods	C
Ctenophora flavibasis	Yellow-banded Cranefly	Tipulidae	Adults only	C
Hexatoma sp.	Black-winged Cranefly	Tipulidae	Adults only	U
Holorusia spp.	Common (Brown) Craneflies	Tipulidae	Adults only	VC
Eristalis spp. *Syrphus* spp. *Xanthogramma* spp.	Hover flies	Syrphidae	Adults & larvae	VC
Dacus cucurbitae	Melon Fly	Tephritidae	Breed in fruits	C
Dacus spp.	Brown Fruit Flies	Tephritidae	Breed in fruits	C
Ceratitis spp.	Mottled-winged Fruit Flies	Tephritidae	Breed in fruits	U
Agromyza spp.	Leaf-miners	Agromyzidae	Larvae mine leaves	C
Phytomyza spp.	Leaf-miners	Agromyzidae	Larvae mine leaves	C
Ophiomyia fici	Fig Leaf-gall Miner	Argomyzidae	Leaf-galls in *Ficus microcarpa*	VC
Drosophila spp.	Vinegar Flies	Drosophilidae	Larvae in overipe fruit	C
Chrysomya megacephala	Bluebottle	Calliphoridae	Larvae saprozoic	C
Calliphora spp.	Bluebottles	Calliphoridae	Larvae saprozoic	VC
Lucilia spp.	Greenbottles	Calliphoridae	Larvae saprozoic	VC

SPECIES	COMMON NAME	FAMILY	REMARKS	ABUNDANCE*
Sarcophaga spp.	Flesh Flies	Sarcophagidae	Larvae saprozoic	VC
Several spp.	Parasitic Flies	Tachinidae	Parasitize caterpillars	VC
Musca spp.	House Flies	Muscidae	Larvae saprozoic	VC
Several spp.	Flies	Muscidae	Larvae saprozoic	C

Order **Hymenoptera**

SPECIES	COMMON NAME	FAMILY	REMARKS	ABUNDANCE*
Arge pagana	Rose Sawfly	Argidae	Larvae on *Rosa* leaves	U
Hemichroa sp.	Camphor Sawfly	Tenthredinidae	Larvae on Camphor	U
Many spp.	Ichneumons	Ichneumonidae	Insect parasites	VC
Many spp.	Braconids	Braconidae	Insect parasites	VC
Saphonecrus sp.	Gall Wasp	Cynipidae	Galls on oak tree flowers	U
Blastophaga spp.	Fig-wasps	Agaonidae	*Ficus* pollinators	VC
Ceratosolen spp.	Fig-wasps	Agaonidae	*Ficus* pollinators	VC
Podagrion spp.	Mantis Wasps	Torymidae	Mantid egg parasite	C
Many spp.	Fig-wasp parasites	Torymidae	Fig-wasp parasites	VC
Ormyrus sp.	Fig Leaf Gall Wasp	Ormyridae	Leaf Galls on *Ficus microcarpa*	C
Ormyrus sp.	Oak Gall Wasp	Ormyridae	Galls on oak flowers	U
Pteromalus puparum	Chrysalis Pteromalid	Pteromalidae	Chrysalis parasite	C
Pachyneuron spp.	Parasitic Wasps	Pteromalidae	Aphid parasites	C
Eudecatoma spp.	Gall Wasps	Eurytomidae	Plant gall-wasps	C
Eurytoma spp.	Parasitic Wasps	Eurytomidae	Insect parasites	C
Sycophila spp.	Fig-wasp 'Parasites'	Eurytomidae	Fig-wasp 'parasites'	C
Tetrastichus spp.	Parasitic Wasps	Eulophidae	Scale parasites	C
Pediobius spp.	Parasitic Wasps	Eulophidae	Aphid parasites	C
Several spp.	Parasitic Wasps	Encyrtidae	Scales & aphids	C
Scolia azurea	—	Scoliidae	Parasitize beetle larvae	U
Trogaspidea oculata	Velvet 'Ant'	Mutillidae	Wingless; parasitic	U
Polyrhachis dives	Black Tree Ant	Formicidae	Aerial nests	VC
Dorylus orientalis	Oriental Driver Ant	Formicidae	Ground; nomadic	C
Oecophylla smaragdina	Red Tree Ant	Formicidae	Aerial leaf nests with silk	C
Solenopsis geminata	Fire Ant	Formicidae	Ground nesting	U
Crematogaster spp.	Cock-tail Ants	Formicidae	Small aerial nests	VC
Pheidologiton spp.	Ants	Formicidae	Terrestrial & ground nesting	VC
Vespa bicolor	Common Wasp	Vespidae	Large aerial nests	VC
Vespa affinis	Large Brown Wasp	Vespidae	Very large aerial nests	VC

SPECIES	COMMON NAME	FAMILY	REMARKS	ABUN-DANCE*
Polistes confusus	Common Paper Wasp	Vespidae	Small flat aerial nests	C
Eumenes pyriformis	Common Potter Wasp	Vespidae	Urn-shaped mud nest	C
Eumenes spp.	Potter Wasps	Vespidae	Spherical mud nests	C
Several spp.	Wasps	Vespidae	Social or solitary	C
Sceliphron spp.	Mud-dauber Wasps	Sphecidae	Lumpy mud nests	C
Ectemnius fossorius	—	Sphecidae	Nest in tree stumps	C
Pompilus spp.	Spider-hunting Wasps	Pompilidae	Adults hunt wasps	U
Leptodialepis bipartitus	Large spider-hunting Wasp	Pompilidae	Large black species	U
Anthophora spp.	Blue-banded Bee	Apidae	Feed from flowers; nest in soil	C
Apis mellifera	Honey Bee	Apidae	Feed from flowers; mostly domesticated	VC
Xylocopa iridipennis	Bamboo Carpenter Bee	Apidae	Nest in bamboo stem; adults at flowers	VC
Xylocopa collaris	Carpenter Bee	Apidae	Nest in dead trees; adults at flowers	U

Order **Coleoptera**

SPECIES	COMMON NAME	FAMILY	REMARKS	ABUN-DANCE*
Cicindela separata	Blue Spotted Tiger Beetle	Cicindelidae	Adults hunt in woods; larvae in earth tunnels	VC
Collyrus spp.	Tiger Beetles	Cicindelidae	Adults arboreal; larvae arboreal	U
Heptodonta sp.	Tiger Beetle	Cicindelidae	Adults hunt in woods	U
Tricondyla pulchripes	Black Tiger Beetle	Cicindelidae	Adults wingless; arboreal	VC
Several spp.	Ground Beetles	Carabidae	Adults only	C
Paederus spp.	Rove Beetles	Staphylinidae	Adults only	C
Several spp.	Rove Beetles	Staphylinidae	Adults predacious	C
Prosopocoilus biplagiasus	Stag Beetle	Lucanidae	Adults	U
Popillia spp.	Flower Beetles	Scarabaeidae	Adults only	U
Protaetia orientalis	Green Rose Chafer	Scarabaeidae	Adults on flowers	VC
Anomala spp.	Flower Beetles	Scarabaeidae	Adults on flowers	VC
Agestreta orichalcea	Large Green Flower Chafer	Scarabaeidae	Adults on wild figs	R
Adoretus convexus	Flower Beetle	Scarabaeidae	Adults on flowers	C
Holotricha geilenkenseri	Cockchafer	Scarabaeidae	Adults on foliage	C
Cyphochilus apicalis	White Cockchafer	Scarabaeidae	Adults on foliage	C

SPECIES	COMMON NAME	FAMILY	REMARKS	ABUN-DANCE*
Xylotrupes gideon	Unicorn Beetle	Scarabaeidae	Adults on foliage	U
Melolontha sp.	Common Brown Cockchafer	Scarabaeidae	Adults on foliage	C
Chalcophora japonica	Large green Jewel Beetle	Buprestidae	Larvae bore timber	U
Chrysobothris sp.	Small Jewel Beetle	Buprestidae	Larvae bore timber	U
Compsosternus auratus	Large Green Click Beetle	Elateridae	Adults only	U
Chenicera sp.	Common Click Beetles	Elateridae	Adults only	C
Megapenthes sp.	Common Click Beetles	Elateridae	Adults only	C
Lychnuris analis	Firefly & Glow-worm	Lampyridae	Sheltered woodland	C
Luciola sp.	Giant Glow-worm	Lampyridae	Only larvae found	U
Several spp.	Soldier Beetles	Cantharidae	Adults only	U
Anobium spp.	Timber Beetles	Anobiidae	Timber borers	U
Bostrychopsis parallela	Black Borer	Bostrychidae	Timber Borer	U
Glischrochilus javanicus	Sap Beetle	Nitidulidae	Live under tree bark	C
Epilachna sparsa	Epilachna Bettle	Coccinellidae	On *Solanum nigrum*	VC
Chilocorus spp.	Ladybird Beetles	Coccinellidae	On foliage; predacious	VC
Coelophora spp.	Ladybird Beetles	Coccinellidae	On foliage; predacious	VC
Rhodolia spp.	Ladybird Beetles	Coccinellidae	On foliage; predacious	VC
Strongylium spp.	Darkling Beetles	Tenebrionidae	On tree trunks	C
Mylabris phalerata	Large yellow-banded Blister Beetle	Meloidae	Adults only	C
Mylabris cinchorii	Small Yellow-banded Blister Beetle	Meloidae	Adults only	C
Epicauta tibialis	Black Blister Beetle	Meloidae	Adults only	C
Epicauta gorhami	Striped Blister Beetle	Meloidae	Adults only	C
Batocera rufomaculata	Red-spotted Longhorn	Cerambycidae	Larvae bore *Sapium discolor* & *Aleurites*	C
Batocera rubus	White-spotted Longhorn	Cerambycidae	Several tree spp.	C
Anoplophora chinensis	Citrus Longhorn	Cerambycidae	*Citrus* mostly	C
Monochamus tesserula	Pine Longhorn	Cerambycidae	*Pinus* bored	C*

SPECIES	COMMON NAME	FAMILY	REMARKS	ABUNDANCE*
Chlorophorus annularis	Bamboo Longhorn	Cerambycidae	Bamboo stems bored	U
Chlorophorus macaonensis	China-berry Longhorn	Cerambycidae	*Melia* and *Prunus*	C
Sp. indet.	Itea Longhorn	Cerambycidae	Bores *Itea chinensis*	C
Apriona germari	Jackfruit Longhorn	Cerambycidae	Jackfruit trees	C
Several spp.	Longhorn Beetles	Cerambycidae	Several tree spp.	C
Metriona circumdata	Green Tortoise Beetle	Chrysomelidae (Cassidinae)	Adults and larvae on *Ipomoea* leaves	VC
Aspidomerpha chinensis	Four-'horned' Tortoise Beetle	Chrysomelidae (Cassidinae)	On *Ipomoea* spp.	C
Aspidomorpha furcata	Two-'horned' Beetle	Chrysomelidae (Cassidinae)	On Ipomoea spp.	C
Sagra purpurea	Purple Bean Beetle	Chrysomelidae (Sagrinae)	*Vigna* & others	U
Oides decempunctata	Ten-spotted Leaf Beetle	Chrysomelidae (Eumolpinae)	*Vitis* plants; defoliate	C
Colasposoma metallicum	Black Leaf Beetle	Chrysomelidae (Eumolpinae)	*Ipomoea* species	U
Aphthona wallacei	Black Flea Beetle	Chrysomelidae (Halticinae)	*Mallotus apelta* leaves	U
Cyrtotrachelus longimanus	Bamboo Weevil	Curculionidae	Larger bamboo species	C
Sipalinus hypocrita	—	Curculionidae	Larvae bore tree trunks	U
Hypomeces squamosus	Gold Dust Weevil	Curculionidae	*Ipomoea, Citrus* leaves	VC
Blosyrus herthus	Sweet Potato Black Weevil	Curculionidae	*Ipomoea* leaves	C
Several spp.	Weevils	Curculionidae	Various plants	C
Several spp.	Bark Beetles	Scolytidae	Bore under tree bark	C

Order **Acarina**

SPECIES	COMMON NAME	FAMILY	REMARKS	ABUNDANCE*
Eriophyes sp.	Schefflera Gall Mite	Eriophyidae	Galls *Schefflera* leaves	VC
Eriophyes spp.	Gall Mites	Eriophyidae	Make erinia on leaves of many different plants	C
Tetranychus spp.	Red Spider Mites	Tetranychidae	Scarify leaves of many plants	C

SPECIES	COMMON NAME	FAMILY	REMARKS	ABUN-DANCE*

(5) FRESHWATER INSECTS

Order **Ephemeroptera**

SPECIES	COMMON NAME	FAMILY	REMARKS	ABUN-DANCE*
Baetis spp.	Small Mayflies	Baetidae	Nymphs in streams	C
Pseudocloen sp.		Baetidae	Nymphs in streams	U
Gen. nov.		Baetidae	Nymphs in streams	U
Caenis spp.		Caenidae	Quiet streams	C
Ephemerella sp.	Swimming Mayfly	Ephemerellidae	Nymphs in streams	U
Ephemera sp.	Burrowing Mayfly	Ephemeridae	Nymphs burrow in gravel	C
Compsoneuria sp.	Common Torrent Mayfly	Heptageniidae	On rocks in torrents	VC
Epeorus sp.	Two-tailed Torrent Mayfly	Heptageniidae	On rocks in torrents	C
Thalerosphyrus sp.		Heptageniidae	Nymphs in streams	U
Choroterpes sp.		Leptophlebiidae	Nymphs in streams	U
Habrophlebiodes gilliesi		Leptophlebiidae	Nymphs in streams	C
Ameletus sp.		Siphlonuridae	Quiet streams	C
Isonychia spp.		Siphlonuridae	Quiet streams	C

Order **Odonata**

SPECIES	COMMON NAME	FAMILY	REMARKS	ABUN-DANCE*
Anax spp.	Darner Dragonfly	Aeschnidae	Freshwater	C
Anaciaeschna jaspidea	Darner Dragonfly	Aeschnidae	Freshwater	C
Heliogomphus sp.	Clubtail Dragonfly	Gomphidae	Streams	C
Onychogomphus sinicus	Clubtail Dragonfly	Gomphidae	Streams	C
Epopthalmia elegans	River Skimmer	Macromiidae	Reservoirs/streams	C
Neurothemis fulvia	Dragonfly	Libellulidae	Freshwater	VC
N. tullia	Dragonfly	Libellulidae	Freshwater	C
Orthetrum spp.	Dragonflies	Libellulidae	Freshwater	C
Orthetrum sabina	Burrowing Dragonfly	Libellulidae	Nymph burrows in detritus	C
Pantala flavescens	Dragonfly	Libellulidae	Widespread; adaptable; migrant	VC
Pseudothemis zonata	Dragonfly	Libellulidae	Freshwater	C
Rhyothemis variegata	Dragonfly	Libellulidae	Freshwater	C
Sympetrum spp.	Dragonflies	Libellulidae	Freshwater	C

SPECIES	COMMON NAME	FAMILY	REMARKS	ABUN-DANCE*
Tramea virginia	Dragonfly	Libellulidae	Ponds	C
Trithemis aurora	Dragonfly	Libellulidae	Freshwater	VC
Zygonyx iris	Dragonfly	Libellulidae	Streams	C
Zyxomma petiolat um	Dragonfly	Libellulidae	Ponds	C
Agriocnemis spp.	Damselflies	Agrionidae	Streams	VC
Ceriagrion latericum	Damselfly	Agrionidae	Streams	C
Ischnura senegalensis	Damselfly	Agrionidae	Ponds	C
(5 other species)	Damselflies	Agrionidae	Freshwater	C
Mnais mneme	} Broad winged Damselflies	Calopterygidae	Freshwater streams	U
Neurobasis chinensis		Calopterygidae	Streams only	U
Rhinocypha perforata	Damselfly	Epallaginidae	Freshwater	C
Euphaea decorata	Damselfly	Euphaeidae	Streams	VC
Euphaea opaca	Damselfly	Euphaeidae	Streams	U
Lestes praemorsus	Spread-winged Damselfly	Lestidae	Freshwater	C
Copera spp.	Damselflies	Platycnemidae	Freshwater	C
(2 other species)	Damselflies	Platycnemidae	Freshwater	C

Order **Plecoptera**

SPECIES	COMMON NAME	FAMILY	REMARKS	ABUN-DANCE*
Perla spp.	Common Stoneflies	Perlidae	Nymphs in streams; carnivorous	VC
Nemoura sp.	Small Stonefly	Nemouridae	Nymphs in streams; herbivorous	U

Order **Orthoptera**

SPECIES	COMMON NAME	FAMILY	REMARKS	ABUN-DANCE*
Oxya sinensis	Small Rice Grasshopper	Acrididae	Paddy field mostly	U
Tetrix subulutus	Aquatic Grouse-locust	Tetrigidae	Streams mostly; subaquatic	VC
Tetrix spp.	Grouse-locusts	Tetrigidae	Mostly waterside; subaquatic	VC

Order **Dictyoptera**

SPECIES	COMMON NAME	FAMILY	REMARKS	ABUN-DANCE*
Opisthoplatia orientalis	Amphibious Cockroach	Blattidae	Streamside; will go under water	VC

Order **Hemiptera**

SPECIES	COMMON NAME	FAMILY	REMARKS		ABUN-DANCE*
Lethocerus indicus	Giant Water Bug	Belostomatidae	Fishponds & ponds		C
Sphaerodema sp.	Small Water Bug	Belostomatidae	Quiet Water	All predacious	
Heleocoris sp.	Small Saucer Bug	Naucoridae	Ponds mostly		U
Nepa sp.	Water Scorpion	Nepidae	Ponds mostly		C
Ranatra sp.	Thin Water Scorpion	Nepidae	Ponds mostly		U
Notonecta spp.	Backswimmers	Notonectidae	Streams & ponds		VC

SPECIES	COMMON NAME	FAMILY	REMARKS	ABUN-DANCE*
Corixa spp.	Water Boatmen	Corixidae	Streams & ponds; herbivorous	VC
Gerris sp.	Pond Skater	Gerridae	On quiet water	VC
Velia sp.	Small Pond Skater	Veliidae	On quiet water	C
Microvelia sp.	Minute Pond Skater	Veliidae	On quiet water	U
Hydrometra sp.	Water Measurer	Hydrometridae	On quiet water	C

(The bracketed remarks "On quiet water" for *Gerris*, *Velia*, *Microvelia* and *Hydrometra* are marked **Predacious**.)

Order **Neuroptera**

SPECIES	COMMON NAME	FAMILY	REMARKS	ABUN-DANCE*
Neochauloides bowringi	Alder-fly	Corydalidae	Larvae aquatic; & predacious	C

Order **Trichoptera**

SPECIES	COMMON NAME	FAMILY	REMARKS	ABUN-DANCE*
Ganonema sp.	—	Calamoceratidae	Flat case of two leaf pieces; slow water	C
Helicopsyche sp.	Snail-case Caddis	Helicopsychidae	Coiled case like snail shell	R
Macronema latum	—	Hydropsychidae	Net like fishtrap on rocks; torrents	VC
Macronema fastosum	—	Hydropsychidae	Net on rocks in torrent	U
Hydropsyche spp.	—	Hydropsychidae	Nets on flat rocks	C
Cheumatopsyche sp.	—	Hydropsychidae	Simple net on rock surface	C
Lepidostoma sp.	—	Leptoceridae	Leaf-case square in section; sand tube initially; slow water	C
Mystacides sp.	—	Leptoceridae	Case leaf-wrapped	C
Triplectides sp.	—	Leptoceridae	Case is bored twig	U
Molanna sp.	—	Molannidae	Flat sand-case	R
Odontocerum sp.	—	Odontoceridae	Heavy sand tube; free larvae carnivorous	C
Genus indet.	—	Odontoceridae	Case of mixed twigs	U
Chimarra sp.	—	Philopotamidae	Small carnivorous free-living larvae	C
Silo sp.	—	Sericostomatidae	Case of small stones	U
Stenopsyche sp.	—	Stenopsychidae	Silk & stone shelter on rocks; swift water; larvae carnivorous	C

Order **Diptera**

SPECIES	COMMON NAME	FAMILY	REMARKS	ABUN-DANCE*
Culex spp.	Mosquitoes	Culicidae	Larvae, pupae acquatic	VC
Anopheles spp.	Malarial Mosquitoes	Culicidae	Larvae, pupae aquatic	C
Aëdes spp.	Mosquitoes	Culicidae	Larvae, pupae aquatic	VC
Mansonia uniformis	Marsh Mosqutio	Culicidae	Swamp species	U

SPECIES	COMMON NAME	FAMILY	REMARKS	ABUN-DANCE*
Chaoborus spp.	'Glassworms'	Culicidae	Larvae planktonic & predatory	U
Psychoda spp.	Moth Flies	Psychodidae	Larvae in wet soil	U
Several spp.	Crane Flies	Tipulidae	Larvae in water	C
Chironomus spp.	'Bloodworms'	Chironomidae	Larvae & pupae benthic	VC
Simulium spp.	Black Flies	Simulidae	Breed in fast streams	U
Several species	Fly larvae	Several families	Larvae & pupae	U
Eristalis sp.	Rat-tailed Maggot	Syrphidae	Larvae in shallow polluted water	U

Order **Coleoptera**

SPECIES	COMMON NAME	FAMILY	REMARKS	ABUN-DANCE*
Cybister limbatus	Giant Water Beetle	Dytiscidae	All stages aquatic	R
Cybister tripunctatus	Large Water Beetle	Dytiscidae	All stages aquatic	VC
Hydraticus sp.	Small Water	Dytiscidae	All stages aquatic	C
Eretes sticticus	Small Brown Water Beetle	Dytiscidae	All stages aquatic	C
Dineutes marginatus	Small Water Beetle	Dytiscidae	All stages aquatic	C
Hydrovatus spp.	Small Water Beetle	Dytiscidae	Only adults recorded	C
Canthydrus spp.	Small Water Beetles	Dytiscidae	Only adults recorded	C
Orectochilus sp.	Common Whirligig Beetle	Gyrinidae	Larvae & adults aquatic	VC
Orectochilus ceylonicus	Large Whirligig Beetle	Gyrinidae	Larvae & adults aquatic	U
Hydrophilus cashmirensis	Scavenging Water Beetle	Hydrophilidae	All stages aquatic	U
Sternolophis sp.	Lesser Scavenging Water Beetle	Hydrophilidae	Only adults found	U
Regimbartia attenuata	Least Scavenging Water Beetle	Hydrophilidae	Only adults found	U
Laccophilus parvulus	Small Red Scavenging Water Beetle	Hydrophilidae	Only adults found	U
Globaria sp.	Small Scavenging Water Beetle	Hydrophilidae	Only adults found	U
Psephenus sp.	Water Penny	Psephenidae	Larvae on rock in streams	VC
Paralichnus sp.	—	Ptilodactylidae	Larvae under silk on rocks	VC
Stenocolus sp.	—	Ptilodactylidae	Larvae in gravel on stream bottom	C

SPECIES	COMMON NAME	FAMILY	REMARKS	ABUN-DANCE*
		Order **Acarina**		
Hydrachna spp.	Red Water Mites	Hydrachnidae	Lentic water usually	C
Atractidea n. nodipalpis	—	(Hygrobatidae)	Small dark species from streams	U
Encentridiphorous hongkongensis	—	(Unionicolidae)	Small black species from pond	U

(6) MARINE INSECTS (including seashore)

SPECIES	COMMON NAME	FAMILY	REMARKS	ABUN-DANCE*
		Order **Collembola**		
Oudemansia eskaii	Seashore Springtail		Intertidal sandy shores	U
Spp. indet.	Seashore Springtails	(various families)	Intertidal sandy shores	C
		Order **Odonata**		
Pantala flavescens	Seashore Dragonfly	Libellulidae	Breed in saline rain pools on rocky shores	U
		Order **Heteroptera**		
Halobates spp.	Sea Skaters	Gerridae	On oceanic sea surface	U
Asclepios sp.	Sea Skater	Gerridae	On surface of coastal waters	U
		Order **Diptera**		
Chironomus spp.	Seashore Bloodworms	Chironomidae	Breed in saline rain pools on the rocky shores	C
Aëdes togoi	Black & White Seashore Mosquito	Culicidae		
Culex sitiens	Brown Seashore Mosquito	Culicidae		
Several species	Seaweed Flies	Ephydridae	Breed in seaweed at H.W.M.	C
Philodicus javanicus	Seashore Robber Fly	Asilidae	Adults hunt on sandy shores	C
		Order **Coleoptera**		
Cicindela anchoralis	Seashore Tiger Beetle	Cicindelidae	On sandy shores at H.W.M.	U
Bryothinusa spp.	Marine Rove Beetles	Staphylinidae	Intertidal sandy shores	U

SPECIES	COMMON NAME	FAMILY	REMARKS	ABUN-DANCE*

INSECTS ON BACK-OF-BEACH PLANTS

<table>
<tr><td colspan="5" align="center">Order Homoptera</td></tr>
<tr><td>Mesohomotoma hibisci</td><td>Hibiscus Psyllid</td><td>Psyllidae</td><td>On young shoots of Hibiscus tiliaceous</td><td>VC</td></tr>
<tr><td>Ptyelus sp.</td><td>Cerbera Spittlebug</td><td>Cercopidae</td><td>Nymphs on Cerebera manghas</td><td>C</td></tr>
<tr><td colspan="5" align="center">Order Diptera</td></tr>
<tr><td>Sp. indet.</td><td>Scaevola Leaf Miner</td><td>Agromyzidae</td><td>Mines in Scaevola leaves</td><td>C</td></tr>
<tr><td colspan="5" align="center">Order Lepidoptera</td></tr>
<tr><td>Sp. indet.</td><td>Scaevola Tussock Moth</td><td>Lymantriidae</td><td>Larvae at Scaevola leaves</td><td>U</td></tr>
<tr><td>Attacus atlas</td><td>Atlas Moth</td><td>Saturniidae</td><td>Larvae on leaves of Sapium</td><td>U</td></tr>
</table>

MANGROVE INSECTS

<table>
<tr><td colspan="5" align="center">Order Homoptera</td></tr>
<tr><td>Ceroplastes rubens</td><td>Pink Waxy Scale</td><td>Coccidae</td><td>On leaves & stems</td><td>VC</td></tr>
<tr><td>Marsipococcus sp.</td><td>Soft Brown Scale</td><td>Coccidae</td><td>On Avicennia leaves</td><td>C</td></tr>
<tr><td colspan="5" align="center">Order Lepidoptera</td></tr>
<tr><td>Hyalareta sp.</td><td>Mangrove Bagworm</td><td>Psychidae</td><td>Long narrow cases on Kanolelia</td><td>C</td></tr>
<tr><td>Sp. indet.</td><td>—</td><td>—</td><td>Round coccoons on leaves</td><td>C</td></tr>
<tr><td>Trabala vishnu</td><td>Rose Myrtle Lippet Moth</td><td>Lasiocampidae</td><td></td><td></td></tr>
<tr><td colspan="5" align="center">Order Diptera</td></tr>
<tr><td>Sp. indet.</td><td>Mangrove Leafminer</td><td>Agromyzidae</td><td>Narrow mines in leaves</td><td>C</td></tr>
<tr><td>Trichocanace sinensis</td><td>Mangrove Fly</td><td>Canacidae</td><td>Swarm in mangroves</td><td>C</td></tr>
<tr><td>Culicoides sp.</td><td>Biting Midge</td><td>Ceratopogonidae</td><td>Breed & swarm in mangroves</td><td>U</td></tr>
</table>

(7) DOMESTIC (URBAN) INSECTS

<table>
<tr><td colspan="5" align="center">Order Thysanura</td></tr>
<tr><td>Lepisma saccharina</td><td>'Silverfish'</td><td>Lepismatidae</td><td>Dry buildings; scavenger</td><td>VC</td></tr>
<tr><td>Thermobia domestica</td><td>Fire-brat</td><td>Lepismatidae</td><td>Warm buildings; scavenger</td><td>U</td></tr>
</table>

SPECIES	COMMON NAME	FAMILY	REMARKS	ABUN-DANCE*

Order **Collembola**

SPECIES	COMMON NAME	FAMILY	REMARKS	ABUNDANCE*
Homidia sp.	Domestic Springtail	Entomobryidae	Inside buildings	U

Order **Orthoptera**

Acheta domesticus	House Cricket	Gryllidae	Dry buildings	U

Order **Dictyoptera**

Periplaneta americana	American Cockroach	Blattidae	Very widespread	VC
Periplaneta australasiae	Australian Cockroach	Blattidae	Status uncertain	U?
Blatta orientalis	Oriental Cockroach	Blattidae	Temperate species	R
Blattella germanica	German Cockroach	Blattidae	Widespread	VC

Order **Isoptera**

Coptotermes formosanus	Wet-wood Termite	Rhinotermitidae	Attacks damp wood	VC
Cryptotermes brevis	Dry-wood Termite	Kalotermitidae	Furniture & dried wood	R

Order **Psocoptera**

Ectopsocus maindroni	House Psocid	Ectopsocidae	On mouldy walls	C
Several species	Stored Products Psocids	—	In stored foods	VC

Order **Hemiptera**

Cimex lectularis	Bedbug	Cimcidae	Nocturnal blood-sucker	U
Triatoma rubrofasciata	Domestic Assassin Bug	Reduviidae	Predacious; old buildings	C

Order **Mallophaga**

Trichodectes canis	Dog Biting Louse	Trichodectidae	On dogs locally	C

Order **Siphunculata**

Pediculus humanus corporis	Human Body Louse	Pediculidae	Usually in body clothing	R
Pediculus h. capitis	Human Head Louse	Pediculidae	In head hair only	U
Pthirus pubis	Crab Louse	Pediculidae	Coarse body hair only	U

SPECIES	COMMON NAME	FAMILY	REMARKS	ABUN-DANCE*
		Order **Lepidoptera**		
Tineola bisselliella	Clothes Moth	Tinaeidae	Stored Clothes	VC
Tinea pellionella	Case-bearing Clothes Moth	Tinaeidae	Cases often on walls	VC
Trichophaga tapetzella	Tapestry Moth	Tinaeidae	Eats coarser fibre	U
Hofmannophila pseudospretella	Brown House Moth	Oecophoridae	Dried vegetable matter and wool	C
Endrosis sarcitrella	White-shouldered House Moth	Oecophoridae		C
Problepsis sp.	Eye Moth	Geometridae	Takes tears from eyes	R
Ephestia eleutella	Warehouse Moth	Pyralidae	Polyphagous pest	VC
Ephestia cautella	Tropical Warehouse Moth	Pyralidae	Dried fruits mostly	C
Ephestia kuehniella	Mediterranean Flour Moth	Pyralidae	Flours mostly	C
Plodia interpunctella	Indian Meal Moth	Pyralidae	Flours, etc.	VC
		Order **Diptera**		
Aëdes albopictus	B/W House Mosquito	Culicidae	Quiet feeder, breeds in tiny pools	VC
Culex fatigans	Brown House Mosquito	Culicidae	Noisy wings	VC
Psychoda spp.	Moth Flies	Psychodidae	Dark humid corners	VC
Calliphora spp.	Bluebottles	Calliphoridae	Larvae saprozoic	VC
Lucilia spp.	Greenbottles	Calliphoridae	Larvae saprozoic	VC
Sarcophaga spp.	Flesh Flies	Sarcophagidae	Larvae saprozoic	VC
Chrysomyia megacephala	—	Calliphoridae	Larvae saprozoic	VC
Drosophila spp.	Vinegar Flies	Drosophilidae	Overripe fruit	VC
Musca domestica	House Fly	Muscidae	Larvae saprozoic	VC
Fannia canicularis	Lesser House Fly	Muscidae	Larvae saprozoic	VC
		Order **Siphonaptera**		
Ctenocephalides felis	Cat Flea	Ceratophilidae	Cats & dogs, tapeworm vector	VC
Leptopsylla signis	Mouse Flea	Leptopsyllidae	On House Mouse usually	C
Xenopsylla cheopis	Tropical Rat Flea	Leptopsyllidae	On rats usually	C
		Order **Hymenoptera**		
Evania spp.	Ensign Wasps	Evaniidae	Parasitize oöthecae of cockroaches	VC

SPECIES	COMMON NAME	FAMILY	REMARKS	ABUN-DANCE*
Ampulex compressa	Cockroach Wasp	Sphecidae	Coackroach predator	C
Vespa bicolor	Common Wasp	Vespidae	Adults seek sugar	VC
Eumenes spp.	Potter Wasps	Vespidae	Round mud nests	VC
Iridomyrmex spp.	Sugar (House) Ant	Formicidae	Most buildings, after foods and sugar	
Tapinoma spp.		Formicidae		C
Monomorium pharaonis	Pharaoh's Ant	Formicidae	Many buildings	C
Anisopteromalus calandrae	Parasitic Wasp	Pteromalidae	Parasitize pupae of *Sitophilus* weevils	C
Choetospila elegans	Parasitic Wasp	Pteromalidae		C

Order **Coleoptera**

SPECIES	COMMON NAME	FAMILY	REMARKS	ABUN-DANCE*
Lasioderma serricorne	Tobacco Beetle	Anobiidae	Stored food stuffs and tobacco	C
Anobium punctatum	Common Furniture Beetle	Anobiidae	Dry timber	U
Stegobium panicum	Drugstore Beetle	Anobiidae	Foodstuffs	C
Tribolium spp.	Flour Beetles	Tenebrionidae	Flours, etc.	VC
Ptinus spp.	Spider Beetles	Ptinidae	Stored foods	VC
Oryzaephilus surinamensis	Saw-toothed Grain Beetle	Silvanidae	Stored foods	VC
Oryzaephilus mercator	Merchant Grain Beetle	Silvanidae	Stored foods	C
Callosobruchus chinensis	Cowpea Bruchids	Bruchidae	Stored pulses	C
Callosobruchus maculatus				
Acanthoscelides obtectus	Bean Bruchid	Bruchidae	Stored pulses	C
Caryedon serratus	Groundnut Borer	Bruchidae	Stored pulses	C
Dermestes maculatus	Leather Beetle	Dermestidae	Dry animal protein	C
Dermestes lardarius	Larder Beetle	Dermestidae	Dry animal protein	C
Anthrenus spp.	Carpet Beetles	Dermestidae	Dry animal material	C
Attagenus piceus	Black Carpet Beetle	Dermestidae	Carpets, furs, etc.	C
Sitophilus zeamais	Maize Weevil	Curculionidae	Maize, etc.	C
Sitophilus oryzae	Rice Weevil	Curculionidae	Rice, etc.	VC
Araecerus fascioulatus	Coffee Bean Weevil	Curculionidae	Many seeds & pulses	VC

Order **Acarina**

SPECIES	COMMON NAME	FAMILY	REMARKS	ABUN-DANCE*
Acarus siro	Flour Mite	Acaridae	Stored flours	VC
Rhizoglyphus spp.	Cheese Mites	Acaridae	Stored foods	C

SPECIES	COMMON NAME	FAMILY	REMARKS	ABUN-DANCE*
Tyrophagus spp.	Cheese Mites	Tyroglyphidae	Stored foods	C
Sarcoptes scabei	Itch & Mange Mites	Psoroptidae	Skin of cats & dogs mostly	C
Rhipicephalus sanguineus	Brown Dog Tick	Ixodidae	Dogs mostly	C

(8) AGRICULTURAL INSECTS

Order **Collembola**

Hypogastrura communis	Springtail	Hypogastruridae	*Brassica* pest in N.T.	C

Order **Orthoptera**

Brachytrupes portentosus	Big-head Cricket	Gryllidae	Polyphagous soil pest	C
Gryllulus spp.	Field Crickets	Gryllidae	Polyphagous pests	C
Teleogryllus spp.	Field Crickets	Gryllidae	Polyphagous pests	C
Gryllotalpa africana	African Mole Cricket	Gryllotalpidae	Polyphagous; fossorial	C
Oxya sinensis	Small Rice Grasshopper	Acrididae	Pest of rice	VC
Chondracris rosea	Large Green Grasshopper	Acrididae	Pest of Gramineae	U

Order **Isoptera**

Macrotermes barneyi	Bark Termite	Termitidae	Polyphagous minor pest	U
Odontotermes formosanus	Large Brown Termite	Termitidae	Polyphagous minor pest	U
Coptotermes formosanus	Wet-wood Termite	Rhinotermitidae	Various trees	C

Order **Homoptera**

Cosmocarta abdominalis	Red/Black Froghopper	Cercopidae	Adults polyphagous	C
Nephotettix spp.	Green Rice Leafhoppers	Cicadellidae	Rice and Gramineae	VC
Typhlocyba spp.	White Leafhoppers	Cicadellidae	Various trees	C
Bothrogonia sp.	Large Red Leafhopper	Cicadellidae	Sunflower & many plants	C
Coelidia spp.	Banded Leafhoppers	Cicadellidae	*Citrus,* & other trees	C
Sogatella furcifera	White-backed Planthopper	Delphacidae	Rice & Gramineae	C

SPECIES	COMMON NAME	FAMILY	REMARKS	ABUN-DANCE*
Inazuma dorsalis	Zig-Zag Winged Planthopper	Delphacidae	Rice & Gramineae	U
Nilaparvata lugens	Brown (Rice) Planthopper	Delphacidae	Serious rice pest in S.E. Asia	U
Perkinsiella saccharicida	Sugarcane Planthopper	Delphacidae	Pest of Gramineae	U
Dictyophara sp.	—	Dictyopharidae	*Citrus* pest	C
Pyrops candelaria	Longan Lantern-fly	Fulgoridae	Lychee, longan pest	C
Pyrops lathburii	Eucalyptus Lantern-fly	Fulgoridae	On *Eucalyptus tereticornis*	U
Pulastya discolorata	Large Moth Bug	Flattidae	*Citrus* pest	C
Diaphorina citri	Citrus Psyllid	Psyllidae	*Citrus* leaves	U
Trioza camphorae	Camphor Psyllid	Psyllidae	Pits on camphor leaves	VC
Several species	Jumping Plant-lice	Psyllidae	Several trees, but not really pests	C
Bemisia tabaci	Tobacco Whitefly	Aleyrodidae	Polyphagous pest	U
Dialurodes citri	Citrus Whitefly	Aleyrodidae	*Citrus* species	U
Aleurocanthus woglumi	Citrus Blackfly	Aleyrodidae	*Citrus* species	C
Aleurocanthus marlatti	Bauhinia Blackfly	Aleyrodidae	Bauhinia leaves	C
Myzus persicae	Green Peach Aphid	Aphididae	Polyphagous pest	VC
Lipaphis erysimi	Turnip Aphid	Aphididae	*Brassica* pest species	VC
Aphis nerii	Oleander Aphid	Aphididae	Oleander & some flowers	C
Aphis gossypii	Melon/Cotton Aphid	Aphididae	Polyphagous pest	VC
Aphis craccivora	Groundnut Aphid	Aphididae	Leguminosae mostly	VC
Toxoptera aurantii	Black Citrus Aphid	Aphididae	*Citrus* species, etc.	VC
Pentalonia nigronervosa	Banana Aphid	Aphididae	Bananas only	R
Chaitoregma sp.	Woolly Aphid	Pemphigidae	Stem galls on privet	VC
Pseudococcus citriculus	Long-tailed Mealybug	Pseudococcidae	Many different hosts	VC
Planococcus citri	Citrus Mealybug	Pseudococcidae	Many different hosts	C
Maconellicoccus hirsutus	Hibiscus Mealybug	Pseudococcidae	*Hibiscus* shoots	VC
Icerya aegyptiaca	Egyptian Fluted Scale	Margarodidae	Polyphagous pest on trees	C
Icerya purchasi	Cottony Cushion Scale	Margarodidae	Several local hosts	VC
Steatococcus assamensis	—	Margarodidae	On croton	C

SPECIES	COMMON NAME	FAMILY	REMARKS	ABUN-DANCE*
Drosicha sp.	Giant Mealybug	Margarodidae	Several local hosts	U
Tachardina sp.	Lac Insect	Lacciferidae	Banana Flower Shrub	C
Saissetia coffeae	Hemispherical Scale	Coccidae	Polyphagous pest	VC
Saissetia oleae	Black Scale	Coccidae	Polyphagous pest	C
Pulvinaria spp.	Brown Scales	Coccidae	Polyphagous pests	VC
Coccus viridis	Soft Green Scale	Coccidae	Polyphagous pest	C
Coccus longulus	Elongate Green Scale	Coccidae	Bauhinia, croton leaves	C
Ceroplastes rubens	Pink Waxy Scale	Coccidae	Polyphagous pest	VC
Ceroplastes sp.	White Waxy Scale	Coccidae	Polyphagous pest	C
Aspidiotus destructor	Coconut Scale	Diaspididae	Polyphagous pest	C
Phenacaspis colkerelli	Oyster Scale	Diaspididae	Oleander leaves	VC
Aonidiella aurantii	California Red Scale	Diaspididae	Pilyphagous pest	C
Aonidiella orientalis	Oriental Red Scale	Diaspididae	Papaya stems	U
Quadraspidiotus perniciosus	San José Scale	Diaspididae	Deciduous fruit, etc.	U
Lepidosaphes beckii	Mussel Scale	Diaspididae	*Citrus* species	C
Chrysomphalus aonidum	Purple Scale	Diaspididae	*Citrus* species	C
Fiorinia japonica	White Scale	Diaspididae	Bamboo palm, etc.	VC
Parlatoria zizyphus	Black Parlatoria	Diaspididae	*Citrus* species	U

Order **Heteroptera**

SPECIES	COMMON NAME	FAMILY	REMARKS	ABUN-DANCE*
Leptocorisa acuta	Rice Bug	Coreidae	Rice and grasses	VC
Leptoglossus australis	Leaf-footed Plant Bug	Coreidae	Polyphagous pest	U
Tessaratoma papillosa	Lychee Stink Bug	Pentatomidae	Lychee and longan	VC
Rhynchocorus humeralis	Citrus Shield Bug	Pentatomidae	*Citrus* species	C
Nezara viridula	Green Shield Bug	Pentatomidae	Polyphagous pest	VC
Scotinophara coarctata	Black Paddy Bug	Pentatomidae	Rice pest	U
Poecilocoris latus	Camellia Shield Bug	Pentatomidae	Tea & camellia bushes in N.T.	U
Cantao ocellatus	Mallotus Shield Bug	Pentatomidae	On Mallotus foliage	U
Catacanthus nigripes	Tea Shield Bug	Pentatomidae	Tea bushes on Tai Mo Shan	U

SPECIES	COMMON NAME	FAMILY	REMARKS	ABUN-DANCE*
		Order **Thysanoptera**		
Gynaikothrips ficorum	Common Fig Thrips	Phlaeothripidae	*F. microcarpa* leaves	VC
Taeniothrips nigricornis	Bean Thrips	Thripidae	Bean flowers	C
Taeniothrips lefroyi	Tea Thrips	Thripidae	Tea flowers	U
Heliothrips haemorrhoidalis	Black Tea Thrips	Thripidae	Polyphagous species	C
		Order **Lepidoptera**		
Conopis sp.	Camphor Clearwing	Sesiidae	Larvae bore camphor trees	U
Pectinophora gossypiella	Pink Bollworm	Gelechiidae	Malvaceae mostly	C
Phyllocnistis citrella	Citrus Leaf Miner	Gracillariidae	Larvae mine *Citrus* leaves	VC
Plutella xylostella	Diamond-back Moth	Yponomeutidae	*Brassica* pest	VC
Zeuzera coffeae	Red Coffee Borer	Cossidae	Many trees attacked	VC
Clania spp.	Bagworms	Psychidae	Several trees	VC
Hyalarcta sp.	Bagworm	Psychidae	Many plants	C
Indarbela disciplaga	Wood-borer Moth	Metarbelidae	Polyphagous bork-eater	VC
Parasa lepida	Stinging Caterpillar	Limacodidae	Various trees	C
Thosea sinensis	Slug Caterpillar	Limacodidae	Various trees	C
Cyclosia papilionaris	Black-veined Moth	Zygaenidae	Tree defoliators	C
Cydia molesta	Oriental Fruit Moth	Tortricidae	Larvae bore fruits	U
Cydia aurantiana	Hibiscus Budworm	Tortricidae	*Hibiscus* buds	C
Cydia pulverula	Fig Budworm	Tortricidae	*F. microcarpa* buds	VC
Homona coffearia	Lichee Leaf-roller	Tortricidae	*Citrus, Litchi,* guava	C
Adoxophyes orana	Summer Fruit Tortrix	Tortricidae	*Citrus, Litchi, Arachis*	C
Rhyaciona cristata	Pink Pine Shoot-Borer	Tortricidae	*Pinus* shoots bored	C
Archips spp.	Flower Tortrix Moths	Tortricidae	Roses, chrysanthemum, etc.	C
Aciptilia sp.	Sweet Potato Vine Borer	Pterophoridae	Larvae bore vines	C
Ochyrotica concursa	Sweet Potato Leaf-Roller	Pterophoridae	Larvae roll leaves	VC

SPECIES	COMMON NAME	FAMILY	REMARKS	ABUN-DANCE*
Cnaphalocrocis medinalis	Rice Leaf Roller	Pyralidae	Larvae roll rice leaves	VC
Chilo suppressalis	Rice Stalk Borer	Pyralidae	Larvae bore rice stalks	VC
Hellula undalis	Oriental Cabbage Webworm	Pyralidae	Larvae attack Cruciferae	C
Tryporyza incertulas	Paddy Stem Borer	Pyralidae	Larvae bore rice stems	VC
Ostrinia furnacalis	Asian Corn Borer	Pyralidae	Maize borer pest	U
Chilo sacchariphagus	Sugarcane Stem Borer	Pyralidae	Sugarcane stalks bored	VC
Nymphula depunctalis	Rice Caseworm	Pyralidae	Rice leaf pest	C
Maruca testulalis	Bean Pod Borer	Pyralidae	Larvae bore legume pods	VC
Sylepta derobata	Cotton Leaf Roller	Pyralidae	*Hibiscus* leaves rolled	C
Palpita indica	Pumpkin Leaf Roller	Pyralidae	Cucurbitaceae leaves	VC
Several other species	—	Pyralidae	Various plants	C
Dehdrolemus punctatus	Lappet Moth	Lasiocampidae	*Pinus* leaves eaten	C
Gastropacha quercifolia	Lappet Moth	Lasiocampidae	Peach leaves eaten	C
Attacus atlas	Atlas Moth	Saturniidae	Camphor, etc. leaves	U
Samia cynthia	Lesser Atlas Moth	Saturniidae	Various trees	C
Arctias selene	Moon Moth	Saturniidae	Camphor, tallow, etc.	VC
Eriogyna pyretorum	Giant Silkmoth	Saturniidae	Camphor leaves	VC
Chilades lajus	Small Blue	Lycaenidae	*Citrus* leaves	C
Pieris rapae	Small White Butterfly	Pieridae	Pest of Cruciferae	VC
Pieris canidia	Small White Butterfly	Pieridae	Cruciferae & other plants	VC
Papilio demoleus	Lemon Butterfly	Papilionidae	*Citrus* & *Fortunella* pest	VC
Papilio polytes	Common Mormon	Papilionidae	*Citrus* pest	VC
Papilio spp.	Swallowtails	Papilionidae	*Citrus* defoliators	C
Erionota torus	Banana Skipper	Hesperiidae	Banana leaf pest	C
Panara buttata	Rice Skipper	Hesperiidae	Larvae eat rice leaves	VC
Calospilos sp.?	Citrus Looper	Geometridae	*Citrus* leaf eater	C
Agrius convoluvli	Convolvulus Hawk Moth	Sphingidae	*Ipomoea* leaves eaten	VC
Hyles lineata	Silver-striped Hawk	Sphingidae	Polyphagous larvae	VC
Hyloicus pinastri	Pine Hawk Moth	Sphingidae	*Pinus* leaves eaten	C
Cephonodes hylas	Bee (Coffee) Hawk Moth	Sphingidae	Larvae defoliate Rubiaceae	U

SPECIES	COMMON NAME	FAMILY	REMARKS	ABUN-DANCE*
Several species	Hawk Moths (Hornworms)	Sphingidae	Larvae defoliate	C
Agrotis ipsilon	Black Cutworm	Noctuidae	Polyphagous cutworm	VC
Agrotis segetum	Cutworm	Noctuidae	Polyphagous cutworm	U
Heliothis armigera	American Bollworm	Noctuidae	Polyphagous leafworm and borer	C
Mythimna loreyi	Rice Armyworm	Noctuidae	Rice pest mostly	VC
Mythimna separata	Rice Armyworm	Noctuidae	Rice pest mostly	VC
Spodoptera litura	Fall Armyworm	Noctuidae	Polyphagous larvae	VC
Sesamia inferens	Pink Rice Borer	Noctuidae	Rice, sugarcane borer	C
Othreis fullonica	Fruit piercing Moth	Noctuidae	Adult pierces ripening fruits	C
Ophiusa tirhaca	Fruit Piercing Moth	Noctuidae	Adult pierces ripening fruits	C
Plusia chalcites	Vegetable Looper	Noctuidae	Pest of Cruciferae	VC
Trichoplusia brassicae	Cabbage Looper	Noctuidae	Pest of Brassicas	C
Anomis flava	Cotton Semi-looper	Noctuidae	Pest of Malvaceae	C
Perina nuda	Banyan Tussock Moth	Lymantriidae	*F. microcarpa* leaves	VC
Euproctis spp.	Yellow-tail Moths	Lymantriidae	Various trees	U
Orgyia postica	Yellow-tuft Caterpillar	Lymantriidae	*Citrus* & peach leaves	U
Lymantria spp.	Tussock Moths	Lymantriidae	Hosts unknown	U

Order **Siphunculata**

SPECIES	COMMON NAME	FAMILY	REMARKS	ABUN-DANCE*
Haematopinus spp.	Animal Lice	Haematopinidae	Domestic animals	U

Order **Mallophaga**

SPECIES	COMMON NAME	FAMILY	REMARKS	ABUN-DANCE*
Menopon gallinae	Chicken Louse	Menoponidae	Pest of chickens	C

Order **Diptera**

SPECIES	COMMON NAME	FAMILY	REMARKS	ABUN-DANCE*
Pachydiplosis oryzae	Rice Stem Gall Midge	Cecidomyiidae	Rice stems galled	U
Dacus cucurbitae	Melon Fly	Tephritidae	Cucurbitaceae fruits	C
Dacus spp.	Fruit Fly	Tephritidae	Guava fruits, etc.	C
Ophiomyia phaseoli	Bean Fly	Agromyzidae	Bean seedling stems & petioles	C
Phytomyza horticola	Pea Leaf Miner	Agromyzidae	Peas, Brassicas, etc.	VC
Hydrellia griseola	Rice Leaf Miner	Ephydridae	Rice leaves mined	C
Notiphila spp.	Rice Leaf Miners	Ephydridae	Rice leaves mined	C
Scatella sp.	Leaf Miner	Ephydridae	*Brassica chinensis*	U
Atherigona orientalis	Rice Seedling Fly	Muscidae	Bores rice seedlings	U

SPECIES	COMMON NAME	FAMILY	REMARKS	ABUN-DANCE*
Stomoxys calcitrans	Stable Fly	Muscidae	Cattle pest in N.T.	VC
? species	Root Flies	Anthomyiidae	Local status not known	?
Pseudolynchia canariensis	Pigeon Louse Fly	Hippoboscidae	Domestic pigeons	VC
Hippobosca equina	Cattle Louse Fly	Hippoboscidae	Cattle ectoparasite	R

Order **Hymenoptera**

SPECIES	COMMON NAME	FAMILY	REMARKS	ABUN-DANCE*
Athalia sp.	Cabbage Sawfly	Tenthredinidae	*Brassica* leaves eaten	U?
Arge pagana	Rose Sawfly	Tenthredinidae	Rose leaves eaten	C
Neodiprion biremis	Pine Sawfly	Tenthredinidae	*Pinus* leaves eaten	U
Solenopsis geminata	Fire Ant	Formicidae	Polyphagous pest	C
Polyrachis dives	Black Tree Ant	Formicidae	Aerial nests in trees	VC
Crematogaster spp.	Cock-tail Ants	Formicidae	Small aerial nests; often with aphids	VC
Oecophylla smaragdina	Red Tree Ant	Formicidae	Arboreal, fierce species	VC
Xylocopa spp.	Carpenter Bees	Apidae	Doubtful as pests	C
Apis mellifera	Honey Bee	Apidae	Reared for honey & wax	VC

Order **Coleoptera**

SPECIES	COMMON NAME	FAMILY	REMARKS	ABUN-DANCE*
Melolontha serrulata	Large Brown Chafer	Scarabaeidae		
Holotrichia geilenkenseri	Large Brown Cockchafer	Scarabaeidae		
Phyllophaga prollaeta	Small Brown Chafer	Scarabaeidae	Larvae found in soil as Chafer Grubs or White Grubs; adults damage various plants	VC
Sophrops cephalotes	Small Black Chafer	Scarabaeidae		
Anomala varicolor	Brown striped Flower Beetle	Scarabaeidae		
Anomala cupripes	Large Green Flower Beetle	Scarabaeidae		
Protaetia orientalis	Green Rose Chafer	Scarabaeidae	Adults eat flowers	VC
Popillia spp.	Flower Beetles	Scarabaeidae	Tree defoliators	U
Popillia histeroides	Black Flower Beetle	Scarabaeidae	Adults eat flowers	C
Chalcophora japonica	Large Green Jewel Beetle	Buprestidae	Larvae bore trees	U
Chrysobothris sp.	Small Jewel Beetle	Buprestidae	Larvae bore trees	U
Bostrychopsis parallela	Black Borer	Bostrychidae	Timber borers (adults)	U
Epilachna sparsa	Epilachna Beetle	Coccinellidae	*Solanum* & cucurbit leaves eaten	VC

SPECIES	COMMON NAME	FAMILY	REMARKS	ABUNDANCE*
Epicauta gorhami	Striped Blister Beetle	Meloidae	Many different flowers	C
Epicauta tibialis	Black Blister Beetle	Meloidae	Many different flowers	C
Mylabris phalerata	Large Yellow-banded Blister Beetle	Meloidae	Many different flowers	C
Mylabris cinchorii	Small Yellow-banded Blister Beetle	Meloidae	Many different flowers	C
Batocera rufomaculata	Red-spotted Longhorn Beetle	Cerambycidae	Larvae bore mountain tallow	C
Batocera rubus	White-spotted Brown Longhorn	Cerambycidae	Fig & mango trees	C
Anoplophora chinensis	Citrus Longhorn	Cerambycidae	*Citrus* mostly	VC
Apriona germari	Jackfruit Longhorn	Cerambycidae	Bore jackfruit trees	C
Chlorophorus annularis	Bamboo Longhorn	Cerambycidae	Bamboo & sugarcane	C
Callosobruchus maculatus	Spotted Cowpea Bruchid	Bruchidae	Ripe pulses bored	C
Callosobruchus chinensis	Oriental Cowpea Bruchid	Bruchidae	Ripe pulses bored	C
Acanthoscelides obtectus	Bean Bruchid	Bruchidae	Ripe bean seeds bored	C
Dicladispa armigera	Paddy Hispid	Chrysomelidae	Rice leaves attacked	C
Colasposoma metallicum	Black Leaf Beetle	Chrysomelidae	*Ipomoea* leaves eaten	VC
Sagra purpurea	Purple Bean Beetle	Chrysomelidae	*Vigna* and others	U
Oides decempunctata	Ten-spotted Leaf Beetle	Chrysomelidae	*Vitis* spp.	C
Aspidomorpha furcata	Two-horned Tortoise Beetle	Chrysomelidae	*Ipomoea* leaves eaten	VC
Lacoptera chinensis	Four-horned Tortoise Beetle	Chrysomelidae	*Ipomoea* leaves	VC
Metriona circumdata	Green Tortoise Beetle	Chrysomelidae	*Ipomoea* leaves	VC
Phyllotreta striolata	Cabbage Flea Beetle	Chrysomelidae	*Brassica* leaves holed	VC
Argopistes spp.	Citrus Flea Beetles	Chrysomelidae	*Citrus* species leaves	U

SPECIES	COMMON NAME	FAMILY	REMARKS	ABUN-DANCE*
Prodagricomela nigricollis	Citrus Flea Beetle	Chrysomelidae	*Citrus* leaves	U
Cylas formicarius	Sweet Potato Weevil	Curculionidae	Sweet Potato tubers	VC
Apion spp.	Flower Weevils	Curculionidae	Flowers of Leguminosae eaten by adults; larvae in soil parasitic on grasshopper eggs	C
Odoiporus longicollis	Banana Stem Weevil	Curculionidae	Banana stems bored	VC
Rhyncophorus sp.	Palm Weevil	Curculionidae	Adults attack palms	R
Cyrtotrachelus longimanus	Bamboo Weevil	Curculionidae	Larger bamboo species	C
Hypomeces squamosus	Gold Dust Weevil	Curculionidae	*Ipomoea* & *Citrus,* etc.	VC
Blosyrus herthus	Sweet Potato Black Weevil	Curculionidae	*Ipomoea* leaves eaten	C
Several species	Weevils	Curculionidae	Various host plants	C

Order **Acarina**

SPECIES	COMMON NAME	FAMILY	REMARKS	ABUN-DANCE*
Tetranychus cinnabarinus	Tropical Red Spider Mite	Tetranychidae	Plant foliage pests	VC
Tetranychus spp.	Red Spider Mite	Tetranychidae	Plant foliage pests	VC
Eutetranychus orientalis	Oriental Mite	Tetranychidae	Frangipani, etc.	C
Eriophyes spp.	Gall Mites	Eriophyidae	Galls on many plants	VC
Aceria spp.	Gall Mites	Eriophyidae	Galls on many plants	VC
Several species	Cattle Ticks	Ixodidae	Cattle pests	U
Dermanyssus gallinae	Red Poultry Mite	Dermanyssidae	Pest of ducks & hens	VC
Argas persicae	Fowl Tick	Argasidae	Ducks & chickens	C

(9) SYMBIOTIC INSECTS

COMMENSALISM

Examples of commensalism are not common in the Insecta. There are a few species of beetles and earwigs that live in the frass in the tunnels made by Longhorn beetles in tree trunks and branches, and this could possibly be regarded as being commensalism. If the association is one whereby a smaller insect (or mite) is transported by a larger host this is termed phoresy; Mallophaga often transport bird skin or feather mites this way, as do Hippoboscidae transport Mallophaga.

INQUILINISM

Order **Dictyoptera**

SPECIES	COMMON NAME	FAMILY	REMARKS	ABUN-DANCE*
Nocticola sinensis	Cockroach	Blattidae	Live in termite nests	U

SPECIES	COMMON NAME	FAMILY	REMARKS	ABUN-DANCE*
		Order Isoptera		
Procapritermes sowerbyi	Termite	Termitidae	Live in other termite nests	R
		Order Lepidoptera		
Several species	Blue Butterflies	Lycaenidae	Larvae in ants' nests	U
		Order Hymenoptera		
Some species	Ants	Formicidae	Live in other ants' nests	U
Saphonecrus sp.	Cynipid Gall Wasp	Cynipidae	Live in galls on oak flowers made by *Ormyrus* sp.	C
Several species	Fig-wasps	Torymidae, etc.	Live in fig syconia	VC

MUTUALISM

SPECIES	COMMON NAME	FAMILY	REMARKS	ABUN-DANCE*
		Order Isoptera		
Coptotermes formosana	Wet-wood Termite	Rhinotermitidae	Micro-organisms in gut digest the plant cellulose material	VC
Cryptotermes brevis	Dry-wood Termite	Kalotermitidae		R
		Order Hymenoptera		
Blastophaga spp.	Fig-wasps	Agaonidae	Develop inside figs (*Ficus* spp.) and later pollinate the flowers	VC
Ceratosolen spp.	Fig-wasps	Agaonidae		VC
Apis mellifera	Honey Bee	Apidae	Adults pollinate flowers	VC
Crematogaster spp.	Cock-tail Ants	Formicidae	Protect aphids and scales for honey-dew	VC
Several species	Ants	Formicidae		VC
		Order Homoptera		
Several species	Scale Insects	Coccidae	Often live associated with ants which protect them & feed on the honey-dew excreted	VC
Several species	Mealybugs	Pseudococcidae		VC
Several species	Aphids	Aphididae		VC
		Order Diptera		
Many species	Hover Flies	Syrphidae	Adults pollinate flowers	VC
Many species	Flies	Muscoidea		VC
		Order Lepidoptera		
Several species	Moths	(various families)	Adults pollinate flowers	VC

SPECIES	COMMON NAME	FAMILY	REMARKS	ABUNDANCE*

PARASITISM

Order **Diptera**

SPECIES	COMMON NAME	FAMILY	REMARKS	ABUNDANCE*
Anopheles spp.	Malarial Mosquitoes	Culicidae	Adults on vertebrates NT	C
Aëdes albopictus	B/W House Mosquito	Culicidae	Around houses, etc.	VC
Culex fatigans	Brown House Mosquito	Culicidae	Around houses	VC
Lasiohelea stimulans	Biting Midge	Ceratopogonidae	Adults on vertebrates	C
Culicoides spp.	Biting Midges	Ceratopogonidae	Adults feed on insects, man & vertebrates	U
Many species	Gall Midges	Cecidomyiidae	Larvae make galls on plants	VC
Tabanus spp.	Horse Flies	Tabanidae	Adults on cattle, etc.	C
Chrysops sp.	Horse Fly	Tabanidae	Adults on cattle, etc.	C
Many species	—	Tachinidae	Larvae in caterpillars	C
Many species	—	Agromyzidae	Larvae make galls on plants	C
Stomoxys calcitrans	Stable Fly	Muscidae	Adults on cattle, etc.	VC
Pseudolynchia canariensis	Pigeon Louse Fly	Hippoboscidae	Adults on pigeons	VC
Hippobosca equina	Cattle Louse Fly	Hippoboscidae	Adults on cattle	R
Leptocyclopodia ferrarii	Bat Louse Fly	Nycteribiidae	On Fruit Bat cave	C

Order **Hemiptera**

SPECIES	COMMON NAME	FAMILY	REMARKS	ABUNDANCE*
Oncocephalus sp.	Assassin Bug	Reduviidae	Micropredator	C
Sycanus croceovittatus	Large Black Assassin Bug	Reduviidae	Micropredator	VC
Triatoma rubrofasciata	Domestic Assassin Bug	Reduviidae	Micropredator	C

Order **Siphunculata**

SPECIES	COMMON NAME	FAMILY	REMARKS	ABUNDANCE*
Pediculus humanus	Human Lice	Pediculidae	Human body & head	C
Pthirus pubis	Crab Louse	Pediculidae	Human coarse hair	U
Haematopinus spp.	Animal Lice	Haematopinidae	Domestic animals	U

Order **Siphonaptera**

SPECIES	COMMON NAME	FAMILY	REMARKS	ABUNDANCE*
Ctenocephalides felis	Cat Flea	Ceratophyllidae	Cats & dogs (man)	VC
Leptopsylla signis	Mouse Flea	Leptopsyllidae	On House Mouse usually	C
Xenopsylla cheopis	Rat (Plague) Flea	Leptopsyllidae	Rats usually (man)	C

SPECIES	COMMON NAME	FAMILY	REMARKS	ABUN-DANCE*
		Order **Mallophaga**		
Menopon gallinae	Chicken Louse	Menoponidae	On chickens locally	VC
Trichodectes canis	Dog Biting Louse	Trichodectidae	On local dogs	C
		Order **Hymenoptera**		
Many species	Ichneumons	Icheumonidae	Larvae in caterpillars, bugs & beetles, etc.	VC
Many species	Braconids	Braconidae		
Many species	Chalcids	Chalcidoidea		
Some species	Gall Wasps	Chalcidoidea	Make galls on plants	C
Philotrypesis spp.	Fig-wasps	Torymidae	'Parasites' inside wild figs	
Sycophila spp.	Fig-wasps	Eurytomidae		
		Order **Lepidoptera**		
Epipyrops anomala	Parasitic Moth	Epipyropidae	Larvae parasitize *Pyrops*	U

All the plant gall-forming insects, leaf-miners, and all the plant-eating insects in general, could be regarded as belonging in this category, but they are mostly omitted from this list as they have already been included in the previous categories.

SPECIES	COMMON NAME	FAMILY	REMARKS	ABUN-DANCE*
		Order **Acarina**		
Dermanyssus gallinae	Red Poultry Mite	Dermanyssidae	Ducks & chickens	VC
Argas persicae	Fowl Tick	Argasidae	Ducks & chickens	C
Rhipicephalus sanguineus	Brown Dog Tick	Ixodidae	Dogs	VC
Several species	Cattle Ticks	Ixodidae	On cattle	U
Sarcoptes scabei	Mange & Itch Mites	Sarcoptidae	Dogs & cats (man)	VC
Tetranychus spp.	Red Spider Mites	Tetranychidae	Many plants	VC
Eutetranychus orientalis	Oriental Mite	Tetranychidae	Frangipani, etc.	VC
Eriophyes spp.	Gall Mites	Eriophyiidae	Many plants	VC

CHAPTER FOUR

KEY TO THE MAJOR GROUPS OF TERRESTRIAL INVERTEBRATES IN HONG KONG

The following key is a modified version of the 'Key to the Major Groups of British Free-living Terrestrial Invertebrates', by Kitty Paviour-Smith and J. B. Whittaker, in J. M. Lambert (ed.), *The Teaching of Ecology: 7th Symposium of the British Ecological Society* (Oxford, Blackwell Scientific Publications, 1967). Permission to include this modified Key (after adaption for the Hong Kong terrestrial fauna) is gratefully acknowledged.

Parasites, pupae and microscopic animals have been omitted. The terrestrial adults of mayflies (Ephemeroptera), dragonflies (Odonata), caddisflies (Trichoptera) and stoneflies (Plecoptera) are included but not their aquatic immature stages as they are sufficiently distinctive to be easily recognized. For further information the textbook by Imms (1964) should be consulted.

Size references are difficult to make at this level but in some cases it is pointed out that the whole group is either 'minute' (i.e. head + body length <1 mm), or else 'small' (<5 mm).

Most of the characters used in the key can be seen if a ×20 magnification is used, but a few characters do require greater magnification, so it is intended that a low power binocular microscope be employed.

KEY TO THE MAJOR GROUPS OF TERRESTRIAL INVERTEBRATES

1   Animals segmented (if not obviously segemented, then either animals worm-like, blunt at both ends and having minute, shining bristles serially arranged along its length, or animals with true jointed legs) 3

1′ Animals not segmented (and not as above) 2

2 Animals soft-bodied and slimy; ventral surface of the body forming a muscular foot .   .   .   .   slugs and snails: **Mollusca**

2' Animals small, worm-like, with a tough shining cuticle
threadworms: **Nematoda**

3 True (jointed) legs present 11

3' True legs absent (annelids and some insect larvae) .   .   . 4

4 Body with many segments, more than 15 (no head capsule)
worms: **Annelida** 5

4' Body with fewer than 15 segments (exclusive of any obvious head) (certain insect larvae)   6

5 With a terminal sucker at each end of the body
leeches: **Hirudinea**

5' Body without suckers   .   earthworms, potworms: **Oligochaeta**

6 With distinct head capsule 7

6' Without distinct head capsule   9

7 Head arranged horizontally so that mouth-parts visible from above (head capsule usually well sclerotized and therefore darker in colour than the rest of the body) 8

7' Head arranged vertically so that mouth-parts not visible from above (head capsule usually scarcely sclerotized so that whitish like the rest of the body)
Larvae of Curculionidae, Coleoptera (part), p. 425

8 Prothorax at least twice the length of, and usually wider than, either of the other two thoracic segments   **Larvae of Buprestidae and some Cerambycidae, Coleoptera** (part), p. 391

8' Prothorax the same length and width as the other 2 thoracic segments   **Larvae of Nematocera** (most), **Diptera**, p. 311

9 Reduced mouth-parts heavily sclerotized (but may need extruding in preserved material)   **Larvae of Higher Diptera,** p. 323

9' Reduced mouth-parts only lightly sclerotized 10

10 Heavily sclerotized, anchor-shaped or toothed 'breast-bone' present
Larvae of Cecidomyiidae, Nematocera (rest), **Diptera,** p. 315

10' 'Breast-bone' absent   **Larvae of Apocrita, Hymenoptera,** p. 337

11   Three pairs of true jointed legs (insects) 19
(also larval mites, see 14; larval pauropods and larval millipedes, see 16)

11' Four pairs of true jointed legs (and with chelicerae and pedipalps)
Arachnida 12

11" More than 4 pairs of true jointed legs 15

12   Segmentation of abdomen distinct (may need to be viewed ventrally) 13

12' Segmentation of abdomen not visible externally . . . 14

13   Pedipalps very large with crab-like pincers (body ending bluntly; animals small) . Chelifers: **Pseudoscorpionidea,** p. 451

13' Pedipalps without crab-like pincers (with pair of eyes, either raised on a tubercle or on a flat, forwardly produced projection)
harvestmen: **Phalangida,** p. 451

14   Abdomen (behind the legs) joined to the rest of the body by a narrow 'waist' . . . . . spiders: **Araneida,** p. 446

14' Abdomen broadly fused to the rest of the body (animals mostly minute, some small) . . mites and ticks: **Acarina,** p. 435
(for larvae, see 11)

15   Appendages behind the head differentiated into 6–7 pairs of walking legs, followed by 5 pairs of small flat plates (pleopods) and 1 pair of uropods ('tail-legs') . woodlice: **Isopoda, Crustacea**

15' Appendages behind the head all (except for the first pair in centipedes) walking legs (N.B. species of millipedes look super-ficially like woodlice) 16

16   With twice-branched antennae (animals minute) . **Paurpoda**
(for larvae, see 11)

16' Antennae unbranched 17

17   Body segments (after the first 3) fused in pairs, giving appearance of 2 pairs of legs on most 'segments' (larvae with fewer than 6 segments visible behind 3 pairs of legs, see 11)
millipedes: **Diplopoda,** p. 455

17' Body segments not fused in pairs, so that only 1 pair of legs per apparent segment 18

18 First pair of legs modified as large jaws (maxillipeds)
                            centipedes: **Chilopoda,** p. 456
18′ First pair of legs not like large jaws (animals small) **Symphyla**

19 Wings absent or extremely reduced 20
19′ Wings present 44

20 With antennae (though these may be much reduced in size) 21
20′ Without antennae but the first pair of legs held out as tactile organs
      (the first 3 abdominal segments each with a pair of minute
      'styliform' appendages; animals minute) . **Protura,** p. 94

21 Always with ventral tube on first abdominal segment; with no more
      than 6 abdominal segments (fourth usually with forked spring
      (furculum); animals minute or small) . **Collembola** 22
21′ Without these characters 23

22 Body elongate with thorax distinct and most or all of the abdominal
      segments visible (ventral tube sometimes greatly reduced);
      furculum sometimes reduced or absent
                    'long' springtails: **Arthropleona (S-O),** p. 95
22′ Body globular, as thorax and the first 4 abdominal segments fused;
      furculum always present
                    'round' springtails: **Symphypleona (S-O),** p. 96

23 Minute appendages (styles) present on at least 2 (usually on many
      more) abdominal segments, in addition to a pair of anal cerci 24
23′ No such styles, except in ♂ long-horned grasshoppers (bush crickets)
      and ♂ cockroaches, and then never more than 1 pair (winged or
      secondarily wingless insects) 25

24 Abdomen ending in segmented median filament between a pair of
      many-joined, filament-like cerci; compound eyes present (ocelli
      may be absent); body covered by scales
                    bristletails: **Thysanura,** p. 93
24′ Abdomen without such a median filament, but with a pair of
      filament-like cerci about as long as the antennae; neither com-
      pound eyes nor ocelli present; body colourless and without scales
      (animals small) . . . . **Diplura,** p. 94

25 With lateral ocelli only (rarely absent, rarely with up to 30 ocelli per side grouped into a pair of 'pseudo-compound' eyes but if so the animal very pale and lightly sclerotized and of the following form); form elongate, without a waist, but with clear thoracic and abdominal segments ... remaining larvae of higher insects . 26

25′ Lateral ocelli absent, but compound eyes (and often dorsal ocelli) present; animal not of the above general larval form. (Rarely the compound eyes reduced to one or a few separate units but if so, the animal as in 36 or 35 respectively) 33

26 Pairs of abdominal false legs (prolegs) present as well as thoracic legs (polypodus larvae)   27

26′ Pairs of abdominal prolegs absent (oligopodus larvae)   .   . 30

27 With 5, or fewer, pairs of unjointed prolegs (these bearing apical rows of minute hooks)
caterpillars: Larvae of Lepidoptera, p. 220
(except Micropterygidae, see 28)

27′ Normally with more than 5 pairs of prolegs (without such hooks)   28

28 Size only a few millimetres; abdominal prolegs (8 pairs) similar to thoracic in being segmented and clawed
Larvae of Micropterygidae, Lepidoptera

28′ Size larger, prolegs not segmented   29

29 Numerous (22–28) ocelli on each side; 8 pairs of prolegs starting with the first abdominal segment; abdomen not annulated
Larvae of Panorpidae, Mecoptera

29′ Only 1 ocellus on each side; 8 or fewer pairs of prolegs starting after the first abdominal segment; abdomen usually annulated
sawfly larvae: Larvae of Symphyta, Hymenoptera, p. 336

30 Sluggish white 'grub-like' larvae   31

30′ Active larvae (or larviform adult) with elongate, somewhat depressed body, often with a well-sclerotized head with obvious forwardly directed mouth-parts and long thoracic legs 32

31 Three simple eyes almost touching in a darkly pigmented patch; retractile legs (3-segmented), appearing conical when retracted, the first pair very close together on either side of mid-ventral line, second and third pairs very wide apart on ventro-lateral margins of body; (head arranged vertically with mouth-parts ventral)   .   .   .   . Larvae of Boreidae, Mecoptera

31′ Ocelli (up to 6) sparsely arranged; non-retractile legs not arranged as above but all three pairs in similar positions on their respective segments; (head often arranged more horizontally with mouth-parts terminal) . . beetle larvae: **Coleoptera** (part), p. 375

32 Larvae fusiform (in 'snake-flies', this shape only behind the heavily sclerotized head and pronotum); (all with very large and obvious 'jaws')   larvae of lacewings, alder-flies, snake-flies etc.:
Neuroptera, p. 213

32′ Larvae (or larviform adult if completely sclerotized) parallel-sided, even if tapering at the ends . . . beetle larvae (rest)
(and larviform ♀ adult glow-worm): **Coleoptera,** p. 375

33 Thorax appearing to consist of only one large segment (mesothorax) since prothorax and metathorax are small but fused with the greatly enlarged mesothorax
adults of wingles flies: **Diptera,** p. 311

33′ Thorax not as above 34

34 Body bearing scales which, if insect very hairy, are visible at least on the abdomen . . . wingless **Lepidoptera,** p. 220

34′ Body without scales 35

35 Mouth-parts modified as long piercing and sucking stylets (usually have triangular-shaped face). (Rarely mouth-parts completely absent, see 46)
nymphal and apterous bugs: **Hemiptera,** p. 145

35′ Mouth-parts present but not modified as stylets . . . 36

36 With a narrow 'waist'; (heavily sclerotized, shining insects; may have club-like vestiges of wings)
ants and other adult wingless **Apocrita, Hymenoptera,** p. 337

36′ Without a 'waist' 37

37 Forelegs greatly expanded, with huge teeth on tibiae and tarsi
nymphal mole-crickets: **Gryllotalpidae, Orthoptera** (part), p. 116
nymphal mantids: **Mantidae, Dictyoptera** (part), p. 128

37′ Forelegs not as above 38

38 Hind-legs (especially their club-shaped femora) very much larger than the first two pairs of legs (and always much longer than the abdomen)   grasshoppers, crickets: **Orthoptera** (part), p. 112

38′ Hind-legs not unusually large compared with the other two pairs of legs 39

39 Head prolonged into a downwardly directed 'rostrum' (beak) bearing mouth-parts at the end   .   .   .   **Mecoptera** (part)

39' Head not as above   40

40 Anal cerci present   41

40' Anal cerci absent 42

41 Dorso-ventrally flattened; cerci short and divergent
cockroaches: **Blattidae, Dictyoptera,** p. 127

41' Body seldom flattened; cerci short and divergent; forelegs raptorial
mantids: **Mantidae, Dictyoptera,** p. 128

41' Body elongate and thin, cerci short and unsegmented, legs usually long and thin   stick insects: **Phasmidae, Phasmida,** p. 120

42 Long thread-like antennae (12–50 segments), prothorax usually small (the front of the head, between antennae, often swollen; insects small   43

42' Short, stubby antennae (6–10 segments), prothorax well developed (insects small-minute)
nymphal and apterous adult thrips: **Thysanoptera,** p. 208

43 No cerci; antennae not moniliform
nymphal and apterous barklice: **Psocoptera,** p. 140

43' No cerci; antennae moniliform; polymorphic, with white soft-bodied workers, soldiers and royal forms
termites: **Isoptera,** p. 133

43" ♀ with swollen fore-tarsi; 2-segmented cerci; ♂ genitalia asymmetrical; antennae moniliform   web-spinners: **Embioptera,** p. 124

44 One pair of wings, the other pair modified as 'halteres' (small club-like organs)   45

44' Two pairs of wings (in a few mayflies and a few beetles, the second pair of wings may be lost completely and, in the latter, hardened forewings (see 49) may be fused together along the mid-line; in one neuropteran sp. the hind wings may be minute)   .   . 47

45 Fore pair of wings modified as 'halteres' (metathorax is the enlarged thoracic segment; insect small)   .   .   .   ♂ **Strepsiptera**

45' Hind pair of wings modified as 'halteres' (mesothorax is the enlarged segment as in 33)   46

46 Mouth-parts present; prothorax indistinct as fused with mesothorax; (tarsi usually with 5 segments but never as few as in 1; wings usually with at least 1 cross vein joining 2 long veins
Diptera, p. 311

46' Mouth-parts completely absent; prothorax quite distinct from mesothorax; (tarsi of 1 segment; wings lacking any cross veins joining long veins; insects small—minute)
adult ♂ **Coccidae, Hemiptera**, p. 200

47 Mouth-parts modified as long sucking stylets (usually have triangular-shaped face) .   .   bugs: **Hemiptera** (rest), p. 145

47' Mouth-parts not thus modified 48

48 Forewings forming usually hardened 'elytra' or wing covers which usually meet in the mid-dorsal line, but never overlap; membranous hind-wings fold longitudinally and transversely under elytra or hind-wings absent 49

48' Forewings at least partly membranous, often overlapping; hind-wings may fold longitudinally but never transversely under elytra 50

49 Elytra usually covering most of the abdomen, but if short, then the end of the body without 'forceps'   beetles: **Coleoptera**, p. 375

49' Elytra always short exposing most of abdomen; and abdomen ending in a pair of more or less converging 'forceps' (unjointed cerci) .   .   .   .   .   earwigs: **Dermaptera**, p. 123

50 Forewings interlocking with hind-wings by means of a ridge on the former and rows of hooks on the latter; (other characters which may be present: very much reduced venation, elbowed antennae; a pigmented spot on the leading edge of forewing)
Hymenoptera, p. 51

50' Wings not interlocking   52

51 Body without a waist .   .   .   sawflies: **Symphyta**, p. 336

51' Body with a waist   ants, bees, wasps, etc: **Apocrita**, p. 337

52 Both pairs of wings strap-like and fringed with long hairs on both leading and trailing edges (antennae and prothorax as in 44; insects small-minute)   .   . thrips: **Thysanoptera**, p. 208

52' Wings otherwise 53

53 Body and wings bearing minute, pigmented scales (sometimes extremely sparse on the wings or mixed with hairs on the body; may be visible only under high magnification)

moths and butterflies: **Lepidoptera,** p. 220

53′ Neither body nor wings with scales 54

54 First vein extending no further than half-way along wing, ending at a thickened joint-like cross vein; thorax distorted to bring all legs almost completely in front of the first pair of wings (body length —3 cm) . . . . dragonflies: **Odonata,** p. 101

54′ Venation otherwise 55

55 First 2 veins extending (unbranched) the whole length of the wing, parallel to each other and more or less to the front edge of the wing; thorax not distorted as above 56

55′ Venation not thus 59

56 At rest, wings held vertically above the body; many 'intercalary' veins present which have lost basal connection to main veins but connected by numerous cross veins; (bearing 2 filamentous cerci and often a third tail filament)

mayflies: **Ephemeroptera,** p. 97

56′ Wings held tentwise over the body; no intercalary veins . 57

56″ Wings held flat over the body 58

57 'Costal veinets' numerous, giving ladder-like effect along front edge of each wing

lacewings, alder flies, snake flies: **Neuroptera** (most)

57′ No such ladder-like effect given by costal veinlets; (small insects covered with whitish, waxy exudation; but this may not be visible in alcohol); hind-wings may be smaller than fore wings

Coniopterygidae, Neuroptera (rest), p. 213

58 Many extra small cells in wings; no cerci; wings capable of being shed by basal fracture . . . . **Isoptera,** p. 133

58′ Most veins poorly developed, few cells; 2-segmented cerci

♂ **Embioptera,** p. 124

59 Forelegs as in 37

mole-crickets: **Gryllotalpidae, Orthoptera,** p. 116

mantids: **Mantidae, Dictyoptera,** p. 128

59′ Forelegs not as above 60

60 Hind-legs as in 38

   grasshoppers, crickets: **Orthoptera** (most), p. 112

60′ Hind-legs not unusually large compared with the other two pairs
    of legs 61

61 Face prolonged downwards into a 'rostrum' (beak) as in 39 (apex
    of ♂ abdomen usually scorpion-like)

   scorpion flies: **Mecoptera** (rest)

61′ Face not thus prolonged 62

62 Prothorax very small compared with either head or mesothorax; at
    rest, wings held tentwise over abdomen 63

62′ Prothorax the same size as, or larger than, the head or mesothorax;
    at rest wings usually held flat (but never tentwise) over
    abdomen 64

63 Front of the head obviously swollen above the upper lip (labrum);
    mandibles (biting jaws) always present (though not always easy
    to see without dissection); usually wing membrane naked between
    the veins but, if not, then hairs minute and colourless (small
    insects) . . . . book lice: **Psocoptera,** p. 140

63′ Front of the head not swollen; mandibles absent; wing membrane
    between the veins always covered more or less densely with
    pigmented hairs . . caddis flies: **Trichoptera,** p. 302

64 Shield-like pronotum much longer than the head or mesothorax or
    metathorax, and almost hiding the head; forewings rather
    thickened and opaque (jaws at rest between the bases of the
    front legs) . . . . cockroaches: **Dictyoptera,** p. 127

64′ Pronotum elongate; large mobile triangular-shaped head with
    conspicuous eyes; forelegs large and raptorial; forewings rather
    thickened and opaque . . mantids: **Dictyoptera,** p. 128

64″ Entire body and legs elongate and usually thin

   stick insects: **Phasmidae, Phasmida,** p. 120

64‴ Prothorax, mesothorax, metathorax and head all of about equal
    size; pronotum not shield-like and not hiding the head; both
    pairs of wings transparent . stoneflies: **Plecoptera,** p. 110

CHAPTER FIVE

INSECTS IN HONG KONG ARRANGED SYSTEMATICALLY

(The following Orders are not included as they have not yet been recorded locally: Grylloblattodea, Zoraptera, Mecoptera and Strepsiptera)

CLASS **INSECTA**

SUB-CLASS **APTERYGOTA**

ORDER **THYSANURA**
(Bristletails)

The bristletails are thought to be the most primitive of all living insects. They are wingless, sometimes without compound eyes, and characteristically with a pair of long tapering antennae, two long many-segmented cerci and an equally long median tail filament. The abdomen bears a variable number of lateral pairs of styliform appendages, on the pregenital segments. Metamorphosis is virtually non-existent. They are generally quite small in size, and cryptic by nature, usually hiding in leaf litter, under stones, and in similar situations.

A few species are found in Hong Kong, but the group has not been studied locally at all. The ubiquitous Silverfish *(Lepisma saccharina)* (Plate 1) occurs in domestic premises and in food stores, feeding on crumbs and fragments of food material or general organic debris. They seem to prefer quiet dry places such as food cupboards. In some buildings in particularly warm situations have been seen occasional specimens of the Fire Brat *(Thermobia domestica)* but it does not appear to be common.

Plate 1.   Silverfish, *Lepisma saccharina* (Thysanura); body length 8 mm.

In forest leaf litter there can be seen several quite large dark species, with large compound eyes and ocelli; they clearly belong to the family Machilidae and are tentatively regarded as being species of *Machilis* and called Woodland Bristletails. Also in this family is the species *Petrobius maritimus* which is quite common on European beaches on the rocks at high tide level; but this species has not been recorded locally.

ORDERS DIPLURA & PROTURA

The next two orders of Apterygota in taxonomic sequence are the Diplura and the Protura; both are quite small groups of primitive wingless insects found in soil and leaf litter. They are widespread as a group and soils in most parts of the world contain representatives of both orders. Because they are so small they are able to penetrate quite deeply in most soils. So far as is known none have been collected locally, although it is presumed that they will occur here.

ORDER **COLLEMBOLA**
(Springtails)

These are also primitive insects, as would be expected by their inclusion in the Apterygota. They are small in size, seldom exceeding 5 mm in length, without wings, and adapted for a life in leaf litter or soil. They are of importance ecologically for they can be found in soil in very large numbers, and are clearly of importance in soil food chains, partly in their role as detritus and fungus feeders and also as prey for the soil predators.

They possess a springing organ on the ventral surface of the abdomen, with which most species can jump distances of several centimetres quite easily. The order is divided into two separate sub-orders according to their body form.

Sub-order **Symphypleona**

The Symphypleona are all globular in shape with the abdominal segments more or less fused together. They are restricted to living on the surface of the soil and in leaf litter, presumably by virtue of their body shape, and have quite well developed eyes, antennae and springing organ. Some species elsewhere are agricultural crop pests for they occur on certain soils in very large numbers and with their biting mouth-parts, they can damage seedling crop plants. There are three large families in this group all of more or less cosmopolitan distribution.

Sub-order **Arthropleona**

The suborder Arthropleona contains about nine families, all of which have attenuate bodies, with the abdominal segments free, and because of their more elongate shape, they tend to be found in the soil rather than just on top. The different families vary in the extent of their development of eyes, antennae and springing organ, and consequently their precise ecological location. The ones that penetrate deepest into the soil tend to have small antenna, vestigial springing organ, and either tiny eyes or else are eyeless. They are also usually white in colour whereas the surface forms are pigmented and may be green, brown, or black and white striped. In other countries there are species of Collembola that are marine and they live on the beaches; others live on the surface of fresh water, and a few live in the nests of termites and ants as inquilines.

It has been observed that locally some species live in rather unusual habitats. In a laboratory sink, in the Northcoate Science Building, there was for several months a flourishing colony of a large black and white species identified as *Homidia* species. Similar colonies have been found on an *Osmunda* board on which an epiphytic orchid is growing on a verandah, and also under the light cover on a domestic freshwater fish-tank.

In temperate regions most springtails are soil and litter species, some descending to considerable depths in the soil. The soils in Hong Kong are generally poor in structure with ill-defined layers and deficient in humus, which might account for the fact that the local springtail soil fauna is sparse, but recent collecting has indicated that there are large numbers of springtails in the forest foliage and also feeding on fungus on dead branches and twigs. Apart from casual collecting, however, the group has not been studied locally.

SUB-CLASS **PTERYGOTA**

DIVISION I **EXOPTERYGOTA (= HEMIMETABOLA)**

ORDER **EPHEMEROPTERA**
(Mayflies)

Several different types of mayfly nymphs can be found in most of the permanent streams, reservoirs, and in the ditches around paddy fields. Generally the body form is correlated with the habitat, for different species prefer different types of freshwater bodies to inhabit. Thus in the swiftly flowing streams the nymphs are broad and flattened (Plate 2) with the legs flattened at the sides of the body; this shape enables them to withstand the effects of the water currents as they cling to the tops and sides of the rocks and stones in the stream. The nymphs with more rounded and cylindrical bodies inhabit in quiet water, either at the sides of streams in the lower reaches, or else in standing water. The few species that burrow in the gravel, such as *Ephemera,* have the abdominal gills transferred from the sides of the abdomen to a dorsal position where they are protected in the burrow by the hump of the thorax in front.

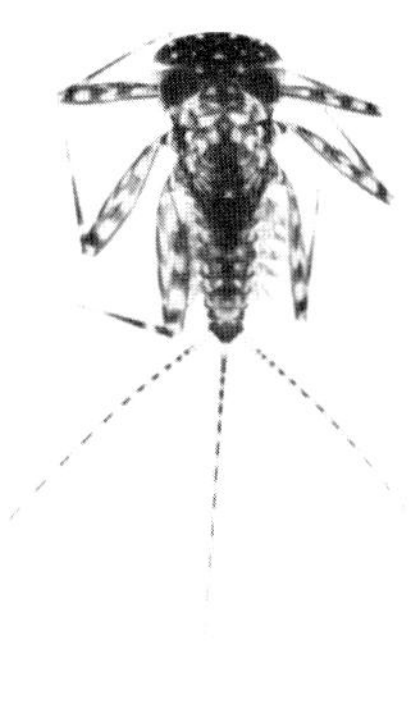

Plate 2. Mayfly nymph, *Compsoneuria* sp. (Ephemeroptera), from fast-flowing stream; body length 8 mm; with three characteristic tail filaments.

Plate 3. Adult mayfly, *Ameletus* sp. (Ephemeroptera), showing two caudal filaments; body length 10 mm.

Most of the nymphs feed on algae and pieces of plant debris, but a few are known to be carnivorous in other countries. They are typically long-lived in that the nymphal life may be a year or more and they may pass through as many as 30 instars during this time. The local species appear to be phytophagous or else detritivorous, although elsewhere *Caenis* species are recorded as being carnivorous.

The nymphs breathe through a series of seven pairs of plate-like lateral abdominal gills. These are tracheal gills and filled with tracheae which are in turn filled with air. The oxygen in the water diffuses across the thin cuticle covering the gills and passes directly into the air in the tracheal system.

Both nymphs and adults have tiny antennae that are scarcely visible to the unaided eye, and both have either two or three long 'tails' at the end of the abdomen. The outer 'tails' are formed from the pair of long segmented cerci, and the central 'tail', if present, is the median dorsal filament, formed by the prolongation of the epiproct (dorsal plate) of the last abdominal segment.

The adults are very short-lived (hence their name), delicate, weak-flying insects with either one or two pairs of wings (Plate 3), and the long tail filaments for balancing while in flight. If the second pair of wings is present it is invariably a small pair. Some species live for only a few hours, others for up to a day or longer but none feed in the adult stage, and in fact they have vestigial mouth-parts and the intestine is filled with air to give them greater buoyancy. At rest the wings are held vertically over the insect's back. These are the insects copied by trout fishermen as 'flies'. The males swarm in a dancing cloud above the water, and a female entering the swarm mates in mid-air with one of the males and then drops to the water surface to lay her eggs. After mating and egg-laying the spent bodies fall on to the water and usually fall prey to the rising fish, or sometimes to predacious insects. Both nymphs and adults play an important role in freshwater food chains.

Up to date about 16 species of local mayflies have been identified tentatively, and these belong to seven families. Classification is usually based largely upon wing venation, which makes the group a difficult one for the non-expert especially as some of the veins are quite faint. The nymphs can be grouped together on the basis of how they live, as (a) flattened torrent stream nymphs, (b) burrowing nymphs, (c) swimming nymphs, and (d) creeping nymphs. On this basis, Hong Kong has *Compsoneuria* and *Epeorus* as flattened torrent stream species; *Ephemera* as a burrower; *Baetis, Pseudocloeon, Choroterpes, Habrophlebiodes, Ameletus* and *Isonychia* as swimmers usually to be found in standing water or in slow–flowing streams; and finally *Caenis* and *Ephemerella* as creepers. The creeping types generally live on a sandy-mud bottom and they cover themselves with debris as they creep around; *Caenis* is elsewhere recorded as being quite carnivorous in feeding habits.

Baetidae

Arranged systematically there are locally *Baetis* spp. and *Pseudocloeon* together with what appears be to a new genus all in the family Baetidae. These are known as Small Mayflies and the nymphs are swimmers to be found in either ponds or slow-moving streams. In streams, such as the Lam Tsuen stream and that in Tai Po Kau Forest, these mayflies will be found in the slower lower reaches, often in the vegetation that grows in the edges of the stream.

Caenidae

The Caenidae is represented by a couple of species of *Caenis;* the adults are night-time swarmers and they are small with the hind-wings lost. The nymphs are creepers in streams with a sandy-mud substrate and often cover their bodies with debris; in some texts they are regarded as being carnivorous but other workers report that they are omnivorous or detritivores. Nymphs are characteristic in that the first pair of lateral abdominal gills are vestigial and the following pair make a cover which shields the remainder and presumably keeps them unclogged from detritus.

Ephemerellidae

Ephemerella is another swimming mayfly, and belongs to the family Ephemerellidae. They are of medium size and also are to be found creeping on the stream bottom and may be covered with debris. However the local species appears to swim more than creep.

Ephemeridae

The Ephemeridae are called Burrowing Mayflies because of their habit of burrowing into the substrate for protection from predators. There appears to be only one species of *Ephemera* occurring locally, but it has sometimes been found in large numbers burrowing in the gravel of the stream bed in several different streams in the New Territories. These are large mayflies and the adults typically swarm in the late evening and at night.

Heptageniidae

The Heptageniidae are the Torrent or Stream Mayflies, and are typically found only in fast-flowing streams. Here in Hong Kong they are confined to the upper reaches of the local streams where the water flows swiftly, and they sit on the sides of rocks and boulders in the streams. Their flattened shape enables them to withstand the dislodging effect of the stream. Although the very common species *Compsoneuria* can be found in large numbers in many different streams as nymphs the adult mayfly of this species has not yet been collected. The Two-tailed Torrent Mayfly is *Epeorus* and can be found in small numbers alongside the *Compsoneuria* in some streams. *Thalerosphyrus* has been collected as an adult but has not yet been found as a nymph. Clearly only a small proportion of the local mayfly fauna has been collected up to date.

Leptophlebiidae

Leptophlebiidae are medium-sized mayflied with swimming nymphs which occur in ponds and quiet streams. *Choroterpes* and *Habrophlebiodes* are both found in New Territories streams in small numbers.

Siphloneuridae

The final family is the Siphloneuridae which has swimming nymphs to be found in quiet water or in aquatic vegetation, and is represented here by *Ameletus* found only as adult, and by *Isonychia* as nymphs in quiet streams.

ORDER **ODONATA**
(Dragonflies and Damsel-flies)

The Odonata are very well represented in Hong Kong by a large number of vividly coloured dragonflies and damsel-flies. They range in size from quite small to very large and are all strong fliers with two large pairs of equally-sized wings.

Sub-order **Anisoptera**

The sub-order Anisoptera comprises the dragonflies proper, and these are stout-bodied insects from 3–10 cm in length with large eyes, tiny tapering antennae and broad wings; they are often brightly coloured, and their three pairs of long bristly legs are positioned anteriorly almost under the head. The fast-flying adults are all predatory and they catch on the wing other insects which are then devoured by their large biting jaws. At rest the wings are held laterally; in some species the wings may be smoky or brown in colour but typically they are transparent. When hunting they hawk up and down streams and paths, returning again and again to a favourite resting place in the sun.

Lai (1971) described 36 species of dragonflies from Hong Kong and 20 species of damsel-flies, but Asahina *(in litt.)* believes that there are probably 100 species in total to be found here.

The newly emerged adult dragonflies spend about two weeks feeding before they start to seek water for breeding. The males then congregate round a suitable stream or pond, and fight each other for the right to defend a certain territory. A female of the same species approaching this territory is immediately seized, and a courtship flight, followed by mating and egg laying, usually occurs. Females tend to stay away from water most of the time, only coming for short periods. Both sexes may mate many times, and sometimes the male remains in attendance while the female lays her eggs. Some species lay their eggs freely in the water, dipping down to the surface as they lay each egg in flight. This behaviour can frequently be observed during the summer months at swimming pools, but presumably in these locations the eggs do not develop successfully. Other species sit on a submerged plant and poke their abdomen down to find the stem; others may even submerge themselves completely during oviposition.

The method of mating is strange; the male bends his abdomen forwards and transfers sperm from the tip to a special storage receptacle situated on the second abdominal segment. The male has special claspers at the end of the abdomen with which he grips the female by the head or anterior thorax. The female then bends her abdomen forward so that the tip touches the male's storage receptacle and she becomes impregnated. After fertilization, the pair may continue to fly about for some time with the male supporting the female by the neck, and both insects using their wings. At various times of the year swarms of dragonflies can be seen soaring and catching other insects in the lee of tall trees or tall buildings sometimes hundreds of metres from the nearest water. Some species are also migratory and there appears to be some movement of dragonflies into and out of the Colony at times during the summer.

The nymphs of dragonflies are aquatic, and some of the larger ones take more than a year to develop. They are predatory and feed on small aquatic animals, sometimes even on small fish, which are captured by their peculiar hinged 'mask' formed from a section of the mouth-parts and bearing a pair of movable claws at the apex. The body shape varies from rather elongate (Plate 4) to quite squat and flattened (Plate 8), and is clearly adapted for clinging to rocks and stones in swiftly flowing streams. The nymphs breathe by drawing water into the rectum whose wall is very thin and covered by many small projections which act as gills. On the occasions when rapid movement is required the dragonfly nymph can propel itself forward rapidly for some centimetres by squirting the water out of its rectum in a muscular contraction. It can perform this movement a number of times in rapid succession, and it makes a very successful escape mechanism from predators. Plates 4–8 show dragonfly nymphs of various species.

When fully-grown the nymphs swim to the water surface and climb up on vegetation where they moult for the last time, usually during the night, and in the morning the newly emerged adults take off on their first flight. Often they fly before their new integument is fully hardened and as they rise the abdomen is soft and sagging, hardening finally in flight.

The classification of dragonflies is a difficult one for the non-specialist as it relies very heavily upon wing venation, but for general studies the separation of dragonflies into their respective families does not confer any great advantage, for many groups are similar in their habits and ecologically. In general the Aeschnidae are usually the largest specimens, and some species prefer to live in running water whereas others prefer ponds and still water. But these habitat preferences tend to apply to individual species rather than to larger taxonimic groups.

Some of the dragonflies show sexual dimorphism which can make field recognition rather difficult. For example Plate 14 shows the male of

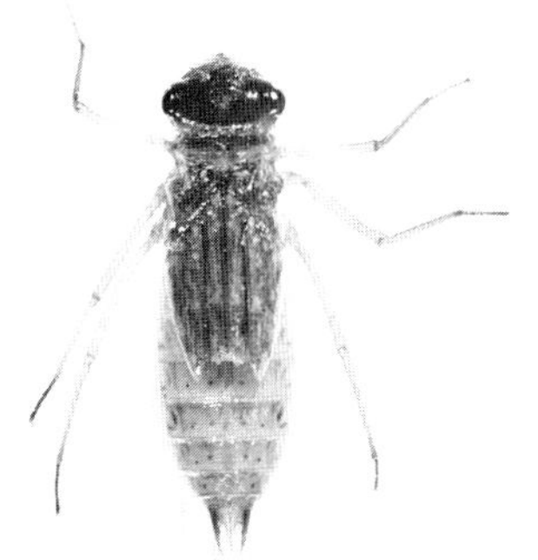

Plate 4. Dragonfly nymph, *Pantala flavescens* (Odonata, Libellulidae); body length 22 mm.

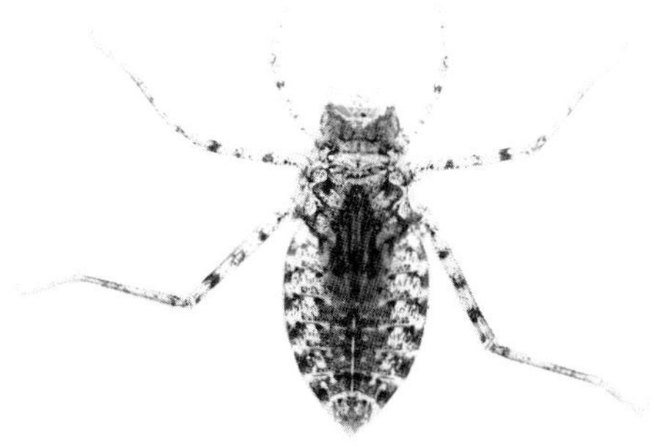

Plate 5. Dragonfly nymph, *Epophthalmia elegans* (Odonata, Macromiidae); body length 37 mm.

Plate 6. Dragonfly nymph, *Heliogomphus* sp. (Odonata, Gomphidae); body length 20 mm.

Plate 7. Dragonfly nymph, *Onychogomphus sinicus* (Odonata, Gomphidae); body length 18 mm.

Plate 8. Dragonfly nymph, *Zygonyx iris* (Odonata, Libellulidae); body length 25mm.

Neurothemis tullia and Plate 15 the female. The dimorphism is so pronounced that they would appear to be two different species. A selection of common Hong Kong adult dragonflies is illustrated in Plates 9–16, consisting of *Tramea virginia*, *Anaciaeschna jarpidea*, *Neurothemis tullia*, *N. fulvia*, *Rhyothemis variegata*, *Trithemis aurora*, and *Pseudothemis zonata*. There is rather a large number of dragonflies with dull coloured bodies and hyaline wings (like *Anaciaeschna jarpidia*) which are difficult to identify, especially in the field.

Plate 9.   Dragonfly adult, *Tramea virginia* (Odonata, Libellulidae); body length 50 mm.

Plate 10.   Dragonfly adult ♀, *Anaciaeschna jaspidae* (Odonata, Aeschnidae); body length 60 mm.

Plate 11. Dragonfly adult ♂, *Pseudothemis zonata* (Odonata, Libellulidae); body length 48 mm.

Plate 12. Dragonfly adult ♀, *Neurothemis fulvia* (Odonata, Libellulidae); body length 37 mm.

Plate 13. Dragonfly adult ♀, *Rhyothemis variegata arria* (Odonata, Libellulidae); body length 35 mm.

Plate 14. Dragonfly adult ♂, *Neurothemis t. tullia* (Odonata, Libellulidae); body length 23 mm.

Plate 15. Dragonfly adult ♀, *Neurothemis t. tullia* (Odonata, Libellulidae); body length 22 mm.

Plate 16. Dragonfly adult ♂, *Trithemis aurora* (Odonata, Libellulidae); with its bright red abdomen; body length 48 mm.

Sub-order **Zygoptera**

The damsel-flies (sub-order Zygoptera) are characterized by being of more slender and delicate build; the wings are narrow at the base (i.e. petiolate) and at rest they are held vertically together, closed, and over the back of the insect. The head is also more transverse and the large eyes thus widely separated. The wings are sometimes brightly coloured and very beautiful. They normally do not stray so far from water as do the dragonflies, but some very small species are sometimes abundant in grassland and low vegetation, especially so in the vicinity of paddy fields.

The nymphs of damsel-flies are also fully aquatic and possess the predatory 'mask' for catching their prey. However, they are very distinctive in that their body form is more slender than dragonfly nymphs and their abdomen terminates in three caudal gills, which in life are gently wafted from side to side to increase ventilation (Plate 17). Although a few species have the squat flattened form which indicates the habit of living in swiftly moving water, damselfly nymphs in general prefer quieter water ponds or slow moving streams.

The common stream species of damsel-fly nymph which is widespread in the New Territories is *Euphaea decorata* (Plate 17), which is quite flattened in shape and is capable of clinging to rocks in torrent regions of the streams. The more slender-bodied forms are to be found lower down the stream courses.

A few common and spectacular damsel-fly species are illustrated in Plates 18–21: *Rhinocypha perforata* (Plate 18), *Neurobasis chinensis* (Plate 19), *Euphaea decorata* (Plate 20), and *Mnais mneme* (Plate 21). Some damselflies have brightly coloured bodies, but again there is rather a large number of species with dull-coloured bodies and hyaline wings which are very difficult to identify. Some species are surprisingly tiny and measure as little as 1.5 cm in body length.

Plate 17. Damsely-fly nymph, *Euphaea decorata* (Odnata, Euphaeidae); body length 15 mm.

Plate 18.   Damsel-fly   adult   ♂,   *Rhinocypha   perforata*   (Odonata,
Epallaginidae); body length 26 mm.

Plate 19.   Damsel-fly adult ♂, *Neurobasis chinensis* (Odnata, Calop-
terygidae); body length 63 mm.

Plate 20. Damsel-fly adult ♂, *Euphaea decorata* (Odonata, Euphaei-dae); body length 40 mm.

Plate 21. Damsel-fly adult ♂, *Mnais mneme* (Odonata, Callopterygi-dae); body length 58 mm.

ORDER **PLECOPTERA**
(Stoneflies)

In some of the permanent streams in the New Territories and on Lantao Island stonefly nymphs can be found. So far three species have been collected and it is thought they belong to the large genus *Perla*. The nymphs rather resemble mayflies but they only have two 'tails' (cerci) for there is no medial caudal filament, and the antennae are quite long and filamentous (Plate 22). The gills are tuft-like and usually situated at the base of the legs on the sides of the thorax.

Plate 22. Stonefly nymph, *Perla* sp. (Plecoptera); body length 18 mm.

Some species are herbivorous and feed largely on algae and diatoms, but most nymphs in the Perlidae are carnivorous and feed on midge larvae and mayfly nymphs. The nymphal life can extend for one or two years during which time there may be 20 or more instars. The last instar nymphs crawl cut of the water and moult on the rocks, so that after the adult stoneflies have flown off, the old nymphal skins (exuviae) remain clinging to the sides of the rocks in the stream bed. The hatching of the adults typically occurs in the spring and early summer in Hong Kong, although a few do emerge at other times during the summer.

The adult stoneflies look very much like the nymphs but possess two pairs of large, clear, gauzy wings which are held down flat over each other along the abdomen at rest (Plate 23). Mating apparently occurs on the ground, and although the adults may live for several weeks, their mouth-parts are rather weak and they feed little. They are poor fliers and are seldom found far from water. Generally stoneflies are only found in clear fast flowing streams and never in polluted water, so their presence can be a useful ecological indcator of low levels of pollution.

Classification schemes differ and a general European system recognizes some nine families whereas a popular American system lists up to a dozen different families. However, it is agreed that the nymphs of the Perlidae are carnivorous, whereas other families such as the Green-winged Stoneflies (Isoperlidae) and the Nemouridae are herbivorous and feed on diatoms, desmids, and other algae. It should however be remembered that for animals that browse or graze in such locations as on rock surfaces under water, there is no real distinction between being herbivorous and being carnivorous—they tend to scrape off all the organic material from the rock surface as most of it is digestible. Many of these species will in fact be omnivorous in feeding habits if one wishes to be punctilious.

It is now known that a species of *Nemoura* (Small Stonefly) can be found in some streams of the New Territories.

Plate 23.   Adult stonefly, *Perla* sp. (Plecoptera); body length 11 mm.

ORDER **ORTHOPTERA**
(Grasshoppers, Crickets, etc.)

The Orthoptera are characterized by having their forewings modified into thickened, protective tegmina. In flight the forewings are held out horizontally and act as aerofoils while the large fan-shaped hind-wings beat and provide the thrust. The base of the hind-wing is often brightly coloured red or yellow which flashes in flight but disappears as soon as the insect alights. All these insects have powerful biting and chewing mouth-parts, and in the larger species the horny, dark mandibles are conspicuous—many large orthopterans will bite if handled carelessly. Most species have strongly developed hind-legs and can jump well. Out of a total of about 14 familes 5 are well represented in Hong Kong.

Tettigoniidae

The Tettigoniidae are long-horned grasshoppers, and have distinctive long tapering antennae, large hind-legs for jumping, and the female has a stout, often curved, ovipositor. As a group they are inhabitants of bushes and trees rather than grassland and are mostly nocturnal in habits. In some species there are alternative colour forms—usually green, but some brown individuals, and some of the cryptic species have a silhouette resembling a folded leaf. Most are vegetarian, and a few are agricultural pests, but some are carnivorous and feed upon other Orthoptera and cockroaches.

A group of common long-horned grasshoppers all have an elongated head ending in a point between the antennae and the jaws which are deflected posteriorly under the head. The commonest form is probably *Conocephalus* sp. and *Euconocephalus* spp. (Plate 24), about 50–60 mm in length, and are usually green but sometimes brown in body colour. These species are attracted to lights and will often be found on buildings in the morning. A similar brown species with a stouter body (Plate 25) is also quite common at times, and has been identified as *Pseudorhynchus* species, and there are several quite small species of *Conocephalus* with the characteristic pointed head. The Large Brown Leaf-grasshopper *(Mecopoda elongata)* (Plate 26) with its characteristic shape and very long hind-legs is a most striking insect to be encountered in bushes or tree foliage. This is one of a group of species with a 'normal' face, as is also the medium-sized green *Elimaea punctifera* (Plate 27) often to be found on Hibiscus bushes where they frequently eat holes in the side of developing flowers. The female Long-horn Grasshoppers use their stout blade-like ovipositor either to make slits in the stem of the preferred

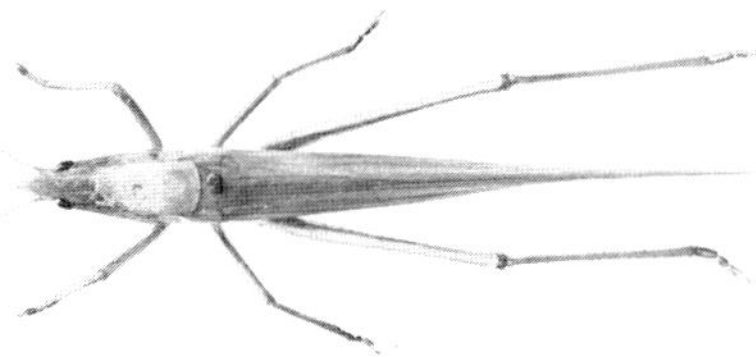

Plate 24.   Long-horned grasshopper, *Euconocephalus* sp. (Orthoptera, Tettigoniidae); body length 35 mm.

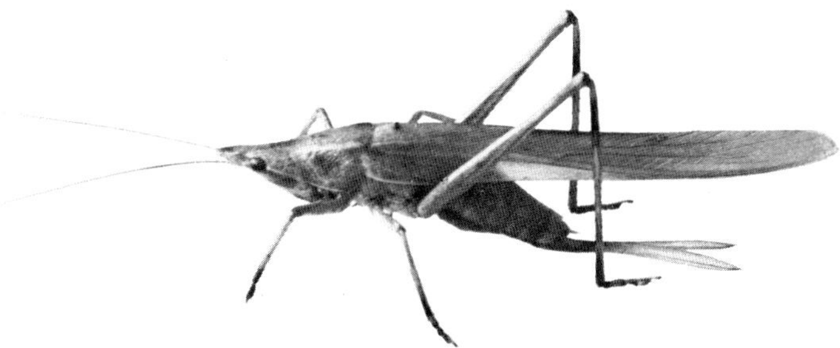

Plate 25.   Long-horned grasshopper, ♀ *Pseudorhynchus* sp. (Orthoptera, Tettigoniidae); body length 45 mm.

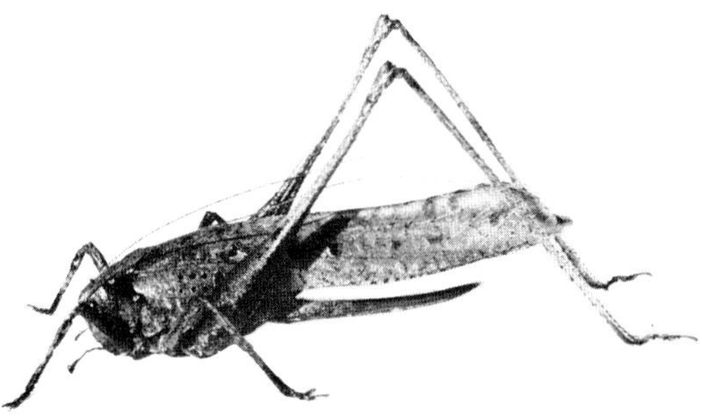

Plate 26.   Large Brown Leaf Grasshopper, ♀ *Mecopoda elongata* (Orthoptera, Tettigoniidae); body length 60 mm.

food plant, or else in soil, where the eggs are laid. The males 'sing' (stridulate) at night and can be heard regularly during the warmer part of the year. The 'song' is produced by rubbing the bases of the two forewings together—the edge of the right forewing rubs on a roughened

Plate 27. Long-horned grasshopper, *Elimaea punctifera* ♀ (Orthoptera, Tettigoniidae); feeding on Hibiscus; body length 25 mm.

patch on the left forewing. The chirping sound produced is intensified by resonance from a smooth circular area (mirror) on the right forewing base. In order to hear these calls long-horned grasshoppers have 'ears' in their 'knees'. The audioreceptory organ is evident as a short slit on the tibia of the forelegs. This group does contain some wingless species and to date only one apterous form has been collected in Hong Kong and this is carnivorous.

Gryllidae

The Gryllidae are crickets, which also have long antennae and a pair of long unsegmented cerci posteriorly. Their hind-legs are stout but they jump only slightly, and the females have long straight ovipositors. Most species are nocturnal and spend the daytime in their subterranean burrows and nests and come out at night to forage; the entrance to the nest hole of the agriculturally important Big-head Cricket *(Brachytrupes portentosus)* (Plate 28) is up to 20 mm in width (Plate 29). This species sometimes does devastating damage in seed beds in the New Territories. Typically, the seedling stems are severed and the plants left on the soil surface for a day in order to wilt, then the following night the limp seedlings are dragged down into the cricket's nest to be eaten. Crickets are omnivorous and some species are pests of some importance. The House Cricket *(Acheta domesticus)* is a minor domestic pest found in the warmer parts of the world, and other species of *Acheta* are minor pests of rice in Hong Kong. A local species of Field Cricket is *Teloegryllus testaceus* (Plate 30), and many tiny crickets can be found, sometimes in very large numbers, in short grassland sward, leaf litter, and sometimes in agricultural fields (e.g. *Tridactylus* spp.).

Plate 28. Big-head Cricket, adult ♂, *Brachytrupes portentosus* (Orthoptera, Gryllidae); body length 38 mm.

Plate 29. Entrance to nest hole of Big-head Cricket; about 20 mm in diameter.

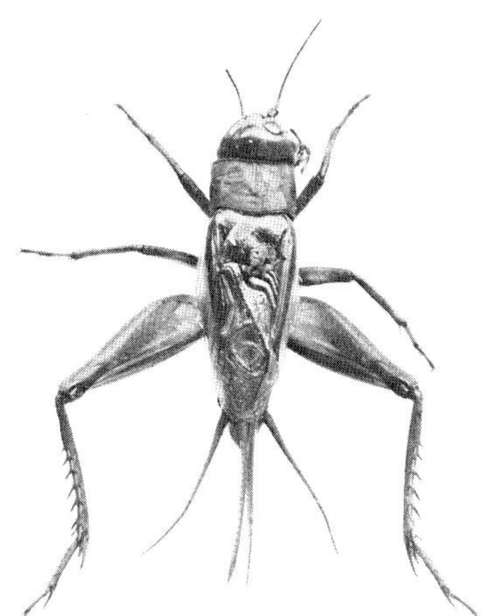

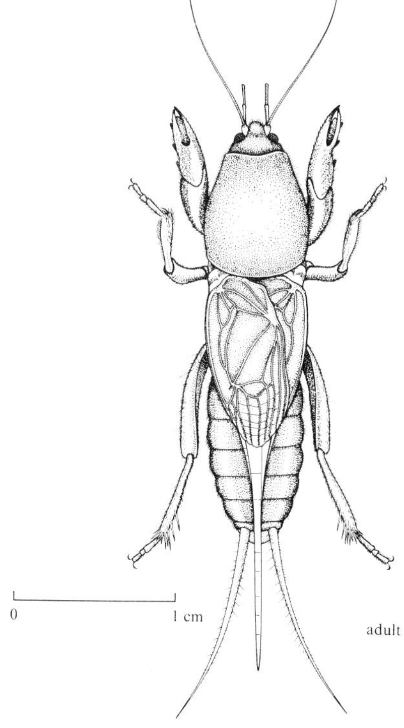

Plate 30. Field Cricket, *Teloegryllus testaceus*, adult ♀ (Orthoptera, Gryllidae); body length 20 mm.

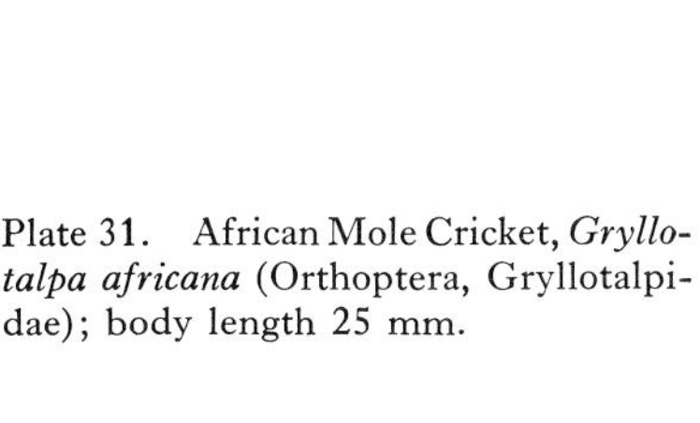

Plate 31. African Mole Cricket, *Gryllotalpa africana* (Orthoptera, Gryllotalpidae); body length 25 mm.

Crickets have sound producing organs similar to those of the long-horned grasshoppers, and the 'song' produced by the males is notoriously loud and piercing. Both right and left forewings possess file, scraper and mirror areas, and as the insect 'sings' the forewings are raised at an angle of about 45° above the abdomen, and moved backwards and forewards one over the other. The song is characteristic of the species, and it is reported that there are distinct aggressive, courting, mating, and recognition components. Most 'singing' takes place at night, when the intensity of sound sometimes rivals that of the cicadas.

Occasionally arboreal crickets can be found, living on the leaves of shrubs and trees, and called respectively Bush and Tree Crickets; they are quite distinctive and belong to different sub-families.

Gryllotalpidae

The third group of importance is the Gryllotalpidae or Mole Crickets. This remarkable group actually contains about 50 species but so far the best known species is *Gryllotalpa africana,* the African Mole cricket, which occurs throughout Africa and Southeast Asia and is quite common in agricultural parts of the New Territories (Plate 31). It spends more or less the entire life underground, digging through the soil easily with its specially enlarged, spade-like forelegs from which it derives its vernacular name. It is a regular minor pest of many agricultural crops throughout its distribution range, eating the roots of the plants as it burrows in the soil. At certain times during the summer the adults fly at night and may be attracted to house lights.

Acrididae

The Acrididae are the short-horned grasshoppers and locusts, with short antennae, very short female ovipositor, and large hind-legs for jumping. They are all vegetarian and many are agricultural pests. The locusts are especially serious pests; in mainland China there is the Asiatic Migratory Locust *(Locusta m. migratoria)* but so far it has not been recorded from Hong Kong. Acridids are very common and wide-spread locally; they are all inhabitants of grassland although occasionally are found in other habitats, and are probably the most serious pests on cereal crops. The eggs are laid in special tubular egg-pods in sandy soil, the top of the tube being sealed off by a specially secreted froth which hardens and dries on exposure to air. Both sexes make chirping noises, although it is more frequent by the males. Each species has a distinctive 'song' used for keeping the population together and for attracting females for mating. The scraping sound is produced by rubbing a row of small

(Left) Plate 32. Large Green Grasshopper, *Chrondracris rosae* (Orthoptera, Acrididae); body length 60–80 mm.

(Above) Plate 33. *Gastrimargus marmoratus,* female; body length 40 mm.

pegs on the hind femur rapidly across a thickened vein on the forewing, although the details of the mechanism vary somewhat in the different groups.

There are ten sub-families in the Acrididae but most are rather difficult to recognize. The largest group is the Acridinae which contains the locusts and the 'typical grasshoppers'. Locally, the range found within this group is enormous; the smallest forms are tiny, stubby grasshoppers no longer than 15 mm and the largest form is the spectacular Large Green Grasshopper *(Chondracris rosae)* (Plate 32), some 60–80 mm in body length, with red basal patches on the hind-wings. This is polyphagous and is a minor pest on many crops locally although usually found on Gramineae. A common species found in grassland on the hills and elsewhere is the common stubby green and brown species *Gastrimargus marmoratus* (Plate 33) with its characteristic crested thoracic ridge ending in a posterior spine and with yellow hind-wings bordered black; this is the common species sold throughout most parts of the year in the local bird shops as food for large insectivorous cage birds. *Ceracris fasciata* is a small species 25–30 mm long, often found in woodland on herbaceous plants, and it has a distinctive broad dorsal stripe (Plate 34). *Pternoscirta caliginosa* (Plate 35) is found near wooded areas, and *Patanga succincta* (Plate 36) is found on hilltops in the grasses. A number of small short-horned grasshoppers have been identified but most are undistinguishable in appearance, and none is very common.

Plate 34. *Ceracris fasciata* (Orthoptera, Acrididae) on sweet potato leaf; body length 25–30 mm.

Plate 35. Short-horned grasshopper *(Pternoscirta caliginosa)* eating on a leaf of *Agave* sp.

Plate 36. Large Brown Hillside Grasshopper (Bombay Locust), *Patanga succincta* (Orthoptera, Acrididae); body length 60–80 mm.

However, one genus is of note: *Oxya* is a small green grasshopper that is semi-aquatic in habits and swims out to growing rice in flooded paddy fields; two species have been recorded in Hong Kong to date.

One sub-family of distinctive appearance is the Pyrgomorphinae where the grasshoppers have pointed heads and short broadly-tapering annulated antennae. A large grassland species of some 60–80 mm long is *Acrida* sp. Some medium-sized similar species, all of similar appearance, called *Attractomorpha* spp., are all very common in grassland and around paddy fields in the New Territories. It is quite likely that in total there will be at least 50 or more species of Acrididae to be found in Hong Kong.

Tetrigidae

The final group of othopterans represented locally is the Tetrigidae or grouse locusts. They are all small and possess a backwardly-pointing extension of the dorsal thoracic skeleton which projects over the wings and abdomen. They have short antennae, and an inconspicuous ovipositor in the female. They are conspicuous on various paths (e.g. Lugard Road), conduit cover paths, and reservoir roads and they fly readily when disturbed. The commonest local forms belong to the genus *Tetrix* (Plate 37). One species is aquatic *(T. subulutus)* and can be found sometimes in streams and paddy fields, but the whole group is more or less associated with water.

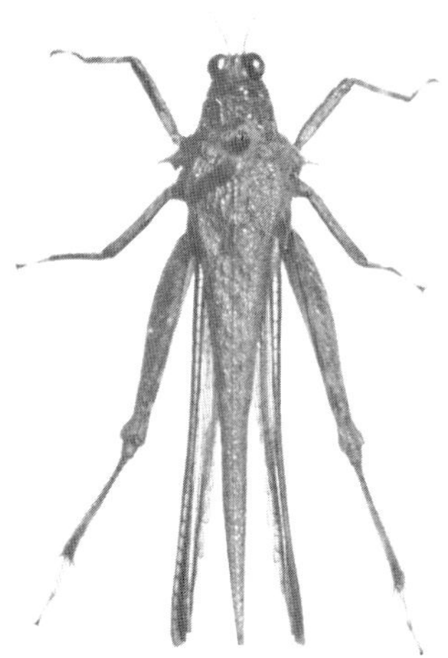

Plate 37.   Grouse Locust, *Tetrix* sp. (Orthoptera, Tetrigidae); body length 12 mm.

ORDER **PHASMIDA**
(Stick and Leaf Insects)

Phasmidae

The Phasmida comprises of two distinct groups; the leaf insects (Phylliidac) which do not occur here at all, and the stick insects (Phasmidae). There are fewer than a dozen species of stick insects in Hong Kong, ranging in size from the very long and thin-bodied *Entoria* (Plate 38 & 39) measuring some 10 cm. in length, down to the shorter, fatter *Datames* (Plate 40), sometimes called twig insect.

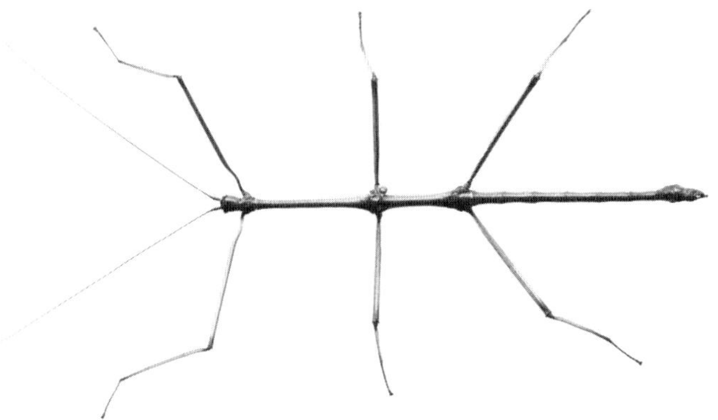

Plate 38.   Stick insect adult ♂, *Entoria* sp. (Phasmida, Phasmidae); body length 100 mm.

Plate 39.   Stick insect, *Entoria* sp., resting in foliage of Bamboo Palm.

In general they are cryptic, sluggish insects which feed on plant leaves at night, and during the day they sit passively amongst the plant foliage (Plate 39). When resting, the forelegs are typically held straight out in front, making the animal even more elongate and stick-like. These insects usually have wings as adults but are slow, clumsy, fliers. The forewings usually make long thin wing covers (tegmina) and the hind-wings are large and fan-shaped, similar to those in grasshoppers. The nymphs are even more elongate and slender than the adults (Plate 41). The adults may live to a year or more in age, and lay 20–30 thick-shelled eggs. *Datames* is usually found on leaf litter rather than in foliage, and its dark brown coloration blends well into the background.

This is essentially a tropical group of insects, and in fact the great majority of the known 2,000 species are inhabitants of the Oriental region. Hong Kong is only on the fringe of this region and the climate here is too cool for part of the year to allow more than a small number of stick insects to survive.

Plate 40.   Stick insect with shorter stubby body, sometimes called twig insect, *Datames* sp. (Phasmida, Phasmidae); body length 50 mm.

Plate 41.   Young nymph of stick insect walking on folage; length 70 mm.

Phylliidae

The Phylliidae (leaf insects) are very well developed in Malaysia, Indonesia and Papua New Guinea, and at times some species are important as defoliators of trees. It appears that the climate of Hong Kong is too cool for them to flourish here.

Plate 42.   Earwig adult, probably *Forficula* sp. (Dermaptera); length 18 mm.

ORDER **DERMAPTERA**
(Earwigs)

The earwigs are a curious group, small in size, and not really abundant as a group anywhere. They are rather primitive, and have weak biting mouth-parts, and the forewings are characteristically modified into short leathery tegmina used for protecting the large and complexly-folded hind-wings with which they fly quite well. They are also characterized in having the abdominal cerci as a pair of large unjointed and heavily sclerotized forceps or pincers.

The origin of the word 'earwig' is obscure. Some etymologists think it refers to the tendency of these insects to crawl into small orifices, being cryptozoic by nature, in which case they readily enter the ear of a person sleeping on the ground. It also is thought to allude to the shape of the hind-wing of Dermaptera; this is ear-shaped, and it is thought that 'earwig' is a corruption of the original 'earwing'.

The forceps have a function that is not really understood, but is more generally thought to be connected with either offence or defence, although some entomologists claim that they are used in the act of copulation.

As a group earwigs are reknowned for showing what is possibly the earliest phylogenetic development of social life and parental case in the Insecta.

In Hong Kong the group is represented by several different species to be found in leaf litter, under loose bark on trees, under the old leaf sheaths of bananas, and under stones. The Holarctic genus *Forficula* (Plate 42) appears to be the commonest here, but in addition, in forest leaf litter can be found the long-horned earwig, which is thought to belong to the genus *Labidura* of the family Labiduridae, and is characterized by having very long antennae.

Within the order are various species to be found living deep inside caves (e.g. Batu caves of Malaya) in total darkness and feeding upon the guano of bats. Other species are ectoparasites of bats in S.E. Asia, and five species of Arixeniina are known to inhabit the bats caves at Hiah, Sarawak, which are inhabited by bats belonging to the family Molossidae (Microchiroptera), colonial roosters which live in hollow trees or caves. *Arixenia esau* is probably the commonest species in the Niah Caves, where it lives in the bat roost and crawls on to the bodies of the bats while they roost, feeding apparently on various body exudates and epidermal debris. The group Hemimerina live on the bodies of two genera of rodents in Africa.

ORDER **EMBIOPTERA**
(Web-spinners)

The oldest group name for the web-spinners is actually Embiidina, which is the name now preferred by the world authority on this group, Dr Edward S. Ross of California. The common name of web-spinners has originated for the obvious reason that these insects spin webs, either on tree trunks with rough bark or on the surface of rocks. The web is distinctive in that it is composed of a series of silken tunnels along the bark crevices and sometimes faecal pellets and frass can be seen along the edges of some tunnels (Plate 43). The family colony lives entirely under the protective silk webbing. The females are wingless, but the males bear two pairs of equally-sized wings (Plate 44), and sometimes at night they fly to lights.

This is a small group of primitive insects, only found in the warmer parts of the world. It is expected by Dr Ross that there is a rich embiid fauna in South China but the group has not been studied in this region. In Hong Kong there are several very common species which according to Ross are 'weed' species, these are *Aposthonia borneënsis, Oligotoma greeniana* and *O. humbertiana.* These species have apparently been carried by man throughout most parts of South and Southeast Asia and are biologically robust and adaptable species. It is typical of these species that breeding may be more or less continuous and in an area adults may be found at all times of the year. The only native species to Hong Kong, *Aposthonia varians,* was found in 1961 by Ross and Hill (Ross, 1978) on the trunks of some large old trees by a cemetary on Route Twisk in the New Territories. It belongs taxonomically to a species group of one of the new genera being established by Ross *(in litt.)* in a major taxonomic study, but for the present time can be referred to as the species above.

The silk glands are situated in the tarsi of the front legs. It appears that generally eggs are laid in early summer and by August many webs can be seen to contain small nymphs. Little is known of their feeding habits, but it is thought that the females and nymphs are herbivorous, possibly feeding on lichens under the webs, and the males might well be carnivorous as they have rather different mouth-parts. The commonest tree species in Hong Kong to act as host for the local web-spinners is Camphor which has ideal bark for such insects, although colonies have been found on the smooth-barked *Ficus microcarpa.*

The web is apparently protective, for it can often be seen that on the same trees as the Embioptera are carnivorous ants in quite large numbers but the ants do not appear to detect the embiids under their webs.

The male insects are peculiar in having genitalia which are quite large and prominant and quite asymmetrical, and the classification of these seven families is based upon the structure of the male genitalia together with the wing venation.

Plate 43.   Web of Embioptera on tree trunk.

Plate 44.   Adult ♂ of *Aposthonia* sp. (Embioptera); body length 10 mm.

ORDER **DICTYOPTERA**
(Cockroaches & Mantids)

This order contains two groups of insects which from the expert point of view are clearly closely related, but superficially they look very different and have quite different habits. These are the families Blattidae (cockroaches) and Mantidae (mantids). The group is rather primitive, with long filamentous antennae, biting mouth-parts and the forewings modified into hardened, protective tegmina, under which are located the large folded hind-wings used for flying.

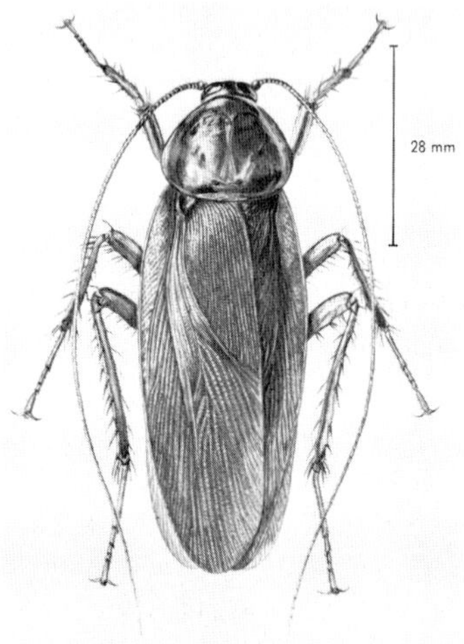

(Left) Plate 45. American Cockroach, *Periplaneta americana* (Dictyoptera, Blattidae); body length 34 mm.

(Below) Plate 46. Oriental Cockroach, *Blatta orientalis* (Dictyoptera, Blattidae); body length 18 mm.

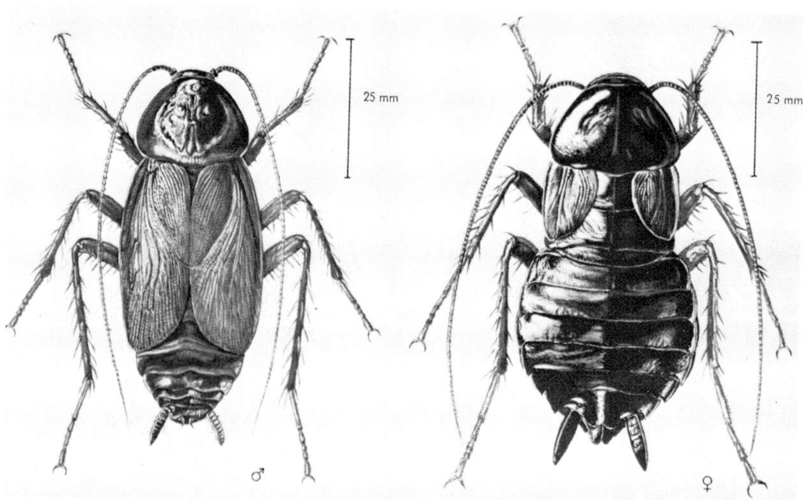

Sub-order **Blattaria**

Blattidae

Cockroaches are omnivorous, and feed on almost anything organic, even the bindings of books. Their flattened shape enables them to hide in small crevices and they will lurk in holes with only the tips of the long antennae showing. Eggs are laid inside a small egg-case, or oötheca, formed from the hardened secretion of the female accessory reproductive glands and the usual number laid per oötheca is 16, packed in two vertical rows. The oöthecae are stuck in dark corners and crevices, under tables, in cupboards and wardrobes. The most obvious species of cockroaches in Hong Kong are the domestic species found in houses, flats, and godowns. The commonest is the large, brown American Cockroach (*Periplaneta americana*, Plate 45); the rare, smaller blackish Oriental Cockroach (*Blatta orientalis*, Plate 46) with the wingless females; and the tiny striped brown German Cockroach (*Blatella germanica*, Plate 47). In forest leaf litter can be found the Small Forest Cockroach (*Blatella lituricollis*) which is externally identical to the German Cockroach (differs only in male genitalia—Romer, *pers. comm.*); but it flies readily, which is quite distinctive. An equally common countryside species is the flattened, large, brown wingless Litter Cockroach (*Opisthoplatia orientalis*, Plate 48), found in leaf litter, under rocks and under loose bark on dead trees. This curious creature has the thoracic dorsum expanded right over the head, so that viewed from above the head is not visible. Occasionally it is arboreal and found in trees, and

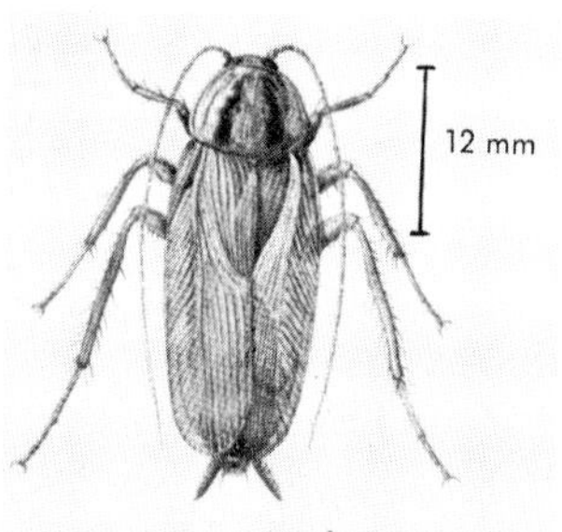

(Above) Plate 47. German Cockroach, *Blatella germanica* (Dictyoptera, Blattidae); body length 13 mm.

(Right) Plate 48. Amphibious Litter Cockroach, *Opisthoplatia orientalis* (Dictyoptera, Blattidae); wingless adult, body length 30 mm.

not infrequently it reveals its amphibious nature by submerging completely under water in freshwater streams. There are several other, less common, ordinary looking small black or brown cockroaches found in the countryside, and occasionally the Australian Cockroach *(Periplaneta australiae)* with the characteristic pale edges to the forewings is found in houses. A fairly common arboreal species is the brown mottled *Onychostylus vilis,* which spends the entire life cycle amongst tree foliage, and a larger brown arboreal forest cockroach is *Rhabdoblatta humeralis.* In soil litter a medium-sized brown species, *Nauphoeta cinerea* and the almost black litter cockroach *Pycnoscelis surinamensis* can be found.

Sometimes during the summer in Hong Kong, a small black wasp can be seen walking around inside houses, with conspicuous long antennae flickering and a tiny triangular stalked abdomen which is flicked up and down as the wasp walks. This is called the Ensign Wasp *(Evannia* sp.) because of the resemblance of the abdomen to a small flag. The *Evannia* parasitize cockroach oöthecae, and so they are welcome guests in any household.

Sub-order **Mantodea**

Mantidae

The Mantidae are known as 'praying mantids' because of the way the raptorial forelegs are held up and folded at the side of the face. They are fierce predatory insects which feed solely upon other insects and they catch their prey using the forelegs in which the spiny tibiae fold back against the femora, rather like a closing pen-knife (see Plate 49). The prey is held very firmly between the opposing sets of spines, like the teeth on the jaws of a gin-trap, and the mantid eats the insect alive, starting at the nearest point. Mantids make interesting, though rather gruesome, pets and are easily kept in captivity—one fresh cockroach or grasshopper per day will suffice to keep them satisfied.

There are at least ten local species of mantid, some of which are quite common. The stout-bodied Large Green Mantid is *Hierodula* sp. (Plates 50 & 51) with body length 60–80 mm, and the equally common but more slender Large Brown Mantid is *Tenodera* sp. (Figs. 52–54). The well known European Mantid, *Mantis religiosa,* is smaller and usually green, about 40–50 mm in length. Another species of similar appearance and brown in colour is *Statilia maculata,* and the species *Acromantis* is only slightly smaller. Some species are very small, e.g. the *Spilomantis occipitalis* is only 20–25 mm in length. Another interesting species is pale green in colour with a flattened body, and it sits on leaves pressed close to the leaf surface (Plate 55), *Leptomantella* species.

Plate 49.   Large Green Mantid in close-up, showing mobile head and spiny fore-limbs with delicate tarsus for walking.

Plate 50.   Large Green Mantid, *Hierodula* sp. (Dictyoptera, Mantidae); body length 60–80 mm.

The eggs of mantids are laid inside large frothy, cream-coloured or brown, egg cases called oöthecae which are stuck on to twigs, tree trunks, walls and rocks (Plate 56). The oötheca is made from the secretion of the female accessory reproductive glands. The liquid froth solidifies upon exposure to air and each oötheca contains up to 200 eggs arranged

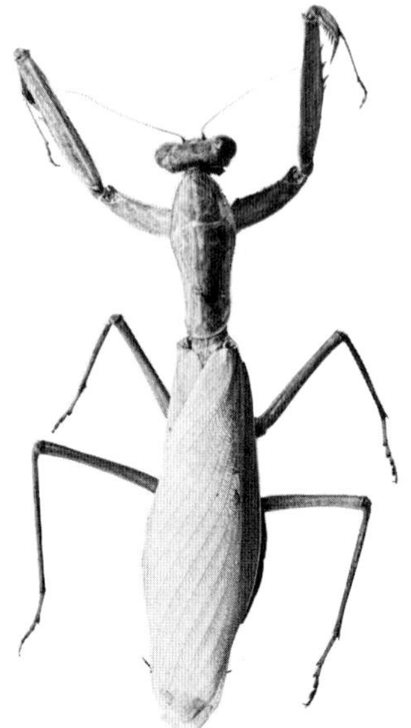

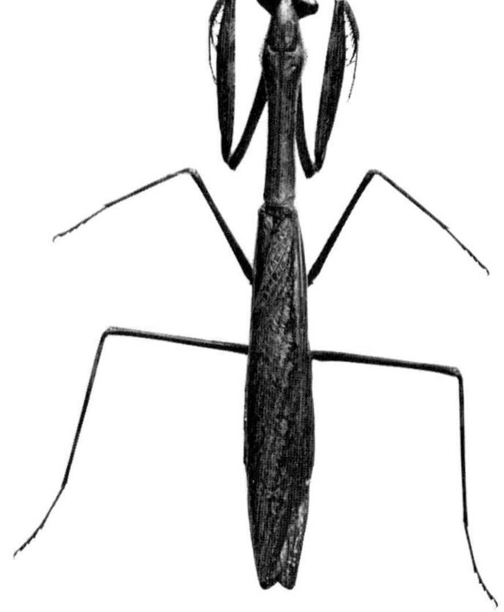

Plate 51. Large Green Mantid, *Hierodula* sp. (Dictyoptera, Mantidae); body length 60–80 mm.

Plate 52. Large Brown Mantid, *Tenodera* sp. (Dictyoptera, Mantidae); body length 60–80 mm.

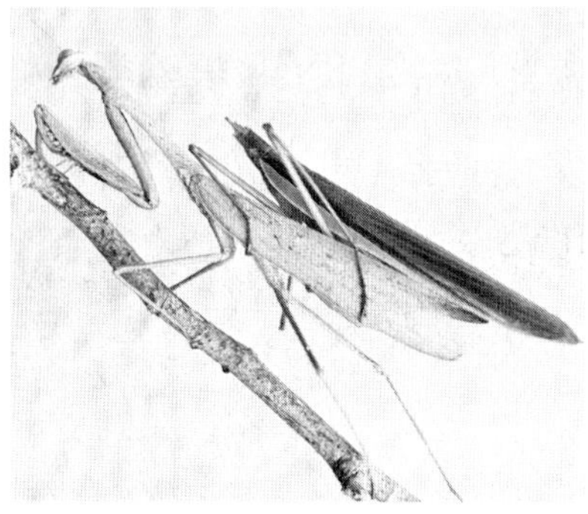

Plate 53. Pair of mating *Tenodera*; the ♀ has eaten the head, prothorax and forelimbs of the ♂, but they remain *in copula*.

Plate 54. Large Brown Mantid hunting in vegetation with poised raptorial forelegs.

in several rows in the centre. A female mantid may lay as many as 20 oöthecae during the warmer part Hof the year in Hong Kong. The different species of mantid have characteristically different shaped and sized oöthecae (Plates 57 & 58).

After a few weeks, the eggs hatch and the mantid nymphs emerge as a swarm of tiny, spindly, miniature replicas of the adults (although

Plate 55.   Green Leaf Mantid (*Lepto-mantella* sp.) on leaf of *Ficus elastica*; body length 40 mm.

Plate 56.   Oötheca of Large Brown Mantid *in situ* on stalk of *Melastoma* bush.

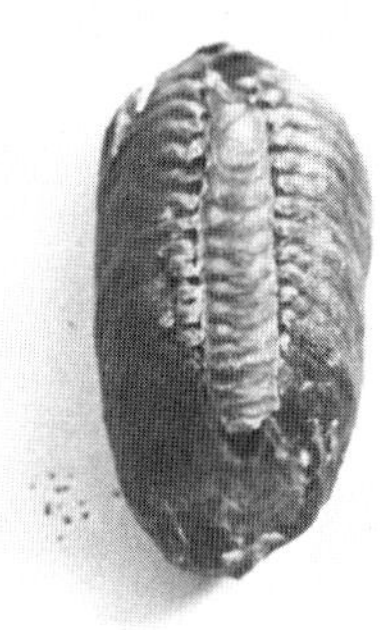

Plate 57.   Oötheca of Large Green Mantid; length 25 mm.

Plate 58.   Oötheca of Large Brown Mantid, Length 35 mm.

wingless; of course). They are equally as voracious as the adults, and if confined together without adequate food they readily eat each other until eventually there are only one or two survivors. The adults will also eat each other, and it is not uncommon for the female to devour the male after, or even during copulation. Oddly enough this does not hinder the process, for she starts to eat at the head, and the removal of the head actually speeds up the mating activity. This is because many actions of an insect work automatically, but are normally inhibited by the brain. A pair of mating *Tenodera* were observed where the female ate the head and prothorax of the male, including his raptorial forelegs (Plate 53). The decapitated male remained *in copula* with the female for 18 hours after 'losing his head' and the complete mating took about 20 hours.

Mantises are well-known in Chinese literature and paintings, and their bodies and oöthecae are also used in Chinese medicine for the treatment of various complaints.

The oötheca normally offers excellent protection for the egg against predators and parasites, but it has been observed that various oöthecae laid on tree trunks in the New Territories have been opened by insectivorous birds. The more remarkable situation is the evolution of a group of chalcid wasps with long ovipositors which parasitize the eggs of mantids. The parasites all belong to the genus *Podagrion* (Torymidae) and it appears that each species of mantid has its own species of parasitizing *Podagrion,* and characteristically the length of the ovipositor of the female wasp is correlated with the thickness of the oötheca wall.

ORDER **ISOPTERA**
(Termites)

The termites are truly tropical insects and thus are not abundant in Hong Kong for the climate here is not hot enough for their liking. Another consequence of local temperatures is that locally the mound-building termites in the family Termitidae do not produce a mound but keep the entire colony underground. A typical *Macrotermes* mound in Malaya is shown in Plate 59. These are primitive but highly specialized insects which live in vast subterranean colonies in a series of distinct polymorphic castes (i.e. different body forms). There are usually one or two castes of workers who provide the physical labour for the maintenance and development of the colony, including collecting and storage of food, building and repairing of the tunnels and galleries, and care and maintenance of the queen and all the eggs and young termites in the nurseries. The royal pair consists of the huge reproductive female (queen) and her normal attendant mate (king) (Plate 60) and they live in a deeply placed central royal chamber. The king has no other function than to fertilize the queen from time to time. The queen becomes little more than a vast egg-laying machine, with a single ovary containing more than 2,000 ovarioles (strings of developing eggs). The eggs develop into individuals of different castes, which are made up of both males and females. The development of an individual's caste is controlled by chemical substances, some of which originate from the queen's body and are circulated throughout the colony by a system of mutual licking. These 'social hormones' are believed to control the relative proportions of the various castes.

Plate 59. Termite mound of *Macrotermes* sp. (Isoptera), in Penang, 1 m high.

Plate 60. Termite royal pair in opened royal chamber, *Macrotermes* sp. (Isoptera); Uganda.

The main castes are wingless, sterile, white-bodied workers of various sizes which begin working before they are fully grown. Other individuals become soldiers, which are also wingless and sterile, but have large, brown, heavy heads with strong jaws which are held straight out in front. For identification purposes it is best to examine the soldiers, for the shape of their head and the size of their mandibles are generally distinctive. The soldiers guard the entrances to the nest and keep out enemies and predators; they use their mandibles to bite and from the frontal gland above the eyes, they squirt a jet of burning formic acid. Some termites have other castes, both winged and wingless, which act as a reservoir of reproductive individuals in case of accident, and when numbers and conditions permit, these also take part in the mating and dispersal of swarms.

The colony nest is made according to a fixed pattern, either in the earth or inside timber, and either soil or wood is used for its construction. Very few animals can digest wood or cellulose directly, and most wood-eaters have bacteria and/or protozoa which live in their intestine and produce enzymes which can break down the cellulose in the wood. Some termites have colonies of flagellate protozoa in their hind gut which produce cellulose-digesting enzymes which break down the complex polysaccharide cellulose into simpler sugars. The termites are then able to absorb and utilize the sugars produced. This is a good example of biological mutual interdependance, called symbiosis or mutualism. Thus, some termites eat wood and with the aid of their symbiotic intestinal protozoa, they are able to derive nourishment from it. Into this category fall the dry-wood (family Kalotermitidae) and wet-wood termites (family Rhinotermitidae).

The mound-building termites (family Termitidae) do not have this intestinal micro-fauna and are not able to digest cellulose. So these termites use the pieces of vegetation and wood collected by the workers and mix them with saliva and construct honeycombed structures called fungus gardens (Plate 61). The gardens are innoculated with a special fungus called *Termitomyces* which the termites go out and collect. The fungus covers the cellulose frame as a fine mycelium and produces series of small round white bodies called brometia upon which the termites feed (Plate 62). Termites are in constant danger of drying up (desiccation) because of their very thin body cuticle, and they always endeavour to keep out of the light and under cover so as to work in a favourable microclimate. When not working in the ground or inside wood but working on the surface, they construct covered passageways out of a mud and saliva mixture over the surface of tree trunks and pathways (Plate 63).

At certain times of the year the young winged reproductive adults

Plate 61. Termite mound, opened to show the fungus gardens (Isoptera; Termitidae).

Plate 62. *Macrotermes* fungus garden, showing small white brometia.

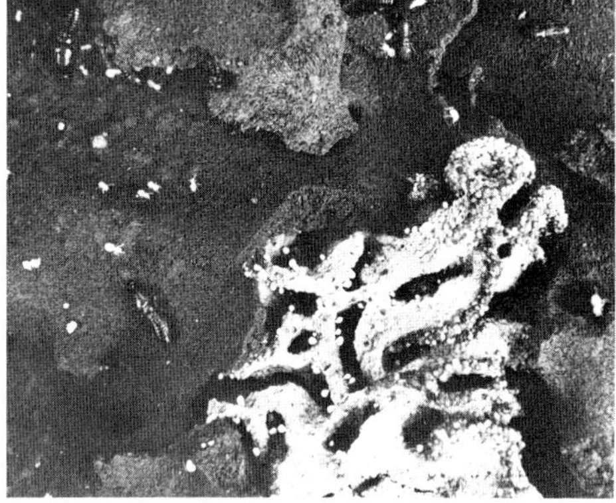

Plate 63. Termite covered pathway on tree trunk; Wu Kwai Sha, N.T.

have their nocturnal mating flights, at which time they are attracted to lights and may invade houses and flats in large numbers. After the nuptial flight when the young royal forms pair off and mate as they fall to the ground, they shed their wings. Those pairs which escape the many predators that collect at these times then find a suitable location and burrow into the ground to make a nest and found a new colony. In the morning after such a nuptial flight when the swarming termites are attracted to house lights, the ground is covered with myriads of discarded wings (Plate 64). The winged forms have two pairs of equally sized wings (hence their name Isoptera) with which they fly in a clumsy and laborious manner.

The termites of Hong Kong were the subject of a paper by Harris (1963) and so the seven local species are well documented.

Rhinotermitidae

There really are only two common species here. The most common is the small Wet-wood Termite *(Coptotermes formosanus)* of the family Rhinotermitidae, which can do considerable damage to buildings and structural timbers. In the wild they live in dead tree stumps where the woody tissues are moist. The winged adults swarm at dusk on humid evenings during May and June. Their body is orange in colour, about 8 mm long, with transparent wings about 11 mm long (Plate 65). They fly to lights at night; flats and houses in parts of the Colony may become innundated on swarming nights if windows are open. As most of the new buildings in Hong Kong are now blocks of flats or offices on a foundation of solid concrete under which is sprayed a heavy dose of the persistent insecticide dieldrin, this pest is now not so serious as it was once. However, given an area of damp wood the flying adults are able

Plate 64. Termites and discarded wings, under house lights, after a nocturnal nuptial swarming.

to start a colony within our modern concrete boxes (flats), and if they establish a nest under wooden parquet flooring or under built-in wardrobes and cupboards, the amount of damage they can do is still considerable. This is a species which produces substitute queens by accelerating the rate of development of certain nymphs so that they are able to take over the duties of the old queen at short notice.

A second species of Wet-wood Termite has been found locally(but it is rare). This is *Reticulitermes fukiensis* which probably makes its nest in tree stumps and buried timber.

Plate 65.   Adult winged Wetwood Termite, *Coptotermes formosanus* (Isoptera, Rhinotermitidae); wingspan 25 mm.

Termitidae

The second most common termite in Hong Kong is the largest species *Macrotermes barneyi* in the family Termitidae (subterranean and mound-building termites). Due probably to the relatively cool conditions prevailing locally, there are no mounds produced in Hong Kong, but *Macrotermes bellicosus* in Africa produces mounds up to two metres or more in height and another species of *Macrotermes* in Malaysia produces mounds about one metre high (Plate 59). It is not common but there are scattered nests around the colony, though the termites are seldom seen, spending most of the time in their subterranean colonies. Colonies under observation have only produced mating swarms once annually on humid or wet evenings in May and June. The winged adult is 20–25 mm long (just body), with a stout dark brown body and pale yellow wings. It has two distinct sizes of worker caste and two soldiers. This group of termites do not have the symbiotic intestinal protozoa and so cannot digest cellulose; they make the honey-comb like fungus gardens and feed on the bromatia produced by the fungus *Termitomyces* (Plate 62). The covered pathways produced on tree trunks and paths locally are usually made by this species (Plate 63).

A similar but slightly smaller species in Hong Kong is *Odontotermes formosanus*. This is also paler than *M. barneyi* but the winged forms have a characteristic T-shaped pale spot on the pronotum of the thorax (first dorsal plate) and the wings are dark brown (Plates 66 & 67). This species is uncommon in Hong Kong, but a few colonies are known and swarming usually occurs in June.

Also belonging to this family are two rare species which might occasionally be found either in small colonies under stones or else living inside the nests of other termites; these are *Capritermes fuscotibialis* and *Procapritermes sowerbyi*.

Plate 66. Adult winged termites, *Odontotermes formosanus* (Isoptera, Termitidae) trapped by rain-water film on concrete path after a nuptial swarming.

Plate 67. Adult winged Subterranean Termite, *Odontotermes formosanus* (Isoptera, Termitidae); wingspan 50 mm.

Katotermitidae

The last local termite species is the Dry-wood Termite (*Cryptotermes brevis*), sometimes called the Furniture Termite, which fortunately does appear to be becoming rather rare now. These termites in the family Kalotermitidae apparently live quite successfully without contact with water or even moisture, and laboratory cultures have lived for years inside a block of dry wood inside a glass jar or tank. The natural habitat for the indigenous species of *Cryptotermes* is in dead branches of standing trees. In this species the head of the soldier is large, black, and cylindrical and is used to plug the narrow outer galleries to prevent the passage of enemies into the wider interior galleries where the community lives. There is no distinct worker caste in the Kalotermitidae; the work of food collecting, maintainance, etc., are carried out by nymphs in various stages of development, which can be arrested permanently or temporarily according to the needs of the community. Swarming apparently takes place at irregular intervals during wet and warm weather, and the small winged adults are brown in colour with the head and body together about 6 mm long and the transparent wings about 11 mm long.

ORDER **PSOCOPTERA**
(Booklice and Barklice)

These are small, soft-bodied insects with long filamentous antennae and a rather bulging head which gives them a characteristic appearance. Some are minute and colourless, but a few species are larger, being up to 7 mm in body length with dark bodies and dark or spotted wings. In addition to their being apterous and winged individuals there are sometimes brachypterous forms with short, stubby, non-functional wings.

The classification is confusing to the non-specialist, and is based upon rather esoteric characters. Different books give different schemes of classification for the Psocoptera. This group was the basis for a series of specialized taxonomic papers by Thornton (1959, etc.) but no general account of the local psocids was ever published. To the general entomologist and to ecologists the two dozen different families of Pscoptera are not particularly significant in that they are difficult to distinguish, especially in the field, and the taxonomic differences do not reflect major ecological differences.

The many local species are usually to be found on trees or on lichen-covered rocks. They apparently feed on algae, fungi and lichens, but will also take small pieces of vegetable matter. Some of the household and stored products species will also eat dried animal material. In fact the greatest damage done to dried insect collections throughout the world is by the psocids; they rate as the major museum pest in most countries. They are so small and colourless that they often escape notice, but in large numbers they invade the bodies of museum specimens and eat out the dried body until just a fragile shell remains.

The arboreal species are generally large with dark coloured wings and up to 7 mm in body length, and they often live gregariously. They can be found in a dense cluster on a rock or tree trunk, of up to 50 or more individuals. If disturbed they immediately disperse, but after a while they gradually reform the group. Occasionally several groups may be found in close proximity on the same tree trunk or rock. These arboreal outdoor species are called bark-lice.

Some species make extensive sheets of silken webbing over the rock or trunk surface under which the insects live and breed. As with the Embioptera it is assumed that the webbing offers the insects a measure of protection from the many predators in their neighbourhood. Psocid webbing is quite distinct from embiid webbing in that it is a broad sheet without discrete tunnels, and more frequently is found over a smooth surface (Plate 68) rather than the rough surface favoured by embiids.

A few species are household pests and can be found in food cupboards, kitchens and amongst books, where they feed on organic debris and pieces of fungus, and occasionally on book bindings, and here they are called book-lice. During the warmer weather in Hong Kong many flats and houses have light patches of mould on some internal walls, and it is very common to find a herd of minute wingless psocids browsing upon this fungus, and sometimes winged or brachypterous individuals, probably of the species *Ectopsocus maindroni,* can be found. Other small species are to be found in stored foodstuffs, particularly rice and especially if the rice is already infested with beetles or other insects.

Plate 68. Web of gegarious species of Psocoptera on a tree trunk.

ORDER **MALLOPHAGA**
(Biting Lice)

These are the biting lice, sometimes called bird lice. Although they are very common on birds, there are a few species to be found on mammals. They are wingless, obligate ectoparasites with biting mouthparts, and they feed on fragments of feathers, hairs, and skin debris, but some also take blood from the host if the opportunity arises, such as from scratches or wounds. Some species show interesting ecological preferences in that they are only found on the head or on the wings, or some other restricted part of the host body. Some birds may carry several different species of Mallophaga but each species being restricted to different parts of the body. The whole life cycle is spent on the host, but some dispersal obviously takes place on young birds in the next, or by bodily contact of adult hosts. One group is confined to elephants only as hosts. The group is well represented throughout the entire world on all types of birds and on some mammals.

Locally, dogs are infested with *Trichodectes canis,* and in other regions there are different species found on cats, horses, and cattle. Domestically, there is the Common Chicken Louse *(Menopon gallinae)* which is often brought home on live chickens from the markets and several species occur on ducks. On domestic pigeons the elongate, slender *Columbicola columbae* is common. Once the host is dead and the body cools, these lice soon leave. There is a particularly spectacular large louse about 7–8 mm in length occurring on the common Black-eared Kite, identified as *Laemobothrion maximum.* Hong Kong was one of the participating countries in a recent survey on bird ectoparasites in Southeast Asia carried out by the U.S. Migratory Animals Pathology Survey (M.A.P.S.) (McClure & Ratanaworabhan, 1972).

Mallophaga are of interest from an evolutionary point of view in that it seems certain there is a relationship between the evolution of the lice and their hosts, so that groups of closely related host species tend to be infested by similar Mallophaga and several scientists have attempted to use this apparent relationship to throw light on the phylogeny of certain groups of birds, particularly the storks and herons (Ciconiidae) and the flamingoes (Phoenicopteridae). Thus it appears on this basis that the flamingoes are more closely related to ducks (Anatidae) than they are to storks, despite previous views to the contrary.

ORDER **SIPHUNCULATA**
(Sucking Lice)

The sucking lice are permanent, obligatory ectoparasites, but confined to mammals as hosts. The piercing and sucking mouth-parts are used to obtain blood from their host. As with the Mallophaga there is striking host-specificity but as a group they are not nearly so common. Virtually any bird collected almost anywhere in the world will be found to be carrying Mallophaga, but the actual number of mammals found with Siphunculata is really quite small, and one can examine a large number of specimens of local mammals without ever finding one louse. One group (family) is confined to seals and other Pinnipedia, and other species occur on mice, squirrels, rabbits, monkeys, elephants, etc., in addition to man and his domestic animals.

On domestic animals the genus *Haematopinus* is found as a series of different species on different hosts. On pigs is *H. suis* (Hog Louse), on buffaloes *H. tuberculatus* (Buffalo Louse), on cattle *H. bovis* (Cattle Louse), and on horses *H. equis* (Horse Louse). Obviously some of these host animals are present in Hong Kong in large enough numbers to sustain a louse population (e.g. pig and buffalo) but so far as is known ectoparasite studies have not been made.

Man is an important host for two cosmopolitan species of lice which are completely host specific. The Human Louse, *Pediculus humanus,* occurs as two distinct varieties; *corporis* is the Human Body Louse which is to be found on the body between the skin and clothing, and the other variety *capitis* is the Human Head Louse which only lives in the fine hair on the head. The Body Louse transmits various diseases when conditions are suitable, such as epidemic fever and trench fever, and have been responsible for several historic epidemics, particularly amongst the soldiers in the trenches during the World War 1. Because of the general improvement in sanitation and hygiene, the Body Louse is no longer common in most parts of the world, but in Victorian days and earlier, it was a general inhabitant of all households in alarming numbers. The Head Louse is traditionally found on children rather than on adults, and in the past most frequently on girls; a few years ago this pest had become quite rare in the Western world, but with the recent change in tonsorial fashions, Head Louse infestations have become alarmingly common amongst schoolboys in Europe and North America, and as recent as 1977, the Head Louse has become a school problem in Hong Kong.

The term 'nits' is used for the eggs which are stuck on to the side of hairs, usually near the base of the hair. They are stuck on to the hair singly although there may be two or three eggs on a single hair at different levels. A 'nit' comb with very fine teeth can be used for combing out the eggs from the hair.

The second species of human louse is the Crab Louse, *Pthirus pubis,* and it inhabits the coarse body hair of man, such as the hairs in the public region and under the arms, the eyebrows, and also occasionally even the eye-lashes. The bites of this species are particularly irritating and infestations are very difficult to eradicate without insecticidal treatment.

ORDER **HEMIPTERA**
(Bugs)

The Hemiptera is a very large order containing some 38 families and 56,000 species, and is divided distinctively into two large sub-orders. The Heteroptera are the true sucking bugs including the predatory animal bugs and aquatic predators as well as some plant-bugs, but the Homoptera are all plant bugs. These two groups are sometimes regarded as being separate orders.

Sub-order **Heteroptera**

Heteroptera are very abundant in Hong Kong, and are characterized by having the forewings membraneous distally and with a thickened basal part. At rest the wings lie flat and overlapping over the abdomen. The mouth-parts are modified into a piercing and sucking proboscis, stylet, or 'beak'; it arises from the front of the head and is sometimes quite short and curved in predatory forms. Many species are sap-sucking but some feed on animal blood, and the aquatic forms are predacious. This sub-order is again divided into two further groups: the Cryptocerata and the Gymnocerata.

Cryptocerata group

The Cryptocerata have very short antennae (shorter than the head length) usually hidden in cavities beneath the eyes, have no ocelli (eyespots) on the head, and they are aquatic in habits.

Naucoridae

The Naucoridae are small, oval, flattened water bugs (called saucer bugs) which swim actively in freshwater. Their forelegs are raptorial and are used for seizing small aquatic insect prey. Several local species are occasionally seen and one has now been identified as *Heleocoris* sp.

Belostomatidae

The Belostomatidae are the Giant Water Bugs, very spectacularly represented locally by the giant *Lethocerus indicus* (Plate 69). This is a far-ranging insect occurring throughout Southeast Asia, India and Africa, and grows up to a length of 90 mm. Several other species also occur in Africa, Australia and the Americas, and in Australia their common name is 'fish-killer'. They feed on small fish, frogs, and other insects, and during the summer they sometimes fly at night in consider-able numbers; occasionally at places near ponds, such as Sek Kong

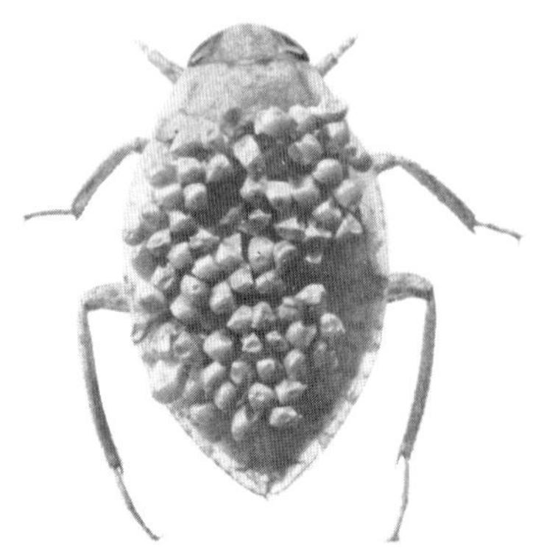

(Left) Plate 69. Giant Water Bug, *Lethocerus indicus* (Heteroptera, Belostomatidae); body length 90 mm.

(Below) Plate 70. Small Water Bug, *Sphaerodema* sp. (Heteroptera, Belostomatidae); body length 12 mm.

Village in the New Territories, they may be seen flying around the street lights and being knocked down by passing vehicles. More than 115 dead or injured bugs were seen on the Sek Kong Village main road one morning in 1964. Because of this inclination to fly to lights at night this insect has sometimes been called the 'Electric Light Bug'. This bug is typical of other predatory heteropteran bugs in that it possesses powerful proteolytic enzymes in its saliva. On capturing a prey the bug 'bites' it with the short curved 'beak' and injects saliva into the wound. The enzymes in the saliva digest and make soluble the body proteins of the prey and the liquid digested food is sucked up by the bug. Thus a carelessly handled Giant Water Bug can inflict a most painful 'bite' on its captor, and would-be collectors of all Heteroptera should treat them with caution!

A second species has recently been identified as *Sphaerodema* sp. (Small Water Bug). It is found in ponds and paddy fields and in lower reaches of streams where the water current is slow. One striking feature is that the male carries the eggs on his back until they hatch (Plate 70). The adult body length is about 10–12 mm.

Nepidae

Water Scorpions constitute the family Nepidae and there are two quite common local species. *Nepa* is the ordinary Water Scorpion (Plate 71) with its prehensile, raptorial forelegs and long respiratory siphon. Generally it swims rather poorly and usually it walks about submerged on pond weeds near the water surface so that the respiratory siphon can be used to bring air down to the submerged insect. The Thin Water Scorpion is *Ranatra* sp., also to be found in parts of the New Territories. Both species fly at night during summer and occasionally end up in swimming pools by mistake. A small species of *Nepa* is sometimes found in paddy fields.

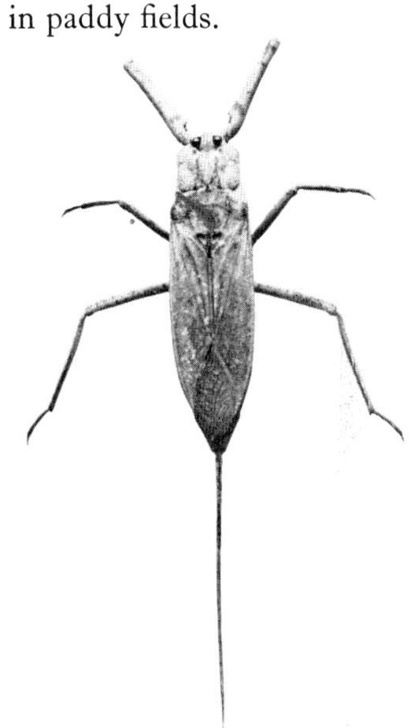

(Left) Plate 71. Water Scorpion, *Nepa* sp. (Heteroptera, Nepidae); length 32 mm.

(Below) Plate 72. Adult Backswimmer, *Notonecta* sp. (Heteroptera, Notonectidae); body length 10 mm.

Notonectidae

The Notonectidae are the fiercely predacious backswimmers with their characteristically keeled backs and mode of swimming. Several species of *Notonecta* (Plate 72) are common here, and the tropical genus of backswimmers *(Anisops)* is probably to be found locally. Both types fly at night; the adults in their search for new bodies of water quite commonly invade swimming pools in very large numbers.

Corixidae

These are the water boatmen and they swim the right way up. *Corixa* is widespread in pools and paddyfields; it is phytophagous and feeds on diatoms and other algae.

Gymnocerata group

The group Gymnocerata contains a larger number of heteropteran families, some of which are aquatic and others predatory, and a number of plant bugs. They all have long obvious antennae, and ocelli on the top of the head between the eyes.

Reduviidae

The Reduviidae are Assassin Bugs, with long slender antennae and legs. They are mobile, predatory, terrestrial bugs with a large but short, curved beak. If handled they can inflict a very painful 'bite', and should be treated with care. Local species include the quite common *Sycanus croceovittatus* (Plate 73) and *Triatoma rubrofasciata* (Domestic Assassin Bug). When young, they are somewhat social and tend to remain together as nymphs, and they can often be found in clusters around a dying snail on a tree trunk or rock wall where their joint salivary secretions dissolved the entire snail body.

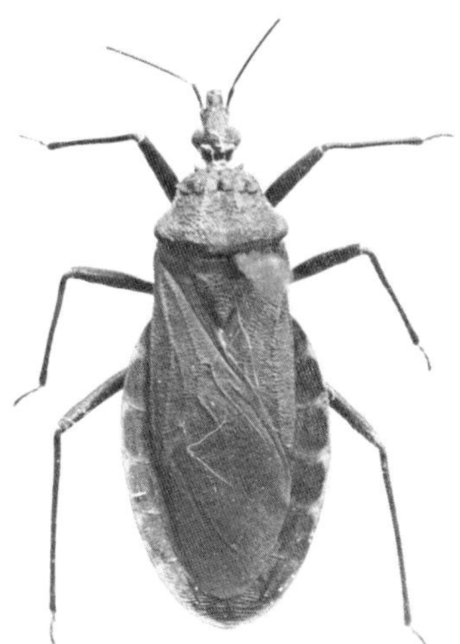

Plate 73. Assassin Bug, *Sycanus* sp. (Heteroptera, Reduviidae); body length 22 mm.

The following two families are predacious water-bugs all of which are to be found living on the surface of water, both as nymphs and adults.

Gerridae

The Gerridae are pond skaters; their extremely long middle and hind-legs with their bristle fringes help to keep the bugs on the surface of the water film. Several species of *Gerris* (Plate 74) are common here, and they move with great rapidity over the water surface of streams and

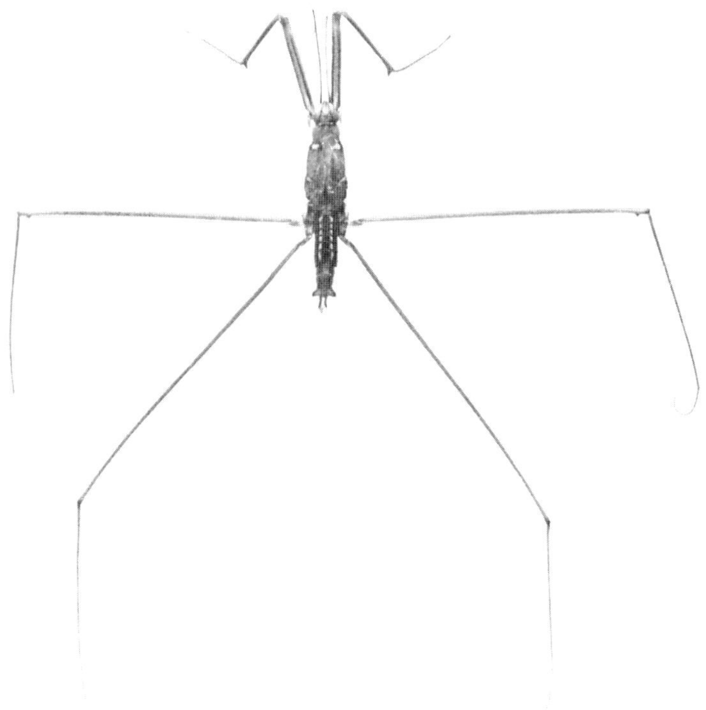

Plate 74.   Adult Water Skater, *Gerris* sp. (Heteroptera, Gerridae); body length 15 mm.

ponds in the New Territories. At certain times of the year the adults make nocturnal dispersal flights. The only species of insect which inhabits the open sea occurs in Southeast Asia and the Indian Ocean; this is a gerrid bug very similar in appearance to *Gerris* which skates on the surface of the sea, and is called *Halobates*. *Halobates* is more abundant off the coasts of Malaysia and Indonesia but it does sometimes occur on the sea off Hong Kong Island. This genus is wingless and does not fly at all. The Coastal Sea Skater, *Asclepios shiranui,* is found in Deep Bay and Three Fathom Cove.

Veliidae

Velia is a very small, stout-bodied (2 mm) pond skater with short legs, found on the surface of quiet streams and ponds in the New Territories. It belongs to the family Veliidae. Another species in this family, *Microvelia,* is even smaller, about 1 mm, and can be found walking on quiet water.

Hydrometridae

In the Hydrometridae are the small, delicate, elongate pond striders or Water Measurers (*Hydrometra* sp.) which are found on the surface of standing waters and slow streams in parts of the New Territories.

Miridae

The Miridae are sometimes called Capsid Bugs; they are small, rather delicate and somewhat flattened bugs that feed on plant sap. A number of odd specimens have been seen but none is very common; *Helopeltis* sp. (Mosquito Bug) is recorded from local *Abutilon* bushes.

Anthocoridae

They are closely related to the Anthocoridae which are small predatory bugs often found on *Ficus microcarpa* where they appear to kill various emerging fig-wasps.

There are several other small families of Heteropteran bugs that are of little consequence, but perhaps a few should be mentioned.

Cimicidae

The Bedbugs *(Cimex lectularis* and *C. rotundatus)* are members of the

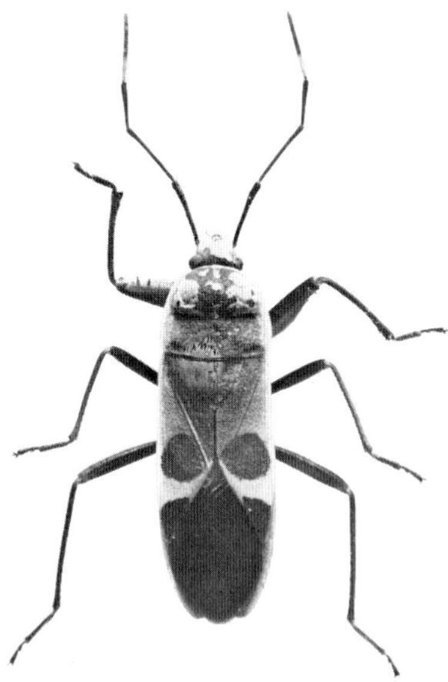

Plate 75. Cotton Stainer (Red Bug), *Dysdercus* sp. (Heteroptera, Pyrrhocoridae); body length 13 mm.

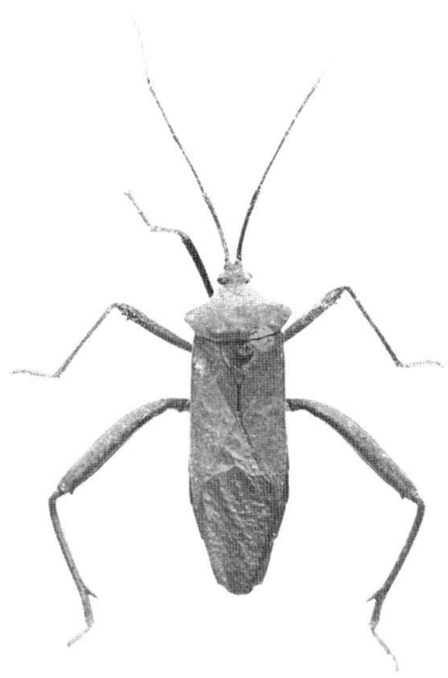

(*Left*) Plate 76. *Mictis* sp., adult (Heteroptera, Coreidae); body length 20 mm.

(*Below*) Plate 77. *Mictis* sp. feeding on, and killing, young shoot of *Cratoxylon ligustrinum.*

Cimicidae. They are ectoparasites of birds and mammals, found mainly in nests and lairs, and are virtually wingless bloodsuckers.

Lygaeidae

The Lygaeidae are small brightly coloured plant bugs some of which are important crop pests in other parts of the world, although a few are predatory on other insects.

Pyrrhocoridae

The Pyrrhocoridae are Red Bugs predominantly red and black in colour, and in this small family is found the Cotton Stainers (*Dysdercus* spp.) of which a couple of species are quite common locally (Plate 75).

Coreidae

The Coreidae have no generally accepted common name, but are abundant in Hong Kong. They are often large in size, usually dull in colour; some species have typically dilated tibiae and antennae, but the latter condition appears to be rare here. Many species have phytotoxic saliva and their feeding kills plant tissues. A very common local species is *Mictis* sp. (Plate 76) frequently found feeding and mating on *Cratoxylon ligustrinum* in August and September (Plate 77); the feeding usually kills the young red shoot. Another large brown species is *Leptoglossus australis* (Plate 78) but this is not often seen, and a slightly smaller brown

0　　5mm

(Left) Plate 78. Leaf-footed Plant Bug, *Leptoglossus australis*, adult (Heteroptera, Coreidae), body length 20 mm.

(Above) Plate 79. *Paradasynus spinosus* (Heteroptera, Coreidae); body length 17 mm.

Plate 80. *Paradasynus spinosus* family group on an unidentified leaf.

and green species is *Paradasynus spinosus* (Plate 79) sometimes found as large groups of nymphs on particular leaves—they appear to stay together until maturity (Plate 80) when they disperse to other plants. The other body form found in the Coreidae is slim and slender, as typified by the very common Rice Bug *(Leptocorisa acuta)* (Plate 81) which in addition to being an important rice pest in southern and eastern Asia, feeds locally on a number of wild grasses. During the warmer months it frequently flies to lights at night in sufficient numbers as to be a nuisance. All these species possess stink-glands which open at the base of the legs and produce a most unpleasant odour when handled or disturbed. Another species of local importance is the Bamboo Bug, *Notobitus meleagris* (Plate 82) which lives on the stems of some large bamboos and their feeding produces long necrotic scars (Plate 83). Some coreids are brightly coloured red and black, sometimes gregarious, and occasionally are seen to kill other insects; Plate 84 shows a *Leptocoris* killing a small shield bug.

Tingidae

The Tingidae are called lace bugs and are attractive little bugs found on plants. They have characteristic reticulate laminate outgrowths from the thorax, and the forewings are also reticulate. The few local specimens obtained have not yet been identified.

Plate 81. Rice Bug, *Leptocorisa acuta* (Heteroptera, Coreidae) on a rice leaf; body length 15 mm.

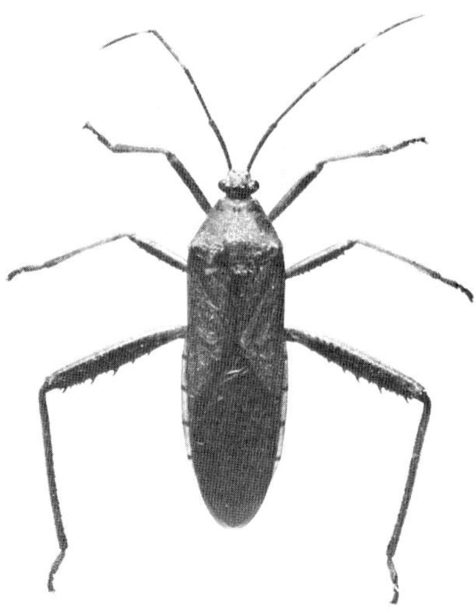

Plate 82. Bamboo Bug, *Notobitus mele-agris* (Heteroptera, Coreidae); body length 21 mm.

Plate 83.   Bamboo stem scarred by Bamboo Bug feeding.

Plate 84.   Coreid bug, *Leptocoris* sp., killing a small shield bug.

Pentatomidae

The last and most important family of heteropteran bugs is the
Pentatomidae, called the shield or stink bugs. There are many local
species, some of which are both striking in appearance and very
abundant. They are mostly large in size with long antennae and a very
extensive scutellum (dorsal skeletal plate), and from their stink glands
they are able to eject a jet of stinking liquid to a distance up to 50 cm,
which will stain clothing and human skin and is extremely irritating if it
contacts eyes. As a group they lay large barrel-shaped eggs either solitary

Plate 85. Shield bug nymphs just emerged from eggs.

Plate 86. Lychee Stink Bug, *Tessaratoma papillosa* (Heteroptera, Pentatomidae); body length 28 mm.

Plate 87. Lychee Stink Bug adult sitting in Litchi foliage.

Plate 88. Nymph of Lychee Stink Bug; body length 19 mm.

or in clusters on the leaves of the host plant. Following eclosion (hatching), the young nymphs often remain together around the empty eggs for some days (Plate 85). One of the most striking local shield bugs is the Lychee Stink Bug *(Tessaratoma papillosa)* found on Litchi and Longan trees, having a large brown body and delicate white waxy dusting underneath (Plate 86 & 87). It is quite remarkable in that all egg clusters of Lychee Stink Bug found have contained precisely 14 eggs. The nymphs are interesting in that they differ considerably in form and

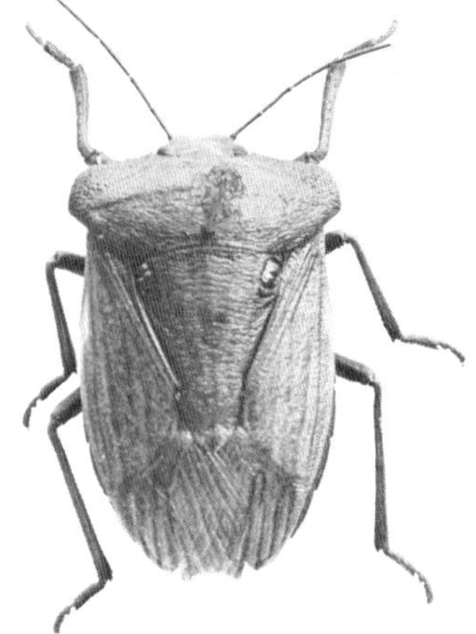

(Left)   Plate 89.   Nymph of Lychee Stink Bug showing waxy dorsal surface, in Longan foliage, Penang, 1974.

(Above)   Plate 90.   *Chalcopsis glandulosa*, adult (Heteroptera, Pentatomidae); length 28 mm.

coloration from the adults (Plate 88). Sometimes the nymphs have the white waxy covering on the dorsal surface in addition to ventrally, but this condition appears to be more frequent in Malaysia than locally (Plate 89).

A similar large brown species of unknown host preference, but distinguished by two dark dorsal makings is *Chalcopsis glandulosa* (Plate 90). A large widespread species of mottled black and white coloration is *Erthesina fullo* (Tallow Stink Bug) (Plate 91) found mainly in wooded areas, but like the other species it does fly to lights sometimes at night. One shield bug of interest is remarkable in being virtually completely cosmopolitan in distribution and is a minor pest on a whole series of different crop plants; this is the Green Shield Bug *(Nezara viridula)* (Plate 92) most frequently encountered in Hong Kong when it flies to lights at night. The Green Shield Bug of 10–20 mm body length has another quite similar relative, slightly larger and with a brown scutum *(Neojurtina typica)* (Plate 93). *Scotinophora coarctata* is the Black Paddy Bug (Plate 94) which is a pest of rice in addition to feeding on many species of grass. Another species found only on grasses is the Elongate Shield Bug, *Megarrhampus hastatus* (Plate 95). There are locally a few very pretty brightly coloured species of Pentatomidae, one of which is *Brachyaulax oblonga* (Plate 96), being predominantly bright blue in colour. Other species of distinctive coloration include the Blue/Green

Shield Bug *Calliphara nobilis* (Plate 97) and the Large White Shield Bug *Chrysocoris grandis* (Plate 98). *Cantao ocellatus* (Plate 99) is interesting in that the scutellum is so enlarged that it projects posteriorly as a

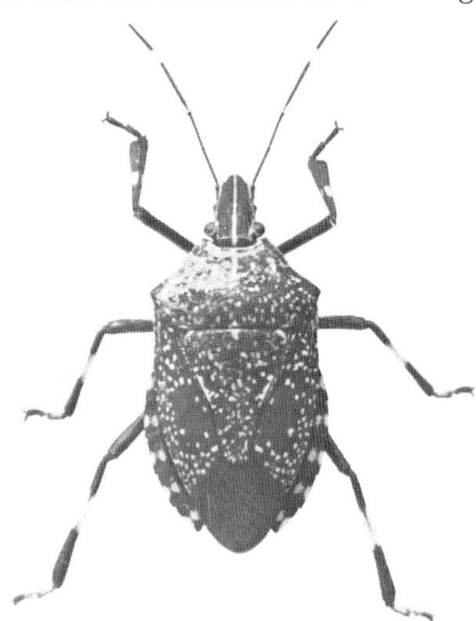

Plate 91. Adult Tallow Stink Bug, *Erthesina fullo* (Heteroptera, Pentatomidae); length 20 mm.

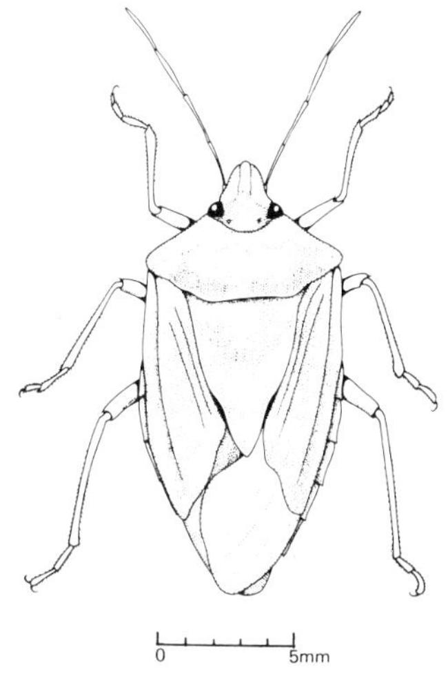

Plate 92. Green Shield Bug, *Nezara viridula* (Heteroptera, Pentatomidae); length 12 mm.

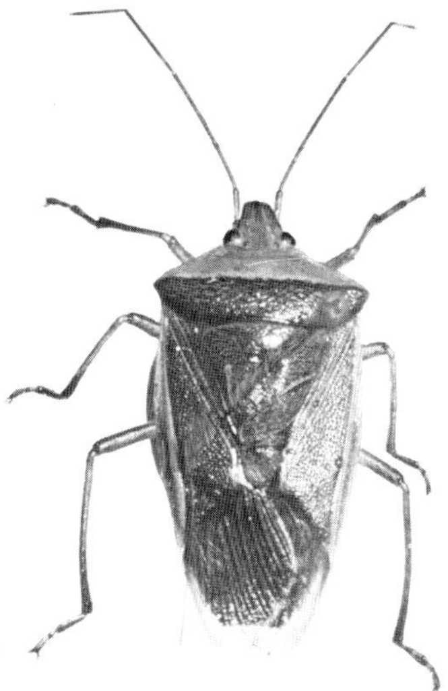

Plate 93. Adult *Neojurtina typica* (Heteroptera, Pentatomidae); length 17 mm.

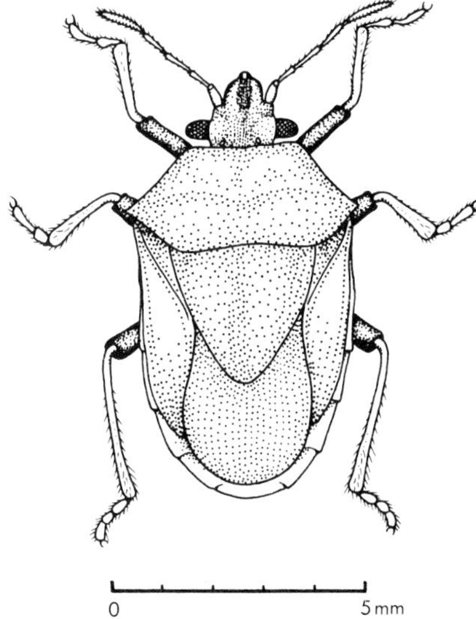

Plate 94. Black Paddy Bug, *Scotinophara coarctata* (Heteroptera, Pentatomidae).

pointed flange overlying the tips of the wings. On tea and camellia bushes in parts of the New Territories are found the beautiful *Poecilocoris latus* (Camellia Shield Bug) with its distinctive blue and orange/red coloration and the equally attractive Tea Shield Bug *(Catacanthus nigripes)* (Plate 100), which appears to be a species of very limited distribution and probably confined to the tea family as host plants. Although most species of Pentatomidae are phytophagous, a few are known to be predatory, but these have not been observed locally.

(Left) Plate 95. Miscanthus Shield Bug, *Megarrhampus hastatus* (Heteroptera, Pentatomidae); body length 21 mm.

(Below) Plate 96. The attractive blue-green *Byachyaulax oblonga* (Heteroptera, Pentatomidae); body length 14 mm.

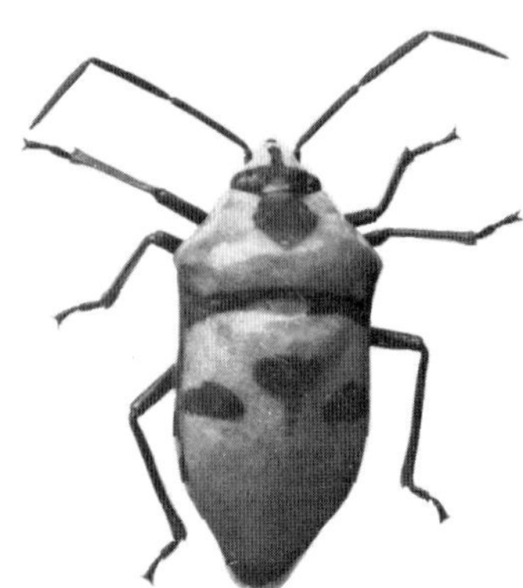

Plate 97. Blue/Green Shield Bug, *Calliphara nobilis* (Heteroptera, Pentatomidae); length 14 mm.

Plate 98. Large White Shield Bug, *Chrysocoris grandis* (Heteroptera, Pentatomidae); length 21 mm.

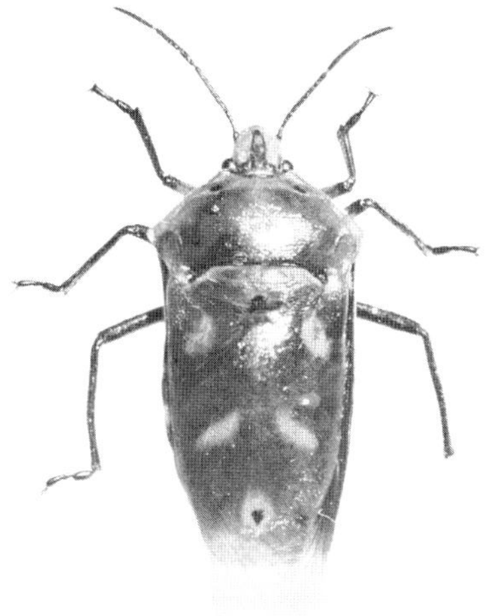 

Plate 99.   Adult *Cantao ocellatus* (Heteroptera, Pentatomidae), found on *Mallotus* foliage; body length 17 mm.

Plate 100.   Tea Shield Bug, *Catacanthus nigripes* (Heteroptera, Pentatomidae); length 16 mm.

Sub-order **Homoptera**

The second sub-order contains the remainder of the plant bugs, the Homoptera, characterized in having the forewing of uniform consistency (albeit either membranous or leathery), with the wings held at rest roof-like over the body, though apterous (wingless) forms are frequent, and the long straight 'beak' arises ventrally from the back of the head and lies between the leg bases. Within the Homoptera are two widely accepted sub-groups, Auchenorrhyncha and Sternorrhyncha.

Auchenorrhyncha

The series Auchenorrhyncha have very short antennae with a terminal arista (bristle), and the rostrum (beak) arises plainly from the head. They are active forms capable of free locomotion.

Super-family **Cicadoidea**

The first four families have several characters uniting them into the superfamily Cicadoidea; they all have their tiny antennae and ocelli situated between the eyes or above.

Cicadidae

The Cicadidae, as would be expected, consists of the cicadas which are a very conspicuous element in the local entomological fauna. At least eight local species of cicadas have been collected, of which several are abundant and widespread at the appropriate time during the spring or summer. The largest and very rare species is the Great Banded Cicada (*Tacua* sp.) which is about 10 cm long, brown coloured and with yellow bands on the abdomen and thorax. The Spotted Black Cicada *(Gaeana maculata)* (Plate 101) with its shiny black body is probably the most common species, and is the first to appear each year, emerging and singing from the end of March through April. A similar species (*Gaeana* sp.) called the Yellow-banded Cicada (Plate 102) is generally rare, but has been recorded as being quite common locally in a few parts of Hong Kong. The other most common species is the Large Brown Cicada

Plate 101.   Spotted Black Cicada, *Gaeana maculata* (Homoptera, Cicadidae); body length 35 mm.

Plate 102.   Yellow-banded Cicada, *Gaeana* sp.; body length 35 mm.

(Cryptotympana mimica) (Plate 103) which is found in very large numbers throughout all wooded parts of the Colony, and it occurs seasonally immediately after the Yellow-spotted Cicada, in May, June and July. It now appears that there are two species of Large Brown Cicada, distinguishable by the length of the sound organ cover.

Three smaller species are also found on trees and bushes in Hong Kong, these being the Green Clearwing Cicada (*Dundubia* sp.) (Plate 104), the Brown Speckled Cicada (*Platypleura hilpa*) (Plate 105), and the distinctive Red-nosed Cicada *(Scieroptera sanguinea)* (Plate 106) with its black thorax, dark forewings, bright red abdomen and a red patch on its face. There are two or possibly three species of the Grass Cicada (*Mogannia* spp.) (Plate 107) and one more similar species (*Pauropsalta* sp.) with a coloured wing base and found mostly on the leaves of *Miscanthus* grass but sometimes on small shrubs in grassy areas. The other species are invariably only found on trees or large bushes, except

Plate 103.   Large Brown Cicada, *Cryptotympana mimica ;* body length 33 mm.

Plate 104.   Green Clearwing Cicada, *Dundubia* sp.; body length 24 mm.

Plate 105.   Brown Speckled Cicada, *Platypleura hilpa*; body length 20 mm.

(Above)   Plate 106.   Red-nosed Cicada, *Scieroptera sanguinea* (Homoptera, Cicadidae); body length 23 mm.

(Above)   Plate 107.   Grass Cicada. *Mogannia* sp., body length 14 mm.

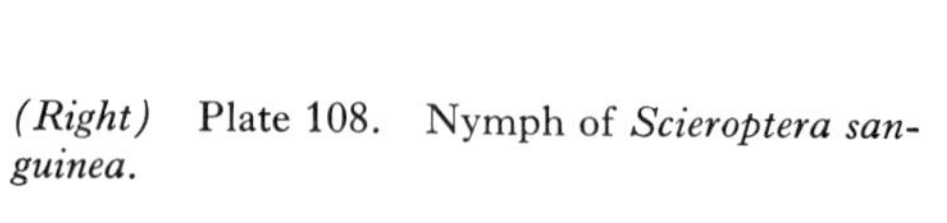

(Right)   Plate 108.   Nymph of *Scieroptera sanguinea*.

for those found on the roadside after being hit and knocked down by passing vehicles.

Whether the larger cicadas show any real preference for certain species of host trees is not at all clear; at times it appears that the thickness of the branch on which they sit, the roughness of bark and other factors are the more important. There probably is a preference for certain trees on the basis of their sap, but these trees are in quite a broad range.

The three smaller aboreal cicadas do not have such a clear-cut-seasonal occurrence and are found in smaller numbers from April through June, although in certain localities (e.g. Cheung Chau Island) they may occur in very large numbers. However, the Red-nosed Cicada is unusual in that it has two emergences each year with a small but definate second emergence in September/October in certain parts of the Colony.

Virtually nothing is known about the life-history of the local cicadas, but it may be deduced from the knowledge of other cicadas in U.S.A. and southern Europe that the females lay their eggs singly or in small groups in twigs which are slit by the ovipositor to make oviposition sites. After several weeks the eggs will hatch and the young nymphs fall to the ground whereupon they start to burrow using their enlarged, mole-like forelegs and embark upon their subterranean nymphal life. While burrowing in the soil, they feed by piercing fine roots with their stylet and sucking the sap. The larger species may well spend several years as subterranean nymphs, but the smaller ones probably only spend one year. Plate 108 shows a nymph of *Scieroptera sanguinea* found burrowing in soil. Upon reaching maturity, the large nymphs crawl out of the earth and crawl up vegetation or, more frequently, tree trunks where they fix themselves securely by gripping with their large spiny forelegs; the skeleton then splits along the back and the new adult cicada crawls out. Plate 109 shows a young adult *Gaeana maculata* emerging from its nymphal skin. The old nymphal skin (exuvium) is left fixed on to the tree trunk or vegetation (Plate 110).

The single most striking feature of cicadas is their 'singing'. The males produce a very shrill penetrating sound from specially developed organs at the base of the abdomen ventrally. The stridulating organ consists basically of a sound-box and a resonating chamber, although some species lack the resonating chamber and produce a sound that is harsher and cannot be controlled. The sound is produced by the tymbal which is an organ on the body surface formed as a ring-like thickening with a crisp, strong membrane stretched across it. A muscle is attached to the membrane and can pull it in suddenly, producing a sound like that made by the distortion of a tin lid. This organ can work for long periods,

Plate 109.   Young adult *Gaeana maculata* emerging from the nymphal skin.

Plate 110.   Exuvium of Large Brown Cicada clinging to a blade of *Miscanthus* grass.

being alternately distorted by the muscle contraction, and returning to its original shape by the elastic nature of the ring as the muscle relaxes. Each of these movements produces a click, and as the 'song' of the cicada is made by a continuous series of little clicks, this mechanism must be very efficient and working at a very high speed. The sound produced is amplified by an air-space acting as a resonating chamber; its hard, rounded walls imparting a more musical quality to the song of some cicadas. The song of each species is quite distinct, and with a little

Plate 111.   Large Brown Cicadas feeding and stridulating collectively on a branch of *Paulowina fortueni*.

practice the local cicadas can be identified by their song. The whole sound organ is usually covered by a hard triangular flap of skeleton, which is sometimes lifted slightly as the cicada sings, producing a difference in the quality of sound emitted. Cicadas also possess special auditory organs called tympana, which lie next to the tymbals, and these are of a similar nature to those of other sound-producing insects. They consist of a tense, sensitive membrane which is distorted by sound waves so that it touches the minute bristles of sensory cells underneath. Apparently, each male cicada has a special muscle which immobilizes the membrane of its own tympanum as it begins to sing, so that it does not deafen itself. The singing is apparently to attract females and as an aid to keep the population together; often groups of cicadas are seen sitting feeding or singing together (Plate 111). It also serves a warning purpose and if a singing male is seized the others in its immediate vicinity usually fly off to another tree. Cicadas are of great interest to the Chinese people. The bodies are used in Chinese medicine; cicadas carved in jade and used to be buried at funerals are obtainable in local jewellery shops as amulets.

Cercopidae

In appearance, the Cercopidae resemble small cicadas but are different in that they jump, hence their common name of froghoppers. Only a few species have been found in Hong Kong but a large red and black species *(Cosmocarta abdominalis)* (Plate 112) is very common. The adult bug feeds on Camphor and other trees, and sometimes several dozen can be seen on a single tree. The frothy spittle-masses in which the nymphs live can be seen in the leaf-axils of *Melastoma sanguineum* and *M. candidum* (Plate 113) in the spring and early summer. It is the production of the

Plate 112. Common Froghooper, *Cosmocarta abdominalis* (Homoptera, Cercopidae); body length 14 mm.

Plate 113. Spittle masses (bugs) of Common Froghopper, on *Melastoma candidum* plant.

Plate 114. Tubicolous Spittle Bug, *Machaerota ccronata* (Homoptera, Cercopidae, Machaerotinae) on stems of *Helicteres augustifolia*.

Plate 115. Green Rice Leafhopper, *Nephotettix nigropictus* (Homoptera, Cicadellidae).

spittle mass, presumably for protection against predators, which earns the nymphs the vernacular name of spittle bugs. Another common spittle-bug is the widespread brown coloured *Ptyelus* sp. to be found both as adults and nymphs on *Cerbera manghas* trees at the back of beaches in the New Territories.

An interesting variation upon this theme is seen in the local tube-building spittle bugs (the Machaerotinae) in which the nymphs of *Machaerota coronata* construct a ridged tube stuck on to the stem of the host plant, exclusively *Helicteres augustifolia* (Plate 114), inside which they live in a bath of spittle. The first instar nymph builds a tiny tube on the stem of the plant, but as it grows, rather than attempting to enlarge this tiny tube, it departs and lower down the stem it builds a new, larger tube in which further development takes place (Marshall & Marshall, 1966).

Cicadellidae

Closely related to the Cicadidae are the Cicadellidae, as is shown by their name. They were formerly called the Jassidae, and they are referred to as leafhoppers, or sometimes as jassids. A number of these tiny leafhoppers are quite common locally on certain host plants. By far

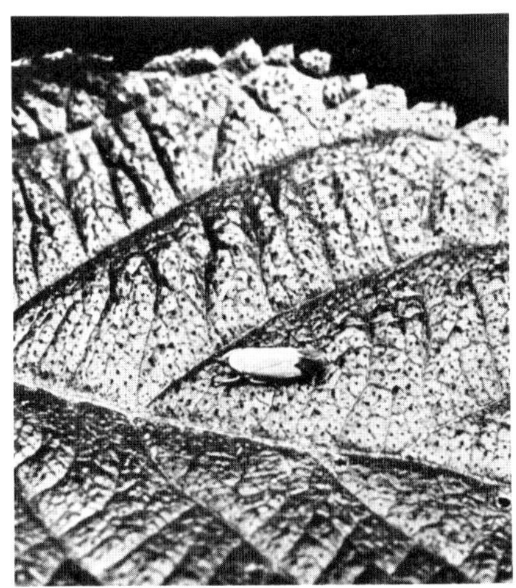

(Left) Plate 116. Green Rice Leaf-hopper, *Nephotettix virescens,* resting on vegation after being attracted to house lights at night.

(Below) Plate 117. Green Rice Leaf-hopper, *Nephotettix* sp. (Homoptera, Cicadellidae); body length 4–5 mm.

the most common and abundant are the Green Rice Leafhoppers *(Nephotettix nigropictus)* (Plate 115) and *N. virescens*(Plate 116), but most likely there are other species to be found locally. *Nephotettix* spp. (Plate 117) feed on a wide range of grasses and several species are pests of economic importance on rice crops throughout Southeast Asia. These tiny green and black bugs fly to lights at night, and several times during the summer period large numbers are attracted to street and house lights (Plate 117). The populations build up during the warm weather and by September and October their neumbers can be enormous. They are important rice pests partly because of the direct damage done by these large numbers of feeding bugs, and also because they are vectors of several plant viruses which cause serious diseases on rice.

Although many leafhoppers are confined to Gramineae (grasses and

(Left)   Plate 118.   White Leafhopper, *Typhlo-cyba* sp., on leaf of *Antidesma bunius*; body length 8 mm.

(Right)   Plate 119.   Citrus Leaf-hopper, *Coelidia* sp.; body length 8 mm.

cereals) as hosts, there are others which are confined to trees (often fruit trees in temperate countries). A fairly common and very widespread genus throughout the world is the white coloured *Typhlocyba* and Plate 118 shows a specimen of *Typhlocyba* sp. on a leaf of *Antidesma bunius* (it has also been observed on many other trees). Another common local species is the tiny yellow and black *Coelidia* sp. (Plate 119) sometimes found in large numbers on *Citrus* bushes and other hosts. The larger species *Bothrogonia* (Plate 120) is found locally on a number of different plants; its reddish brown coloured forewings and its size are very distinctive.

Membracidae

The final group of bugs related to cicadas is the Membracidae which are characterized by their bizarre body form, usually with the thoracic pronotum extended posteriorly as a large spine. The common local species recorded is *Tricentrus* sp. found on several different trees, including *Antidesma bunius* and *Bauhinia*. It has small thoracic 'horns' which give it a buffalo-type appearance.

Super-family **Fulgoroidea**

The superfamily Fulgoroidea is a large group of diverse forms, predominantly tropical in distribution, all with antennae and ocelli situated below the eye level. They are collectively called 'planthoppers' because most are jumping in their habits. Most of them are not of particular importance economically, ecologically, or even faunistically.

Fulgoridae

The most spectacular local species is *Pyrops candelaria,* the lantern-fly (Fulgoridae) (Plates 121 & 122) with its distinctive long 'snout', mottled green forewings and bright yellow and black hind-wings. It is common in parts of the New Territories where it is found on Litchi and Longan trees, usually sitting on the trunks (Plate 123); it has also been observed in the Hong Kong Island on Mango and *Acacia* tree trunks. Usually adults are found from June to November, and nymphs from January to June. A second species, *P. lathburii,* is occasionally found on *Eucalyptus* trees in some parts of the New Territories; but this is smaller and is predominantly brown in colour. The purpose of the 'snout' is not known, but it was once erroneously thought to be luminous, hence the common name lantern-fly. *Pyrops* is unusual in that it possesses a most remarkable parasite—a small caterpillar. The adult is a small brown moth, *Epipyrops,* with no mouth-parts and so does not feed. It lays its eggs on the branches of the tree; the young, active caterpillar attaches itself to the wings of the bug and it feeds on blood derived from a wing vein. As it grows, the caterpillar moves on to the upper surface of the insect's abdomen, where it bites a hole in the cuticle with its long thin jaws, and scoops up blood and underlying tissue with its mouth-parts. Only one large *Epipyrops* larva is found on any single lantern-fly. The

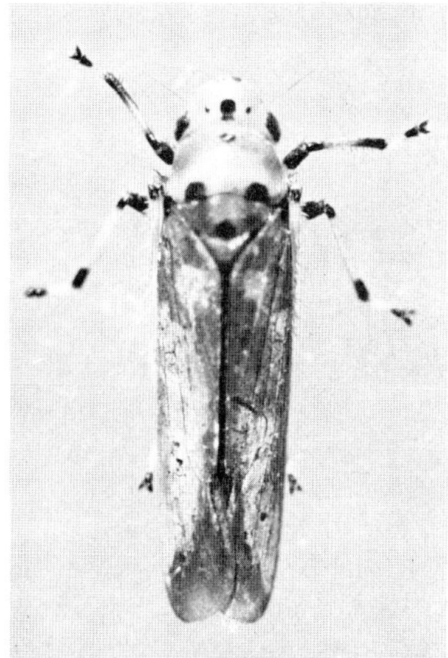

Plate 120. Large Brown Leafhopper, *Bothrogonia* sp.; body length 14 mm.

Plate 121. Lantern-fly, *Pyrops candelaria,* in natural position.

(*Above*)   Plate 122.   Lantern-fly, *Pyrops candelaria* (Homoptera, Fulgoridae); body length 40–50 mm.

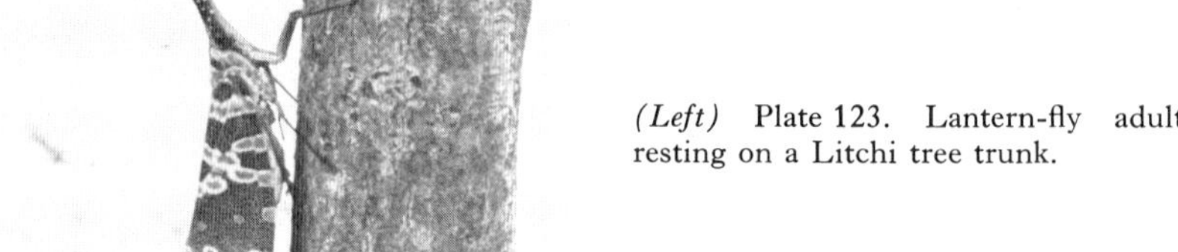

(*Left*)   Plate 123.   Lantern-fly adult resting on a Litchi tree trunk.

mature caterpillar has a remarkable covering of white, silk-like material over the body surface. The moth larva at full maturity leaves the bug and settles on a branch where it pupates in a fluffy pupal case.

Delphacidae

The true planthoppers are in the family Delphacidae, which are typically small brown bugs distinguished by the possession of a movable, serrulate, spur on the hind tibiae. *Sogatella furcifera* (Plate 124) is the White-backed Planthopper; the Sugarcane Planthopper *(Perkinsiella saccharicida)* (Plate 125) is reported to be a major pest of sugarcane in South China and is expected to occur in Hong Kong. Planthoppers are mostly found on grasses and graminaceous crops, and many species can be seen locally, including *Nilaparvata lugens* (Brown Planthopper of Rice) and *Paurohita fuscovenosa* (Bamboo Planthopper).

Dictyopharidae

Dictyopharidae are smallish bugs, often with an exterior head process, looking like a 'snout'. One common local species is *Dictyophara* sp. frequently to be seen on *Citrus* foliage.

Flattidae

Flattidae are called moth bugs, or sometimes (in U.S.A.) flattid planthoppers and several local species are found in small numbers. In the New Territories, the large, white, pale-green or yellowish species *Pulastya discolorata* (Plate 126) is not uncommon, especially on *Citrus* plants. The smaller green *Salurnis marginellus kershawi* (Plate 127) is found often on *Antidesma bunius* (Plate 128) and on other plants; another small white species is also found occasionally. The nymphs of flattids tend to be gregarious in habits and are covered with long white waxy filaments.

Ricaniidae

A very similar group of bugs is found in the Ricaniidae, but their wings tend to be held less vertically and more laterally, and there are a couple

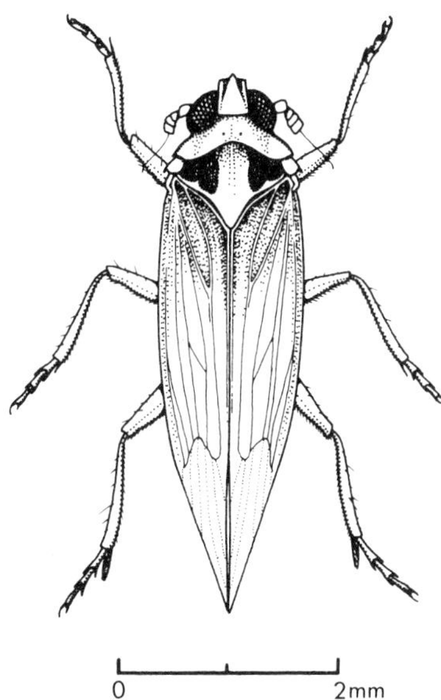

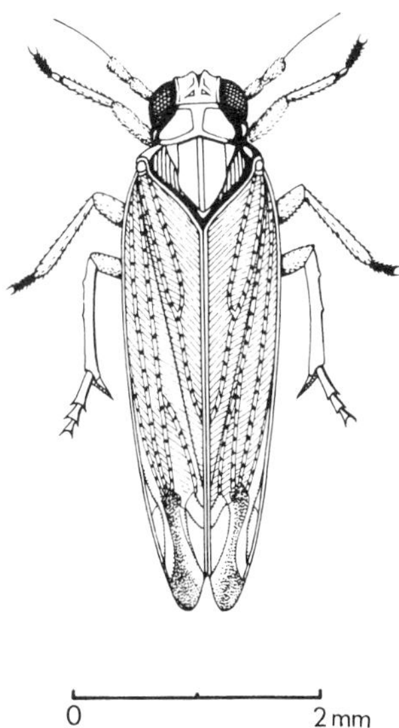

Plate 124. White-backed Planthopper, *Sogatella furcifera* (Homoptera, Delphacidae); body length

Plate 125. Sugarcane Planthopper, *Perkinsiella saccharicida*; adult female; body length

(*Above*) Plate 126. Moth bug, *Púlastya discolorata* (Homoptera, Flattidae); body length 20 mm.

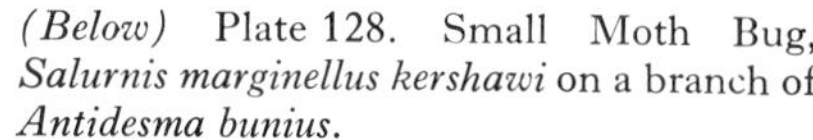

(*Left*) Plate 127. Small Moth Bug (Homoptera, Flattidae) resting on grass.

(*Below*) Plate 128. Small Moth Bug, *Salurnis marginellus kershawi* on a branch of *Antidesma bunius*.

of local species of interest. The brown and white species *Ricania* sp. (Plate 129) appears to be specific to *Miscanthus* grass, on which very large populations sometimes build up during the summer. The nymphs are again white with long waxy filaments. The distinctive black and white species *Ricania speculum* (Plate 130) occurs regularly in small numbers on different local plants, and has been recorded as a minor pest of oil palm in Malaysia. A green species of *Ricania* is sometimes found on citrus.

Other families

There are several other small families within the Fulgoroidae, all of tropical distribution but the only species common in Hong Kong is the pale green, flattened and pointed *Kallitaxila macaona* in the family Tropiduchidae found on the leaves of *Antidesma bunius* and other plants.

(Above) Plate 129. *Ricania* sp. (Homoptera, Ricaniidae) on *Miscanthus* grass; length 8 mm.

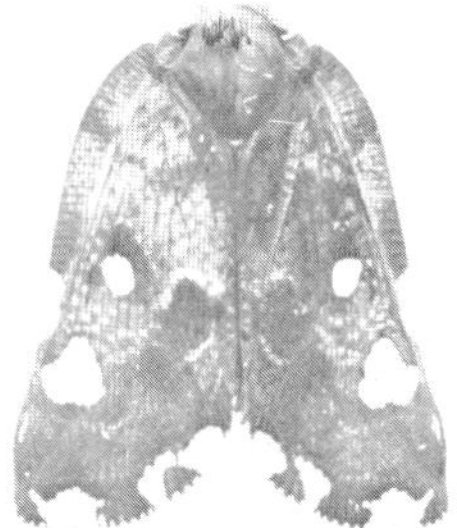

(Left) Plate 130. *Ricania speculum* (Homoptera, Ricaniidae); body length 10 mm.

Sternorrhyncha

The series Sternorrhyncha are characterized by having well-developed antennae, without a terminal arista (bristle), although sometimes the antennae are atrophied. The rostrum is either wanting or else apparently arising between the anterior coxae (leg bases) and the female form is often inactive or incapable of locomotion.

Psyllidae

The Psyllidae are small bugs that look much like tiny cicadas or aphids, and are sometimes called, because of this resemblance, jumping plant lice; if disturbed, they jump readily for considerable distances. Locally, they are an interesting group in that there are a number of species here that are quite similar in appearance as adults but the nymphs live in a series of quite different habitats and some produce spectacular effects on their host plants. *Homotoma radiatum* can be seen just sitting on the leaves of *Ficus superba* var. *japonica* where its feeding produces no apparent reaction on the plant body. *Paurocephala* (new species) infestations on *Ficus hispida* however cause leaf-curling (Plate 131) and a heavy infestation can cause serious distortion of young leaves. *Trioza camphorea* infests the leaves of Camphor and when the nymphs are settled on the under-surface of young leaves, they cause the leaf tissue to proliferate down (up) underneath them so that the nymph is eventually sitting at the bottom of a small pit and the infestation is evident as lumps on the upper surfaces and open pits on the under-surface of the leaf in which the nymphs can be seen sitting (Plate 132). *Trioza erytrea* is an African species confined to Citrus, but other species of *Trioza* are widespread in the tropics on several different hosts.

Plate 131. Leaf-curl on *Ficus hispida* produced by *Paurocephala (n. sp.)* (Homoptera, Psyllidae).

Plate 132. Leaf-pits on Camphor produced by nymphs of *Trioza camphorea* (Homoptera Psyllidae).

Plate 133. Leaf-pits on *Syzygium* leaves produced by nymphs of *Megatrioza* sp. nr. *vitiensis* (Homoptera, Psyllidae).

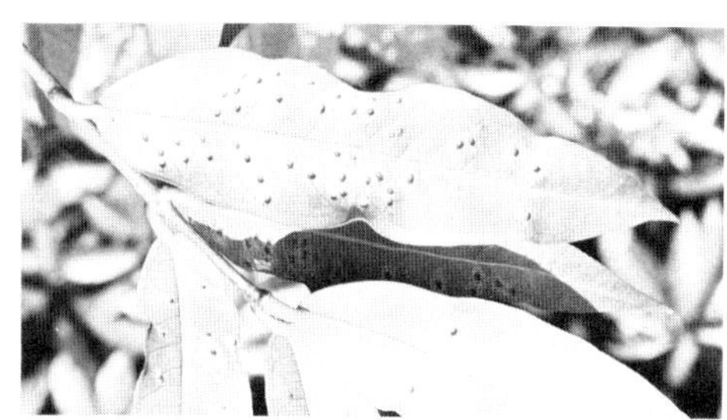

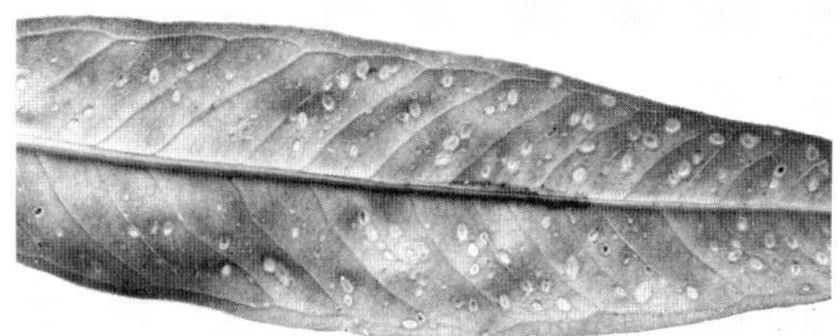

Plate 134. *Megatrioza* nymphs seen *in situ* in their pits on *Syzygium* leaves.

A similar psyllid nymph is seen on *Syzygium* leaves—this is *Megatrioza* sp. nr *vitienses* (again, probably a new species) (Plates 133 & 134). A species producing more effect on the host plant is *Macrohomotoma striata* (Plate 135) which is found in the terminal buds of *Ficus microcarpa* where the nymphs produce large quantities of waxy white filaments and their feeding produces shortened, distorted shoots, usually resulting in the death of the shoot (Plate 136 & 137). The most spectacular effects produced are the large, spherical leaf-galls on *Ficus variegata* var. *chlorocarpa* (Plate 138). Occasionally, the leaf galls are so numerous as to completely cover the upper leaf surface (Plate 139). There are usually three broods of this psyllid *(Pauropsylla udei)* in Hong Kong each

Plate 135. Adult *Macrohomoto-ma striata* (Homoptera, Psyllidae); body length 4 mm.

(Right) Plate 136. Shoot of *Ficus microcarpa* with infestation by nymphs of *Macrohomotoma striata* (Homoptera, Psyllidae).

(Below) Plate 137. Nymphs of *M. striata* on leaf of *Ficus microcarpa*.

Plate 138. Leaf-galls on *Ficus variegata* var. *chlorocarpa* produced by nymphs of *Pauropsylla udei* (Homoptera, Psyllidae).

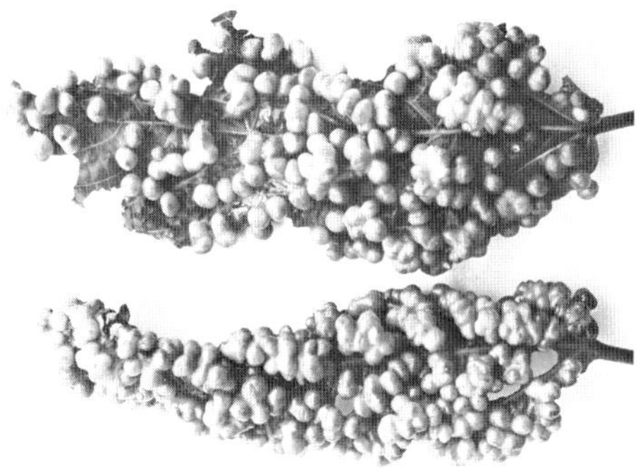

Plate 139. Leaf-galls on *Ficus variegata* var. *chlorocarpa* produced by nymphs of *Pauropsylla udei* (Homptera, Psyllidae).

summer. Most psyllid nymphs are similar in appearance, being flattened and rounded with large flattened wing buds at the sides of the body, but the *Trioza* group which makes leaf-pits are generally immobile and more scale-like (Plate 134).

Aleyrodidae

The Aleyrodidae are the minute whiteflies and blackflies, some species of which are important as crop pests, particularly the cosmopolitan and polyphagous *Bemesia tabaci* (Plate 140) found on a wide range of host plants in warmer countries. On citrus plants locally there are two different aleyrodids to be found, although neither appears to be particularly abundant here. *Aleurocanthus woglumi* is the Citrus Blackfly (Plate 141) found in many parts of the tropics and sometimes on hosts other than Citrus. Similar species are to be found here on the under-surface of guava and camphor leaves (Plate 142), and another *(Pealius fici)* occurs on the upper surface of leaves of *Ficus microcarpa* where it causes the leaf edges to roll in longitudinally (Plate 143). *Dialurodes citri* is the Citrus Whitefly (Plate 144) found also on many other plants in tropical Asia and the New World.

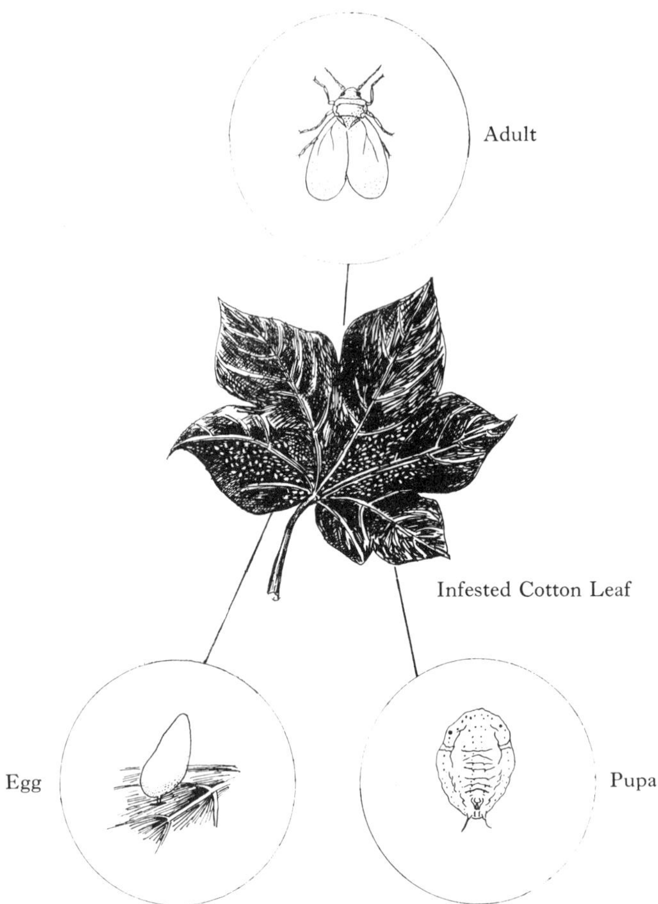

Plate 140.   Tobacco Whitefly, *Bemisia tabaci* (Homoptera, Aleyrodidae).

Local infestations in Hong Kong are not frequent but occasionally, severe cases occur. In 1976 there were very heavy infestations on the leaves of *Bauhinia* by the blackfly *Aleurolobus marlatti* (Plate 145) and in the spring of 1977 the leaves were covered with small dark, mottled-winged adults. In the autumn of 1974 and 1975, very severe infestations of *Aleyrodes lonicerae* (Plates 146 & 147) were observed on *Oxalis corymbosa* growing as a natural undercover in verandah plant pots. In this restricted habitat, the whitefly population grew rapidly and spread from pot to pot, eventually (after a few weeks) killing the *Oxalis* plants completely. At the height of each infestation, virtually all leaves were

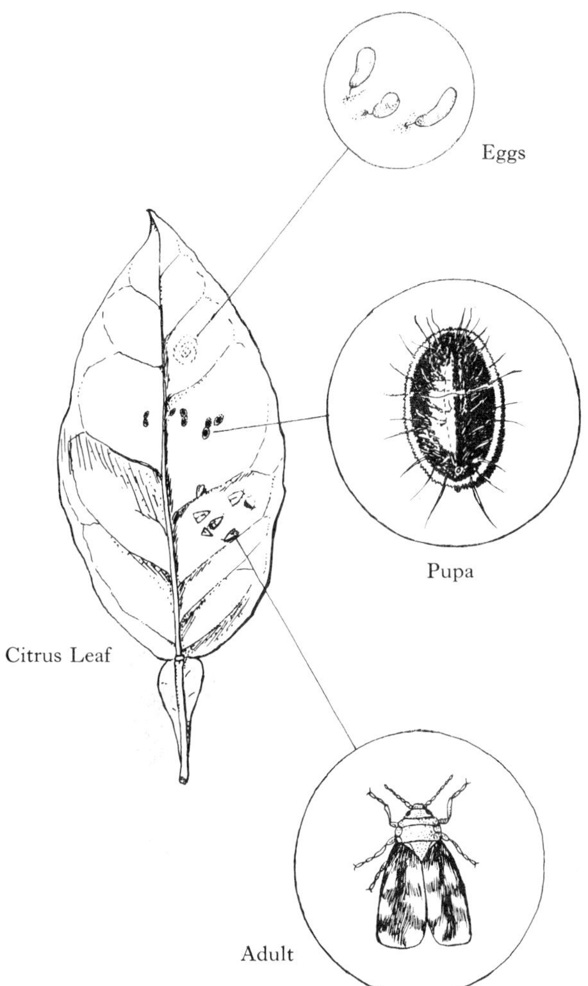

Plate 141. Citrus Blackfly, *Aleurocanthus woglumi* (Homoptera, Aleyrodidae).

densely covered on the under-surface with nymphs, white wax, and adult bugs (Plate 145), and when disturbed the tiny adults flew up as a fine white cloud.

Super-family **Aphidoidea**

Aphididae

The Aphidoidea contains the aphids, the great majority of which

Insects of Hong Kong

Plate 142.   Blackfly nymphs, *Aleurocanthus* sp., under Camphor leaves.

Plate 143.   Blackfly nymphs, *Pealius fici*, rolling leaves of *Ficus microcarpa*.

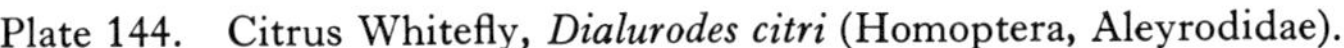

Plate 144.   Citrus Whitefly, *Dialurodes citri* (Homoptera, Aleyrodidae).

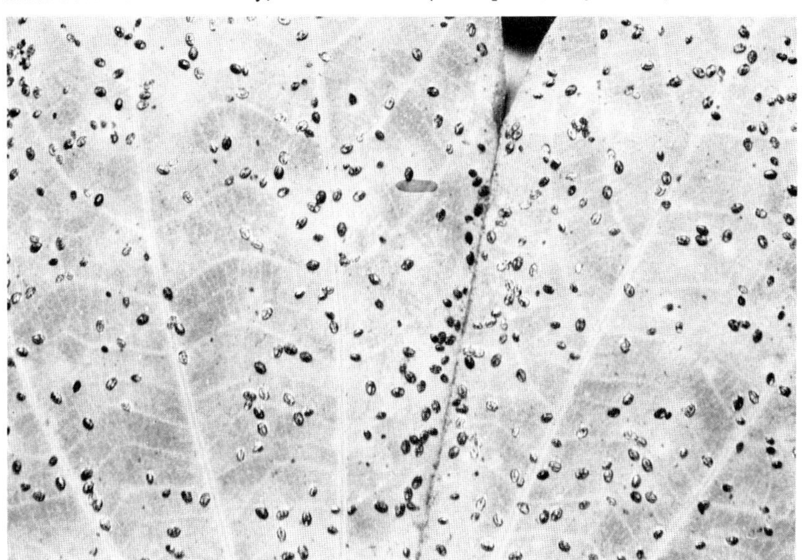

Plate 145.   Blackfly nymphs, *Aleurolobus marlatti* on leaves of *Bauhinia* (Homoptera, Aleurodidae).

(Above)   Plate 146.   *Oxalis corymbosa* leaves heavily infested with the Whitefly *Aleyrodes lonicerae.*

(Left)   Plate 147.   Close-up of adult Oxalis White-fly, *Aleyrodes lonicerae* (Homoptera, Aleyrodidae); body length 1.5 mm.

belong to the family Aphididae. Aphids are small plant bugs called erroneously either 'greenfly' or 'plant-lice', and in temperate parts of the world they are one of the most important groups of plant pests as well as being one of the most abundant groups of insects. Although there are some aphid species which are essentially tropical in distribution, the group generally is temperate and they are best represented in the Palaearctic and Nearctic regions. In the tropics, reproduction is invariably by parthenogenetic (i.e. reproduction without fertilization) viviparity (born alive and active, as opposed to hatching from eggs) and males are seldom recorded.

In Hong Kong, there appears to be a few tropical species, which are abundant during the summer period. The important cosmopolitan temperate species (e.g. *Myzus persicae,* etc.) that are major agricultural pests, are numerous and damaging only during the cooler winter months, and are extremely difficult to find during the hotter summer period. On house plants aphid infestations usually commence in November/December and gradually build up to a peak in January/February but tend to disappear after April. However, they may have killed their host plants long before then.

A heavy aphid infestation on young shoots can be very damaging for not only the host plant may succumb and die, but more important, especially from an agricultural point of view, is the fact that most aphids are vectors of viruses causing many different serious diseases on agricultural plants. *Myzus persicae* for example is known to be a vector of more than 100 different plant viruses. Locally Turnip Mosaic Virus causes extensive damage to *Brassica* crops in the New Territories, where it is distributed by *Lipaphis erysimi* (Turnip Aphid), and *M. persicae.*

Observation of infested plants reveals that most aphid colonies are to be found either on very young leaves and shoots, or else on old senescent leaves. This is primarily because an active plant has free sugars in the sap in the young tissues where growth is taking place, but in mature leaves the photosynthesized carbohydrates are rapidly converted into polysaccharides such as starch and the amount of free simple sugars in the sap is not large. As a leaf senesces, the stored insoluble starches are reconverted into soluble sugars for translocation to other parts of the plant body before leaf abscission takes place. Thus the sap of very young tissues and of senescent leaves has a high content of dissolved simple sugars. During the course of evolution, many aphids have utilized this situation and become attracted to yellow colours rather than dark greens. Simple field experiments reveal this preference for the yellow colour. The end result is that certain species of aphid show preference for young tissues and will be found on the shoots and buds, other species prefer senescent yellowing leaves, and some are found in both locations.

Another aspect of aphid infestations is that as some other plant sap-sucking bugs, the aphids imbibe large quantities of plant sap in order to find sufficient dietary proteins, and as a consequence, they have to excrete large quantities of both water and soluble sugars. This excess sugar is excreted in the form of concentrated drops called 'honey-dew' from the anus of the insect. With a large aphid population, the amount of honey-dew produced is considerable and in Europe, certain tree-inhabiting species (e.g. Sycamore Aphid) produce a fine 'rain' of honey-dew under large infested trees. The honey-dew tends to cover the leaves and twigs and then drips on to lower leaves until the whole plant is covered. Usually, the honey-dew is colonized by Sooty Moulds which cover the foliage with a black film and is most unsightly on ornamentals. Not all aphid species produce large quantities of honey-dew, for example, *Myzus persicae* produces little.

In natural situations, it is usual to find that aphid colonies are attended by ants which feed on the excreted honey-dew, and it appears that the

Plate 148. Mummified aphid nymphs on *Brassica* stem, some showing parasite emergence holes.

ants protect the aphids by keeping away predators and parasites. In the absence of ant guardians, it is usual to find that many of the aphids nymphs are immobile, brown and swollen. These are parasitized nymphs and are called 'mummies' (Plate 148), and in due course, tiny parasitic wasps will emerge through a small circular hole. A regular pest control

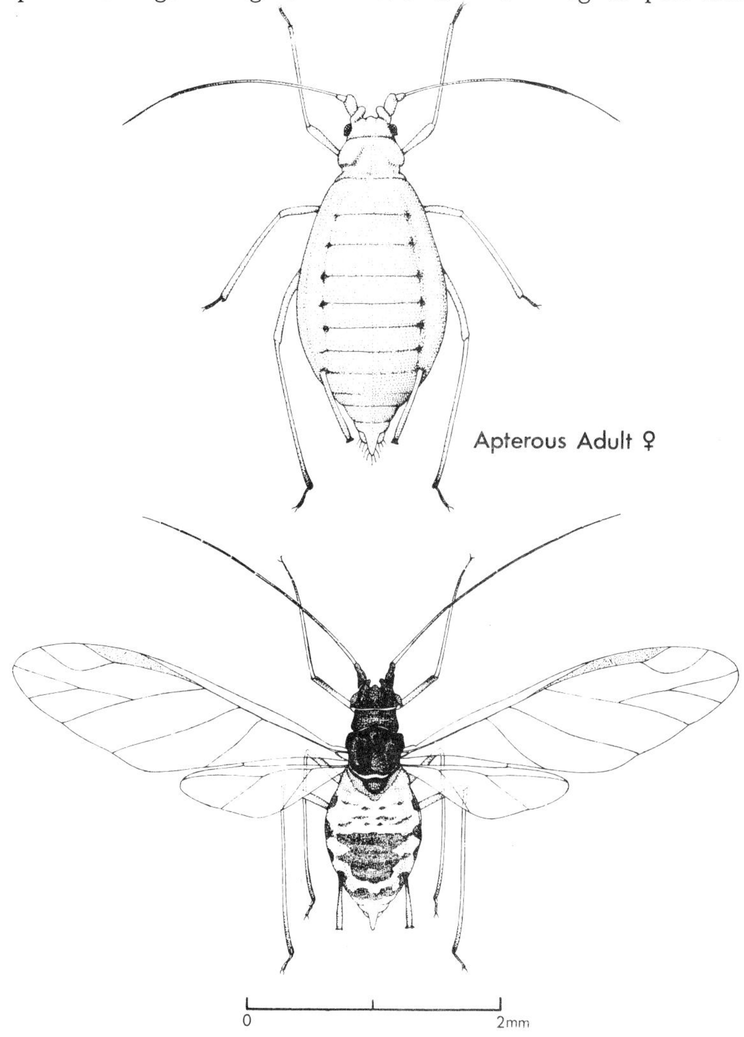

Plate 149. Green Peach Aphid, *Myzus persicae* (Homoptera, Aphididae), possibly the commonest aphid species occuring in Hong Kong.

measure used against arboreal infestations of aphids, mealybugs and some scales is to spray-band the trunk with dieldrin to kill the ants and to prevent more ants from climbing the trunk. In the absence of protecting ants, the bugs are usually quite quickly attacked by their many natural enemies.

Aphids are found on all types of plants, from trees, bushes, shrubs and herbs down to grasses, and they also vary considerably in their host specificity. Some species such as Banana Aphid are almost monophagous and more or less confined to *Musa* as hosts. Others are oligophagous,

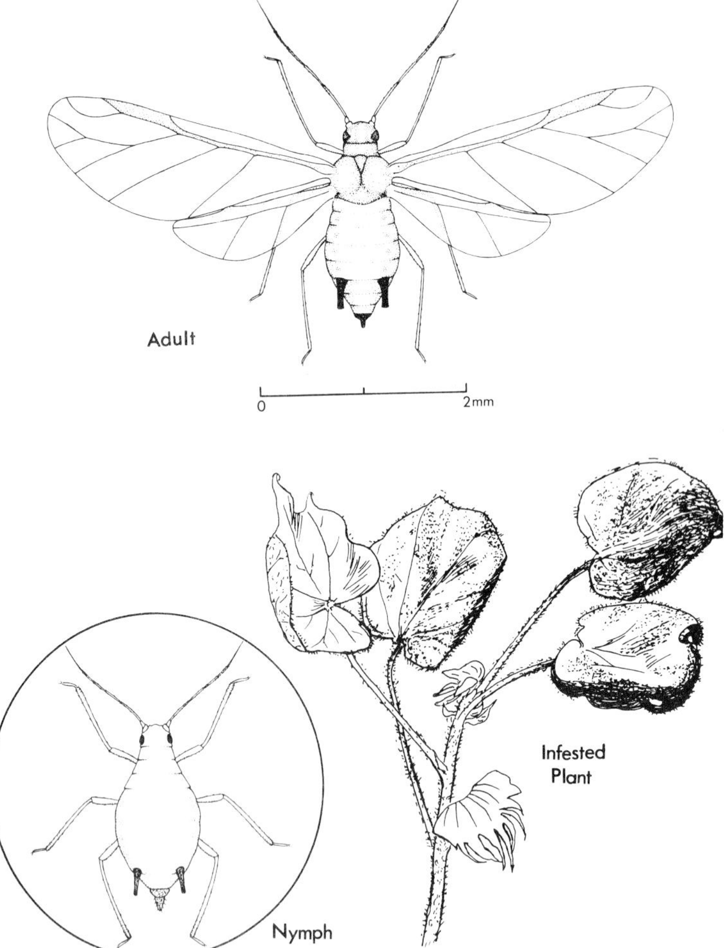

Plate 150.   Cotton (Melon) Aphid, *Aphis gossypii* (Homoptera, Aphididae).

e.g. Groundnut Aphid is generally restricted to Leguminosae, and Wheat Aphid to temperate cereals and grasses. At the other extreme there are *Myzus persicae* and *Aphis gossypii,* both of which are polyphagous and feed on a very wide range of host plants.

Some of the more important local species are Green Peach Aphid *(Myzus persicae)* (Plate 149), Cotton (Melon) Aphid *(Aphis gossypii)* (Plate 150–152), Groundnut Aphid *(Aphis craccivora)* (Plate 153), Turnip Aphid *(Lipaphis erysimi)* (Plate 154 & 155), Citrus Aphids *(Toxoptera* spp.) (Plate 156), Black Bean Aphid *(Aphis fabae solanella)* on *Solanum nigrum* (Plate 157), Maize Aphid *(Rhopalosiphum maidis),* Banana Aphid *(Pentalonia nigronervosa),* and Black Pine Aphids *(Cinara* spp.). An additional spectacular local aphid is *Greenidea ficicola* (Plate 158) occurring on *Ficus microcarpa* and *F. chlorocarpa,* which has enormously long and bristly siphunculi and consequently is very distinctive in appearance. The species *Toxoptera aurantii* (Plate 156) has also been recorded infesting the leaves and young shoots of *F. microcarpa.* A very distinctive species of aphid *(Aphis nerii)* having yellow nymphs with black siphunculae is common in the winter on shoots and leaves of Oleander and some other flowering shrubs (Plate 159).

Plate 151.   Cotton (Melon) Aphid, *Aphis gossypii,* on *Hibiscus* flower bud.

Plate 152.   Cotton (Melon) Aphid, *Aphis gossypii,* on young *Hibiscus* leaves.

Pemphigidae

The family Pemphigidae is a small group of woolly aphids, characterized by the nymphs often producing long white waxy filaments and the adults having no visible siphunculae. On *Miscanthus* grass can be found colonies of Sugarcane Woolly Aphid *(Ceratovacuna lanigera)* (Plate 160 & 161) and another species can be seen on twigs and leaves of *Dalhbergia balansae.* An unknown species of woolly aphid belonging to the genus *Chaitoregma* is common locally and produces very large spectacular galls on the twigs of Chinese privet *(Ligustrum sinense)* (Plate 162). A large grey species producing little wax is locally common on bamboo stems (Plates 163–164), infestations usually starting under

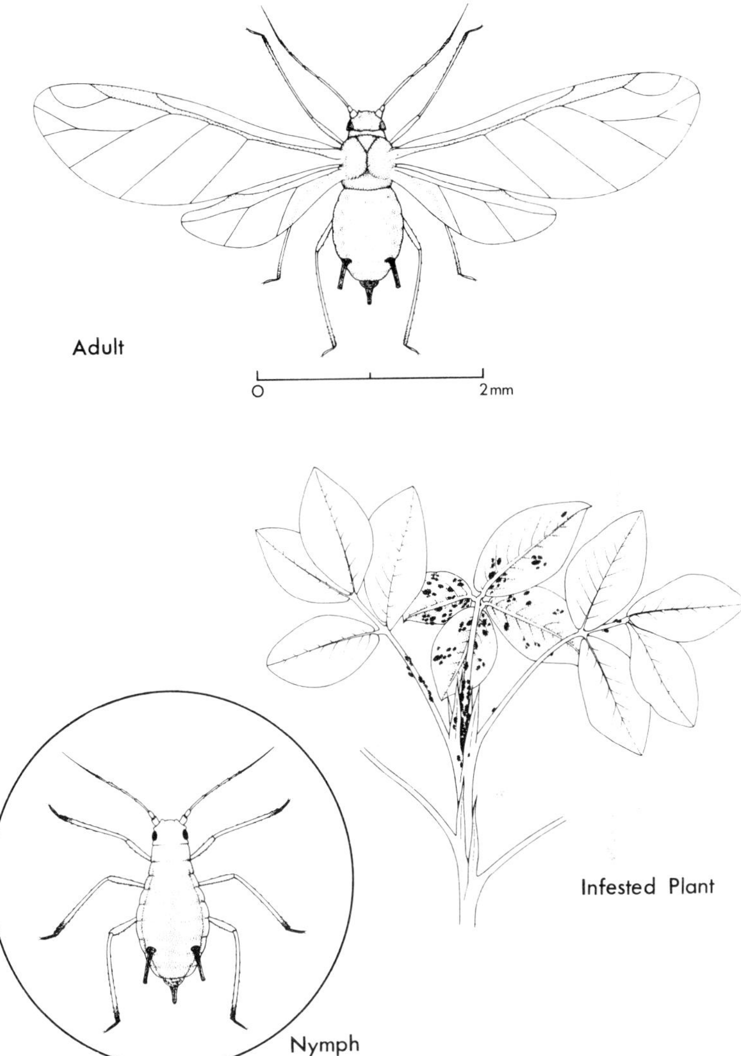

Plate 153. Groundnut Aphid, *Aphis craccivora* (Homoptera, Aphididae).

the leaf-sheathes and often extending over the entire internode. This species has been identified as *Pseudoregma bambusicola*. Some woolly aphids produce little honey-dew and are not usually attended by ants, but the Bamboo Woolly Aphids does produce copious honey-dew and is attended by ants, and wasps and flies are also attracted by the sugar.

  Insects of Hong Kong

Plate 154.   Turnip Aphid, adults and nymphs, on a *Brassica* leaf.

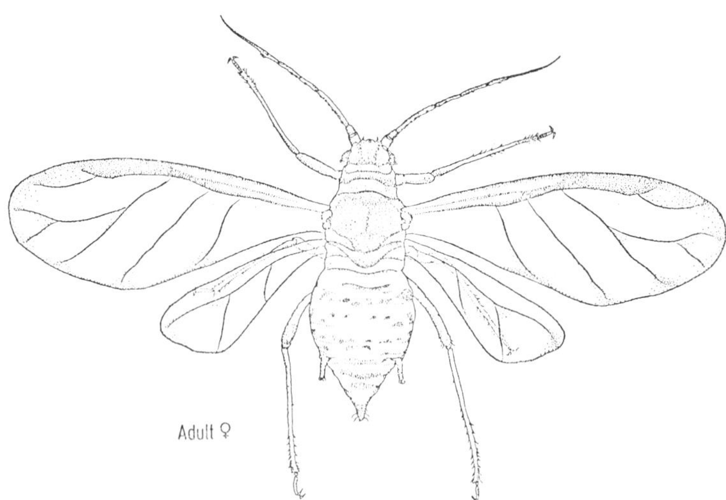

Plate 155.   Turnip Aphid, *Lipaphis erysimi* (Homoptera, Aphididae).

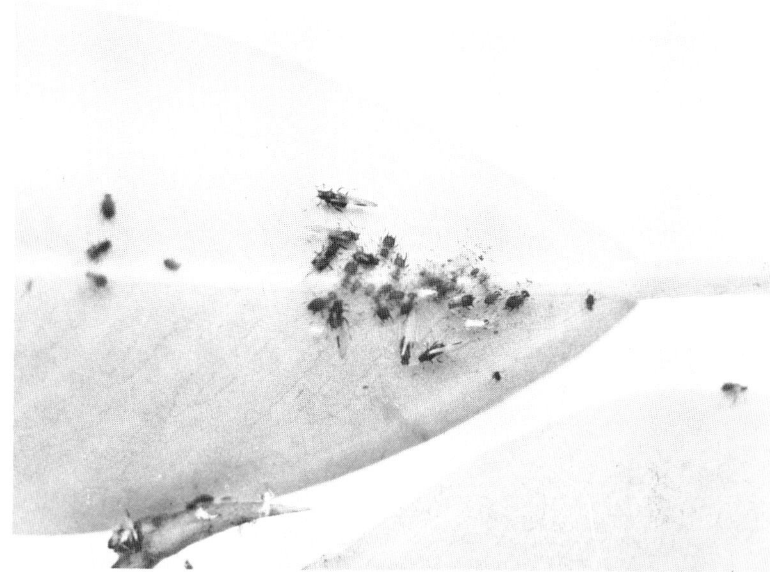

Plate 156. Black Citrus Aphid *Toxoptera aurantii* (Homoptera, Aphididae) on *Ficus microcarpa*.

Plate 157. Curled leaves of *Solanum nigrum* indicating infestation by *Aphis fabae solanella* (Homoptera, Aphididae).

Plate 158.   Fig Aphid, *Greenidea ficicola*, with ants, on *Ficus microcarpa*.

Plate 159.   Aphid nymphs, *Aphis nerii*, on Oleander.

Plate 160.   Sugarcane Woolly Aphid, *Ceratovacuna lanigera* (Homoptera, Pemphigidae) on *Miscanthus* grass.

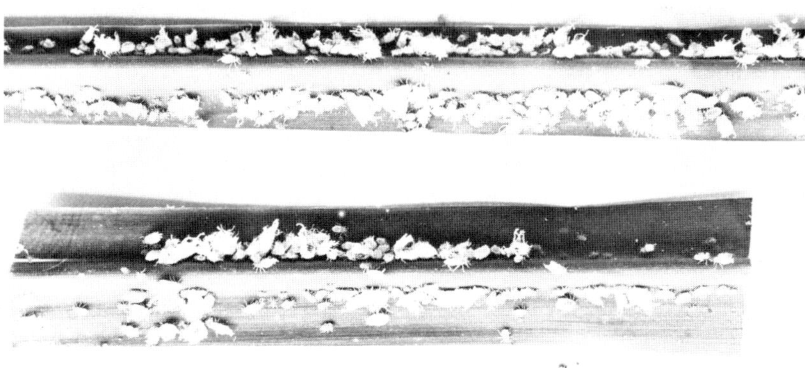

Plate 161.   Sugarcane Woolly Aphid, *Ceratovacuna lanigera* (Homoptera, Pemphigidae) on *Miscanthus* grass.

Plate 162.   Twig galls on Chinese Privet, caused by Woolly Aphid *Chaitoregma* sp. (Homoptera, Pemphigidae).

Plate 163.   Bamboo Woolly Aphid, *Pseudoregma bambusicola* (Homoptera, Pemphigidae) on stem of bamboo.

Plate 164. Bamboo Woolly Aphid, *Pseudoregma bambusicola* (Homoptera, Pemphigidae) on a lateral shoot of bamboo.

Phylloxeridae & Adelgidae

The family Phylloxeridae is very small, and contains the well known Vine Phylloxera *(Viteus vitifolii)* ; the family Adelgidae is equally small and confined to conifers. Neither groups have been recorded in Hong Kong.

Super-family **Coccidoidea**

The next group of families is the Coccidoidea where the females are scale-like or covered with a waxy exudation, apterous and sometimes legless. The tiny males are dipterous (two-winged) (Plate 181) with the hind-wings reduced and the mouth-parts atrophied. The first instar nymphs are minuscule and very active and called 'crawlers'—this is the dispersive stage of the insects.

Margarodidae

The Margarodidae contains the well-known fluted scales, and the females have well-developed legs and antennae and are always mobile. *Icerya purchasi* (Plate 165) is the Cottony Cushion Scale common on *Cassia* trees, *Citrus, Casuarina,* and other plants. These insects are interesting in that each adult is hermaphrodite and viviparous, and the males are seldom found. The nymphs tend to settle initially on the leaves alongside the main vein, but as they mature they return to the twigs where their survival is more likely. *Icerya* in Hong Kong is being

Plate 165.   Cottony Cushion Scale, *Icerya purchasi* (Homoptera, Margarodidae) on *Cassia surattensis*.

Plate 166.   Egyptian Fluted Scale, *Icerya aegyptica* on Croton leaves.

controlled biologically quite succesfully by the ladybird (Coccinellid beetles, *Rodolia* spp.) imported from India by the Department of Agriculture & Fisheries. A heavy infestation of *Icerya* has been observed killing branches of *Cassia surattensis*. Other species of Margarodidae are often seen heavily infesting *Croton* bushes, and have been identified as *Icerya aegyptica* (Plate 166), and *Steatococcus assamensis*.

Lacciferidae

The Lacciferidae (lac insects) have the female body covered by a thick resinous scale. The Indian Lac Insect *(Laccifera lacca)* scales are the source of stick-lac from which commercial shellac is produced. One local lac insect is *Tachardina* sp. found in dense incrustations on the twigs of *Michelia figo* (Banana Flower Shrub) (Plate 167); and *T. aurantica* is found on twigs of Lime in Penang (Plate 168).

Plate 167.   Lac Insect, *Tachardina* sp. (Homoptera, Lacciferidae) infesting stems of Banana Flower Shrub, *Michelia figo*.

Plate 168.   Lac Insect, *Tachardina aurantica* (Homoptera, Lacciferidae) on twigs of Lime, Penang, showing parasite emergence holes in the dead female scales.

Pseudococcidae

Mealybugs constitute the family Pseudococcidae, and as a group are not so common here as might be expected in a tropical country. A selection of typical, widespread mealybugs is shown in Plates 169–171.

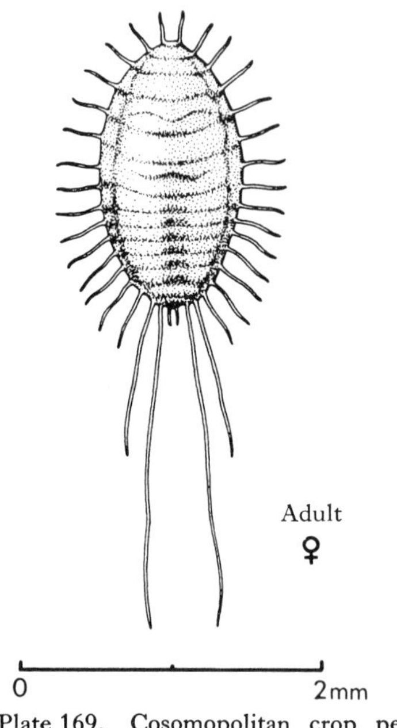

Plate 169. Cosomopolitan crop pest, Long-tailed Mealybug, *Pseudococcus adonidum* (Homoptera, Pseudococcidae).

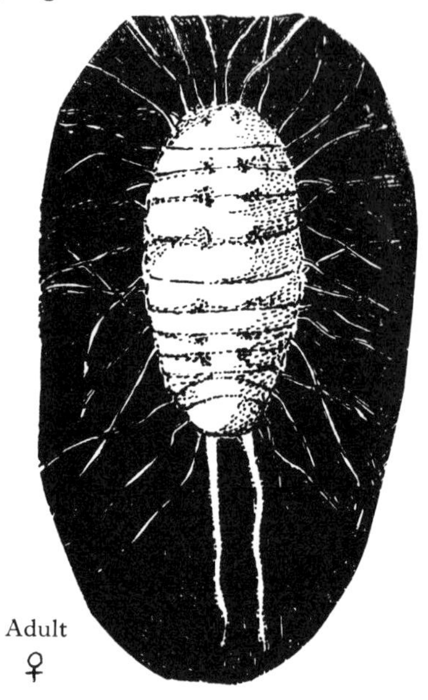

Plate 170. Striped Mealybug, *Ferrisia virgata*.

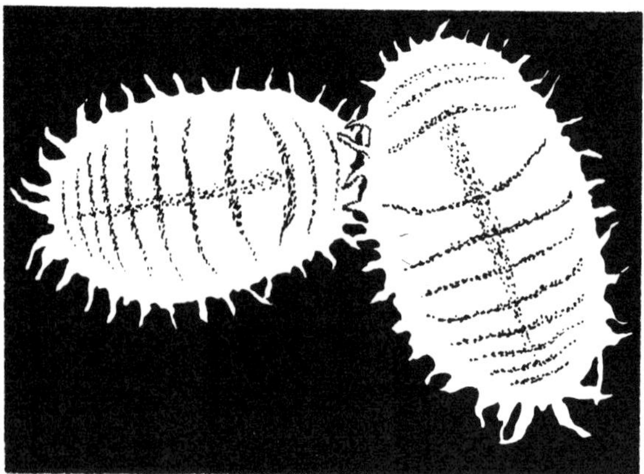

Plate 171. Root or Citrus Melaybug, *Planococcus citri*.

Plate 172.   Long-tailed Mealybug, *Pseudococcus* sp. nr *citriculus* (Homoptera, Pseudococcidae) on Kumquat fruit.

Plate 173.   Hibiscus Mealybug, *Maconellicoccus hirsutus* (Homoptera, Pseudococcidae) infestation on shoot of *Hibiscus,* causing 'bushy-top'.

One polyphagous Long-tailed Mealybug which is common locally is *Pseudococcus* sp.nr *citriculus,* which has been recorded as common on Citrus spp. and other Rutaceae (Plate 172), *Ficus microcarpa, F. elastica, Asparagus lucidus,* Bamboo Palm, and other plants. In appearance it is somewhat similar to *Maconellicoccus hirsutus* that attacks *Hibiscus* shoots, causing shoot distortion, stunting, and sometimes death of the shoot (Plate 173); this species is the only mealybug known to have such a toxic effect on its host plant.

The adult female mealybugs are quite mobile and are not so physically reduced as the scale insects; they Fill walk from plant to plant quite readily. Most mealybugs produce quite a copious flow of honey-dew and so it is usual for their infestations to be associated with sooty moulds and ants. In other countries some species of mealybug (Plate 171) are found subterraneously on the roots of the host plants (as are also some aphids) but this has not yet been recorded in Hong Kong, although ground-level infestations on *Oxalis* roots have been recorded.

Coccidae

There are many small families in the Coccoidea, but they are mostly of limited distribution and do not appear to be represented locally. The Coccidae is, however, one of the most important families and is well represented in Hong Kong. They are called the 'soft scales' and are characterized by having the 'scale' as part of the insect body—in fact it is an extension of the dorsal skeletal shield. If one lifts up a scale with a pair of fine forceps, one can find on the underside the tiny legs and antennae of the female insect. The younger instars are capable of limited locomotion and the quite large scales apparently have the power to move should their location become untenable. They are found on leaves, fruits, and stems of dicotyledenous plants but seldom on grasses. One of the commonest local brown scales is *Saissetia coffeae,* the Helmet (Hemispherical) Scale (Plate 174). As an important coffee pest it is

Plate 174. Helmet Scale *Saissetia coffeae* (Homoptera, Coccidae) on leaf of *Dicladenia boliviensis* vine.

cosmopolitan in distribution and is found locally on Frangipani leaves, *Cycas revoluta* (Sago Palm, Plate 175), and very abundantly on stems and leaves of the creeping vine *Dipladenia boliviensis* (Plates 174 & 176). The brown *Pulvinaria* spp. are also found on *Cycas* leaves (Plate 175) and on Grapefruit leaves. On a number of plants can be found the Soft Green Scales (*Coccus* spp.) of which the cosmopolitan and polyphagous *Coccus viridis* (Plate 177) is probably the commonest. Other locally occuring widespread polyphagous pest species are the Pink Waxy Scale (*Ceroplastes rubens*) (Plates 178 & 179) and Black (Olive) Scale (*Saissetia oleae*) (Plate 180) found on *Ficus microcarpa* twigs. A few large flat brown scales up to 5 mm. in diameter are sometimes found on leaves of *Ficus microcarpa* and *F. elastica* and are thought to be *Paralecanium* sp.

The extent of honey-dew excretion in the Coccidae varies considerably from copious to rather little, but most species are typically found with ants in attendance. The most extreme case of ant/scale interdependence was seen in the New Territories where *Crematogaster* workers have built a series of small arboreal sub-nests on branch forks enclosing completely a small colony of large female *Saissetia* scales. Apparently this pheno-menon is well known, and the species observed locally is probably *Saissetia formicarius*. Soft scales are usually, but not always, found on the undersurface of leaves alongside the main veins or sometimes around the periphery of the leaf lamina, or found on twigs. Probably the most frequently infested host plant for scale insects in Hong Kong are the

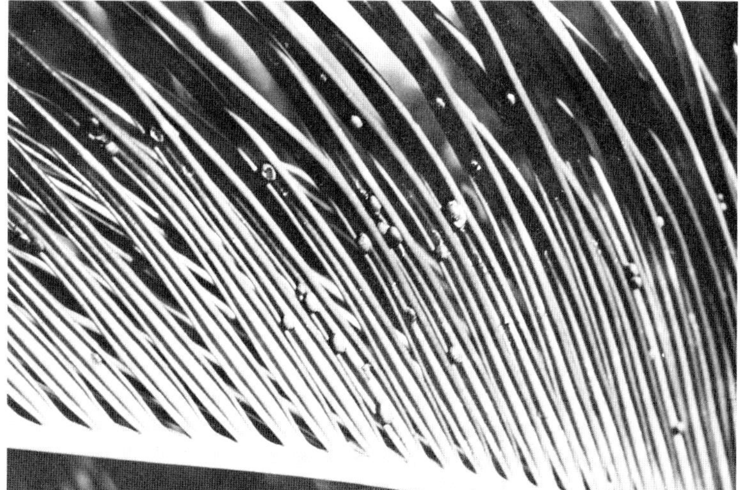

Plate 175.　Mixed infestation of *Saissetia coffeae* and *Pulvinaria psidii* on leaf of *Cycas* palm.

Citrus bushes on which several different soft scales, mealybugs and armoured scales are common.

The very widespread pest species *Ceroplastes rubens* (Pink Waxy Scale) is interesting in that it is commonly found on the leaves and twigs of the mangrove pioneer plants *(Avicennia marina)* (Plates 178 & 179) and most abundantly at the seaward edge of the mangrove community.

(Above) Plate 176. Helmet (Hemispherical) Scale, *Saissetia coffeae* (Homoptera, Coccidae) on stem and leaves of *Dicladenia boloviensis* vine.

(Left) Plate 177. Soft Green Scale, *Coccus viridis* (Homoptera, Coccidae) on *Hibiscus* stem.

Plate 178.   Pink Waxy Scale, *Ceroplastes rubens* (Homoptera, Coccidae) on leaves of the mangrove pioneer shrub *Avicennia marina*.

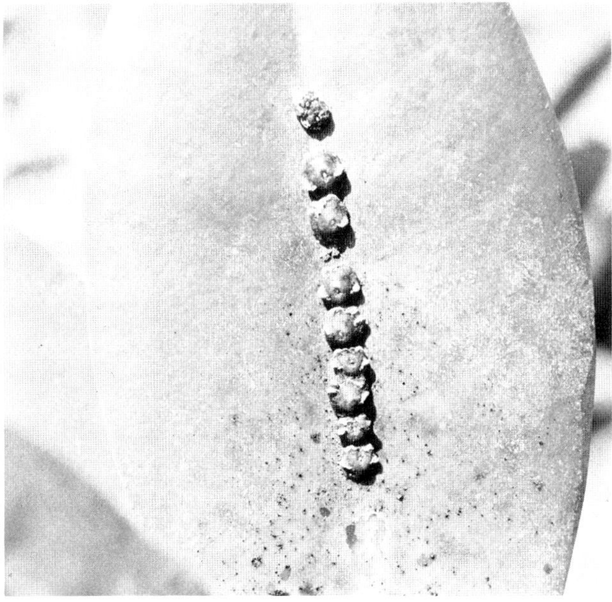

Plate 179.   Pink Waxy Scale, *Ceroplastes rubens* (Homoptera Coccidae) on leaves of the mangrove pioneer shrub *Avicennia marina*.

Plate 180.　Black (Olive) Scale, *Saissetia oleae* (Homoptera, Coccidae) on *Ficus microcarpa*.

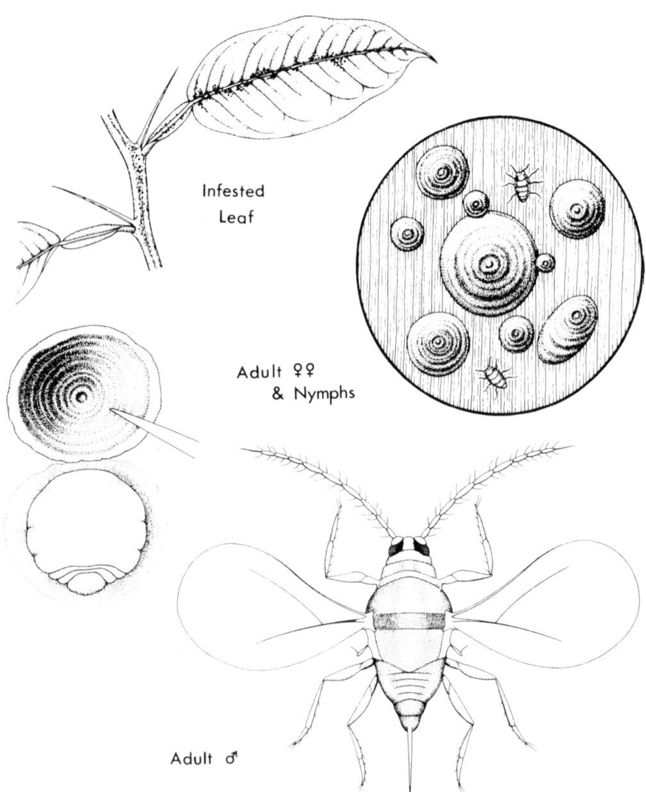

Plate 181.　San José Scale, *Quadraspidiotus perniciosus* (Homoptera, Diaspididae).

Diaspididae

The diaspididae are the armoured (hard) scales, characterized by producing no honey-dew, having greater morphological reduction, and having the 'scale' separate from the insect body (Plate 181). Frequently the scale of the first nymph is seen as a distinct part of the final scale, this being particularly noticeable in the case of the 'mussel-type' scales. Very often the armoured scales are found on the upper surface of leaves as well as on twigs, in contrast to the soft scales which are typically found on the undersurface of leaves. A number of widespread and polyphagous species are common in Hong Kong, including California Red Scale *(Aonidiella aurantii)* (Plate 182), Coconut (Transparent) Scale *(Aspidiotus destructor)*, Mussel Scale *(Lepidosaphas beckii)* on *Citrus* spp., and Purple Scale *(Chysomphalus aonidum)* (Plate 183) also on *Citrus*. A few other species are quite abundant here, including Oyster Scale *(Phenacaspis cockerelli)* on Oleander leaves (Plate 184), and *Fiorinia* sp. nr *japonica* on leaves of Bamboo Palm (Plate 185) and *Cupressus,* and *Parlatoria proteus* on leaves of *Ficus tinctoria* spp. *gibbosa.*

Scale insects have modified their life histories in order to compensate for more or less immobility of the adult female, because when an animal species loses its powers of dispersal it is very likely to be well along the road to extinction. Most of the soft scales (Coccidae) are oviparous and lay eggs, and it appears that some are parthenogenetic in that the female does not require mating prior to egg-laying. The eggs are laid under the scale of the female body, and as they are laid the body of the female

Plate 182. California Red Scale, *Aonidiella aurantii* (Homorptera, Diaspididae) on a small orange fruit.

Plate 183.   Purple Scale, *Chrysomphalus aonidum* (Homoptera, Diaspididae) on *Citrus* leaf.

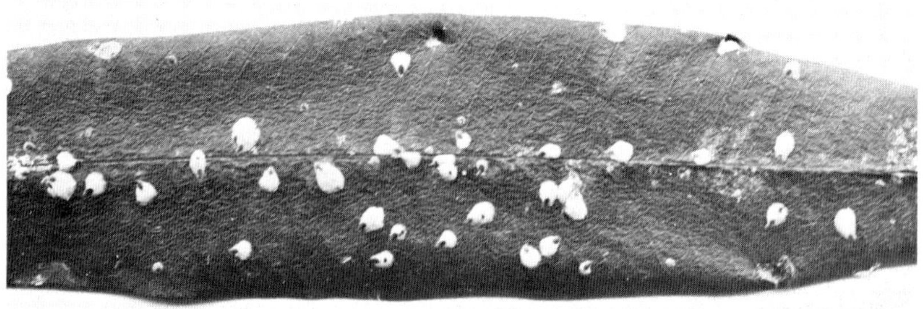

Plate 184.   Oyster Scale, *Phenacaspis colkerelli* (Homoptera, Diaspididae) on Oleander leaves.

Plate 185.   *Fiorinia* sp. nr. *japonica* (Homoptera, Diaspididae) on leaf of Bamboo Palm.

shrinks, so more eggs are laid in the space provided. Finally the female scale dies and the 'scale' remains fixed on to the plant surface quite firmly, shielding the egg mass underneath.

From the eggs emerge the first instar larvae which have well developed legs and sense organs, are very active, and are called 'crawlers'. They do not feed and their function is that of dispersal. They actively crawl all over the host plant, and if it touches an adjacent tree they pass on to this too. They can be spread from tree to tree (host to host) by active crawling, by falling, by air currents, and on the bodies of birds, other animals, and other insects. Experiments have shown that crawlers of Sugarcane Scale in East Africa were carried as far as 1,000 m by air currents through the cane field. After a short time, if the crawlers are on a suitable host plant they moult and turn into a small 'scale' whereupon they insert their proboscis into the plant tissues, start to feed, and become more or less sessile. This type of life history is found in all the various families of the Coccoidea.

Many of the hard scales (Diaspididae) are also oviparous and lay eggs under their scale, but a number are viviparous and the female gives birth to living young (crawlers).

ORDER **THYSANOPTERA**
(Thrips)

Thrips are minute insects, seldom longer than 4 mm, having two pairs of narrow strap-like wings with long fringes of setae. The adults are usually black or brown in colour but the nymphs are more often red, orange, or yellow. The order is divided into two sub-orders, the Terebrantia in which the female has a short, saw-like ovipositor tucked under the end of her abdomen, and the Tubulifera where the female has the terminal segments drawn out into a short tube-like ovipositor. The former group make incisions into the leaf tissue when laying eggs, whereas the latter just stick the eggs on to the leaf epidermis.

In Europe, the various grass and cereal thrips (and others) swarm at times during the summer on sultry humid days, when they are popularly known as 'thunder-flies', and are very obvious because of the extent of the aerial swarms. However, in Hong Kong, such populations do not build up and swarming has not been recorded up to date. To the inexperienced eye the local thrips would scarcely be noticeable except for the damage they cause. They are found in small numbers in grass flower heads, and in the flowers of some Leguminosae and Compositae, and in some other plants. Some cause noticeable damage to the young leaves of local plants.

Gynaikothrips ficorum, the Common Fig Thrips (called the Cuban

Plate 186.   Common Fig Thrips, *Gynaikothrips ficorum* (Thysanoptera, Phlaeothripidae); causing leaf folding on the Chinese Banyan.

Laurel Thrips in U.S.A.), is very common on the Chinese Banyan *(Ficus microcarpa)* causing the young leaves to fold over and become red coloured and lumpy in appearance (Plate 186). Inside the folded leaf will be found a pair of black adult thrips and either a cluster of white eggs (Plate 187) or a cluster of young yellowish nymphs and the empty egg-shells. Thrips have strange asymmetrical mouth-parts which rasp

Plate 187. Folded leaf of *Ficus microcarpa* showing adult thrips and some eggs inside.

Plate 188. *Gynaikothrips kuwani* causing leaf distortion on *Aporusa chinensis.*

the surface of leaves and flowers, rupturing the epidermal cells an sucking up the exuding sap. This surface cell rupturing usually takes place on young leaves and flowers, and results in distortion of subsequent growth. Presumably it is the feeding of the adult pair of thrips which causes the young fig leaf to fold over. A very closely related thrips, *Gynaikothrips kuwani*, causes a different type of leaf distortion on *Aporusa chinensis* (Plate 188).

Another interesting species of thrips in the Tubulifera is *Gigantothrips elegans* which is found on mature leaves of both *Ficus microcarpa* (Plate 189) and *Ficus variegata* var. *chlorocarpa* where it feeds on the undersurface and lays small numbers of eggs; it watches over the young pale reddish nymphs after they emerge. Its size is large for a thrips and measures at least 4 mm, and is black in colour, with a very obvious tubular ovipositor. The specimens usually encountered are single females.

Studying arboreal thrips infestations are puzzling at first for pupation takes place in the soil, and the aerial infestations of large nymphs may disappear overnight as they drop off the leaf or flower on to the soil below to pupate. Two species of the widely occurring *Taeniothrips* are found in the flowers of beans and tea bushes in the New Territories, and other local species of importance are *Chloethrips oryzae* (Plate 190) (Rice Thrips) to be found on rice; *Frankliniella* sp. (Plate 191); and *Heliothrips haemorrhoidalis* (Black Tea Thrips) (Plate 192) on a number of local plants.

The best-known and most noticable thrips are those to be found in

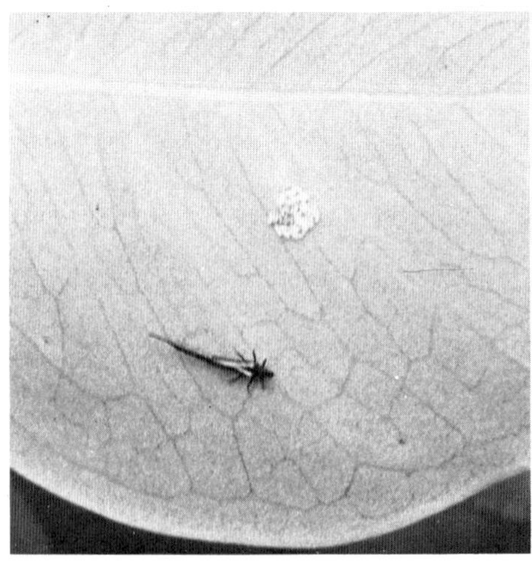

Plate 189. Adult ♀ thrips *(Gigantothrips elegans)* with egg cluster on underside of leaf of Chinese Banyan *(Ficus microcarpa)*; body length 4 mm.

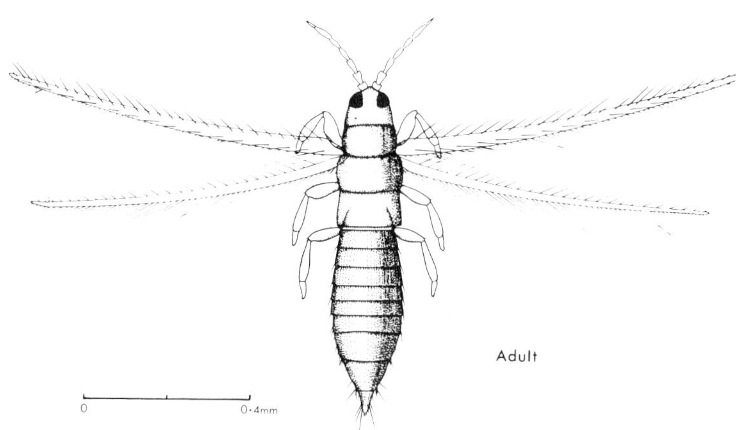

Plate 190. Rice Thrips, *Chloethrips oryzae* (Thysanoptera); length 1 mm.

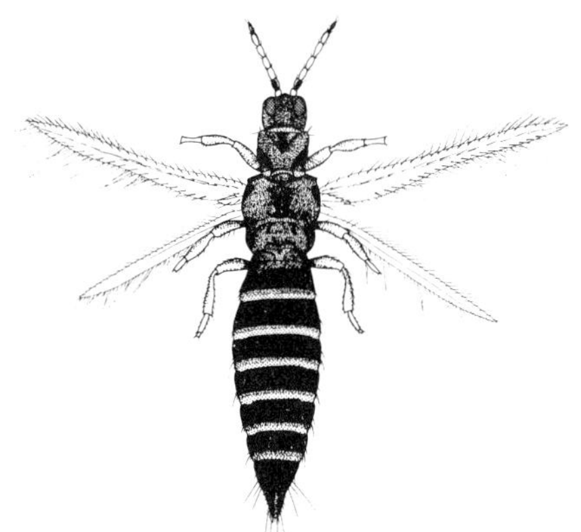

Plate 191.   Adult *Frankliniella* sp. (Thysanoptera).

flower heads or those causing damage to leaves of flowering plants, but appearently by far the most abundant are the tiny fungus feeding species which occur in leaf litter and on dead twigs. These species are seldom encountered however, unless especially looked for with beating trays. Another large group of thrips inhabits the basal region of grass tufts, usually amongst the dead leaves as these are also fungus-feeders. Most of these species can only be identified by taxonomic experts, and to the casual observer they are just small species of thrips.

A few species of thrips are predacious and they feed on very small nymphs of scale insects and mealybugs (Coccoidea), and some predacious species can be found locally in grass tussocks where they appear to be preying on small mealybugs.

    Insects of Hong Kong

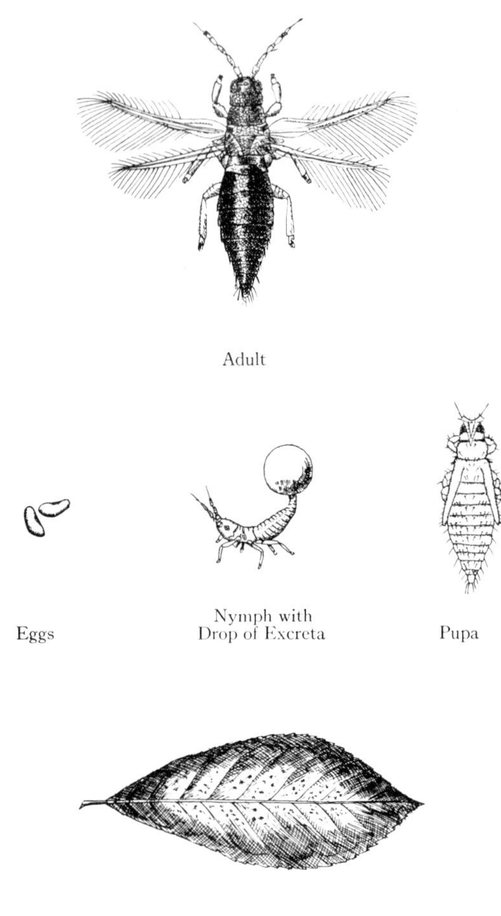

Adult

Eggs

Nymph with
Drop of Excreta

Pupa

Damaged Leaf

Plate 192.   Black Tea Thrips, *Heliothrips haemorrhoidalis* (Thysanoptera); showing immature stages, adult, and damaged leaf of host plant.

SUB-CLASS **PTERYGOTA**

DIVISION II **ENDOPTERYGOTA (= HOLOMETABOLA)**

ORDER **NEUROPTERA**
(Lacewings, Alder-flies, Ant-lions, etc.)

The Neuroptera is a heterogenous assemblage of predatory insects particularly well-represented in the tropical parts of the world. It is divided into two sub-orders, Megaloptera and Planipennia.

Sub-order **Megaloptera**

The Megaloptera are regarded as the more primative, and the larvae have normal biting-type mouth-parts.

Raphidiidae

The snake-flies (Raphidiidae) have terrestrial predatory larvae often found under loose tree bark, but none has been recorded from Hong Kong as yet.

Corydalidae

The alder-flies are larger insects, with two pairs of large, equally-sized wings, often spotted; their aquatic predatory larvae have lateral abdominal gills. In some of the freshwater streams in the New Territories and Lantao Island such larvae are regularly found (Plate 193). Adult alder-flies are sometimes also collected from areas near streams (Plate 194),

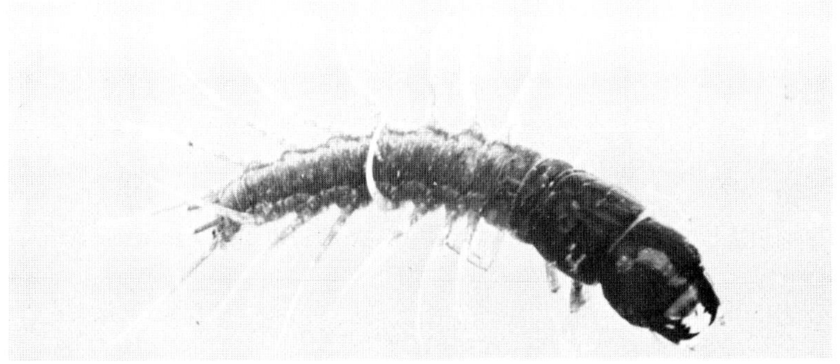

Plate 193. Larval Alder-fly from local freshwater streams in New Territories, *Neochauloides bowringi* (Neuroptera, Corydalidae); body length 30 mm.

Plate 194. Adult Alder-fly, *Neochauloides bowringi* (Neuroptera, Corydalidae); body length 18 mm.

Plate 195. Adult Green Lacewing, *Chrysopa* sp. (Neuroptera, Chrysopidae); wingspan 28 mm.

and it is thought that these must belong to the same species, which has recently been identified as *Neochaulioides bowringi* (Reik, *in litt.*). The larvae are fierce and predatory; they will kill and eat most smaller aquatic insects, worms and crustacea.

Sub-order **Planipennia**

The majority of species are found in the Planipennia and the adults have wings with much secondary vein branching. The larvae have a peculiar type of exerted piercing mouth-parts which are used for siezing insect prey and sucking out the body juices through a specially formed suction canal. Two of the largest families are the Chrysopidae (green lacewings) and the Hemerobiidae (brown lacewings). Both are common locally, and both are of importance in that their larvae kill large numbers of aphids, small bugs, and small caterpillars found on plant foliage.

Chrysopidae

The green lacewing adults are as the name suggests, green in colour, with long tapering antennae and two pairs of large membranous wings. (Plate 195). Their larvae are active, fusiform in shape or broader, and with many body tubercules and spines (Plate 196). Sometimes they carry the empty shells of their devoured prey impaled upon their spines, presumably as a form of camouflage. The eggs of Chrysopidae are characteristic in that they are laid as a group on long delicate pedicels on plant foliage (Plate 197).

Plate 196.   Larvae of Green Lacewing showing spiny body and large jaws.

Plate 197. Eggs of Green Lacewing, *Chrysopa* sp. (Neuroptera, Chrysopidae) on *Cupressus* foliage.

Hemerobiidae

The brown lacewings (Hemerobiidae) are often a little smaller than their green counterparts but otherwise very similar. Their larvae are more slender and fusiform, smooth and with no lateral tubercules. Lacewing larvae generally have jaws that are delicate and needle-like but have no internal teeth along their inner margins.

Mantispidae

The last family in this group is called the Mantispidae and the adults look just like small mantids (*c.* 20 mm long), complete with elongate raptorial forelegs, but with typical neuropteran wings. They have not yet been recorded locally.

Ascalaphidae

The next group contains the ant-lions and the family Ascalaphidae. In the Ascalaphidae there is one local species *Hybris subjacens* which is found quite regularly as the adult, and is characterized by the long knobbed antennae (Plate 198), without which it would look remarkably like a dragon-fly. The larvae are said to lurk in plant foliage.

Myrmeleontidae

The Myrmeleontidae contains the ant-lions which are very well-

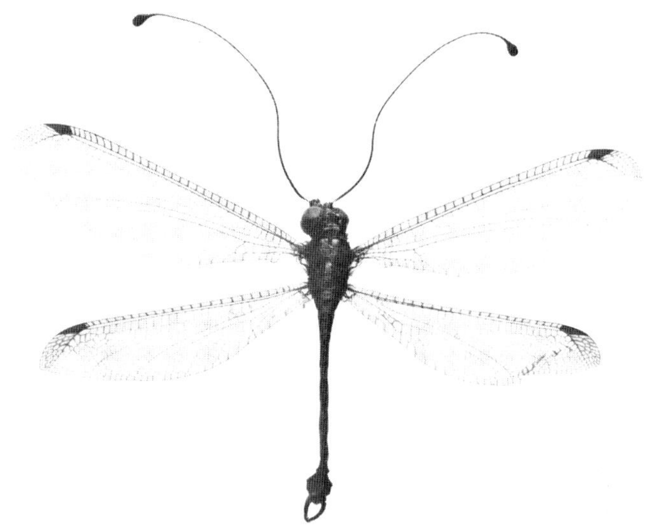

Plate 198.   *Hybris subjacens,* adult (Neuroptera, Ascalaphidae); wingspan 86 mm.

represented in Hong Kong. The two species belong to the genus *Myrmeleon* which again resemble a dragonfly but are characterized by their short, thick antennae (Plate 199) (in dragonflies, the antennae are minute and tapering). It is the larvae that are known as 'ant-lions' for they are squat, stout-bodied insects with huge, hooked and toothed mandibles (Plates 200 & 201) and they lurk in the bottom of pits made in sandy soil (Plate 202). Here they wait for small walking insects and other arthropods to fall over the edge of the pit; any luckless creature that falls into the pit is immediately seized by the huge jaws and when firmly impaled, its body juices are sucked out; later the dried shell is discarded over the side of the pit. Because small ants are probably the commonest walking insects in these dry areas, they form the most usual prey, hence the common name. On one occasion, a large ant was seen to slip into an ant-lion pit where the enthusiastic predator seized the ant by one leg; however, the ant was too large for it to handle and it scrambled out of the pit and was last seen disappearing into the vegetation at the side of the path, with the ant-lion still firmly clinging to one leg.

The ant-lions are nocturnal and so they are normally not observed making their pits during daylight, but careful observation will reveal their mandibles showing at the bottom of the pit while the insect body is completely buried in the sand. They are quite common in certain

Plate 199.   Ant-lion adult, *Myrmeleon* sp. (Neuroptera, Myrmeleontidae); wingspan 60 mm.

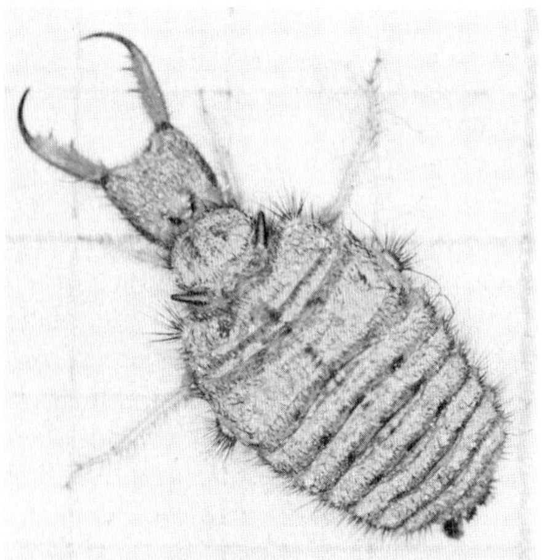

Plate 200.   Larvae of species (A) of Ant-lion (*Myrmeleon* spp.) found in Hong Kong, removed from their pits in the soil.

Plate 201.   Larvae of species (B) of anti-lion (*Myrmeleon* spp.) found in Hong Kong, removed from their pits in the soil.

parts of Hong Kong, especially on various paths that go around the hillsides above the reservoirs on both Hong Kong Island and in the New Territories. They are also common in the dry, rain-sheltered soil under bridges over conduit paths. The major requirement by the larvae is a light dry soil; pupation is said to occur in the soil, and the weakly-flying

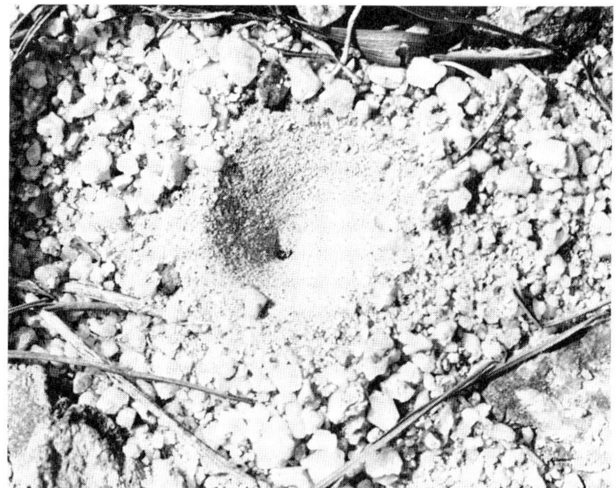

Plate 202. Ant-lion pit in sandy soil, with the insect mandibles just evident at the base of the pit; pit diameter about 20–30 mm.

adults can be found during the summer months. They are nocturnal and sometimes fly to lights and so either fly into rooms at night or else can be found sitting on buildings in the mornings. There appears to be two very similar species occurring in Hong Kong.

Coniopterygidae

The final group of neuropterans to be found locally are in the family Coniopterygidae, which are uncommon, small, pale brown, and lacewing-like in appearance when sitting, about 4–6 mm long, usually with mealy (waxy) wings. They also look like aphids and are often found in association with aphids, on which they presumably prey.

ORDER **LEPIDOPTERA**
(Moths and Butterflies)

The vast majority of Lepidoptera are moths (although to the casual observer butterflies may appear to be more numerous), most of which are small, with a drab brownish colour, rather inconspicuous, and also nocturnal in habits. Thus most moth species go almost completely unnoticed. The entire body and wings are covered in tiny tile-like scales and sometimes hairs. The scales are easily dislodged and rub off during capture and handling, and also come off naturally during the insect's life. They have two pairs of wings, primitively equal-sized, which function as a large single pair by overlapping. Some moths which are fast fliers have an arrangement of special bristles (a frenulum) on the hind-wing which fit into a fold on the forewing and hold the wings together.

The typical biting mouth-parts of the primitive insects have been supplanted in the Lepidoptera by modified structures adapted for sucking nectar from flowers. The long, coiled proboscis typical of this group is formed by elongation of the galea of the paired maxillae. It is coiled neatly under the head when not in use, and is extended by special muscles and hydrostatic blood pressure. The tip of the proboscis can be moved quickly and sensitively in and out of flowers, while the insect either balances on the flower or hovers near it. The longest proboscis belongs inevitably to the hawk moths (Sphingidae) and may be over 15 cm in length.

The antennae can be used as a character to separate moths from butterflies. Most moths have either thin filiform antennae or else large feathery ones as in the Saturniidae, whereas butterflies have distinctly clubbed antennae. Butterflies also hold their wings vertically over the body at rest, whereas the moths tend to hold their wings at rest in a more lateral and horizontal position. However skippers (Hesperiidae) tend to be somewhat intermediate in this respect, and they are generally accepted as butterflies.

The moths mostly use scent for sex location and so the male moths have antennae that are more feathery (pectinate) than the females. Being nocturnal in habits they are not able to use sight (as do the butterflies) for this purpose. This behaviour is most highly developed in the Emperor Moths (Saturniidae) where the males with their huge feathery antennae are reputed to be able to locate virgin females by their scent at distances of up to three miles at night in the dark.

The larval stages of Lepidoptera are very similar in general pattern of structure, but vary enormously in detail. They are all termed caterpillars and have a long segmented soft body with a well defined head capsule. There are three pairs of thoracic legs and a varying number of soft fleshy legs on the abdominal segments, terminating in a pair of claspers on the last body segment. Typically there is a pair of lateral spiracles per body segment, which are quite obvious. Most caterpillars are herbivorous and feed on plant foliage, but sometimes they may be cannibalistic, and some bore into tree trunks and branches where their development is prolonged and slow. Most caterpillars are voracious feeders and grow rapidly and after some weeks they pupate. Butterflies pupate inside a chrysalis, the shape of which varies according to the group of butterflies concerned. Usually they are not hidden, and may be cryptically coloured and shaped, but sometimes they are flamboyant in their beauty and conspicousness, such is the case with the golden chrysalis of *Euploea midamus* (Plate 264). The skippers are again intermediate in this respect for they tend to hide their rather drab pupae in folds of leaf material. Moths generally produce smooth, brown, symmetrical pupae which are either formed in soil, in plant litter, or in woven coccoons spun in vegetation. The woven coccoons are made from silk produced by special glands opening on the jaws of the caterpillars. The group Lepidoptera is very large, containing an estimated 22 families and 140,000 species throughout the world.

The most primitive Lepidoptera bear vestiges of the biting mouthparts that the early ancestors must have possessed, and another primitive feature is to have the two pairs of elongate wings not attached or overlapping, and functioning inefficiently as separate wings. Very primitive species have not been recorded in Hong Kong up to date, for example, no Swift Moths (Hepialidae) have been collected here, either as the low-flying crepuscular adults or as the long white (black-headed) soil inhabiting caterpillars.

Sesiidae

The Sessiidae are Clearwing Moths, not to be confused with the local members of the Amatidae who are sometimes called Wasp Moths, but are related to tiger moths (see p. 287). The Clearwings spend their larval life boring in tree trunks, branches or roots, although some species prefer bushes or vines (e.g. Current Clearwing Moth of Europe and N. America, and Sweet Potato Clearwings of Africa). The only local species found up to date is *Conopis* sp. (Plate 203) from the base of the trunk and surface roots of Camphor trees and as mentioned later is of interest ecologically. The larvae bore in the wood for at least a year, and when

Plate 203.   Camphor Clearwing Moth, *Conopis* sp. (Lepidoptera, Sesiidae); body length 11 mm.

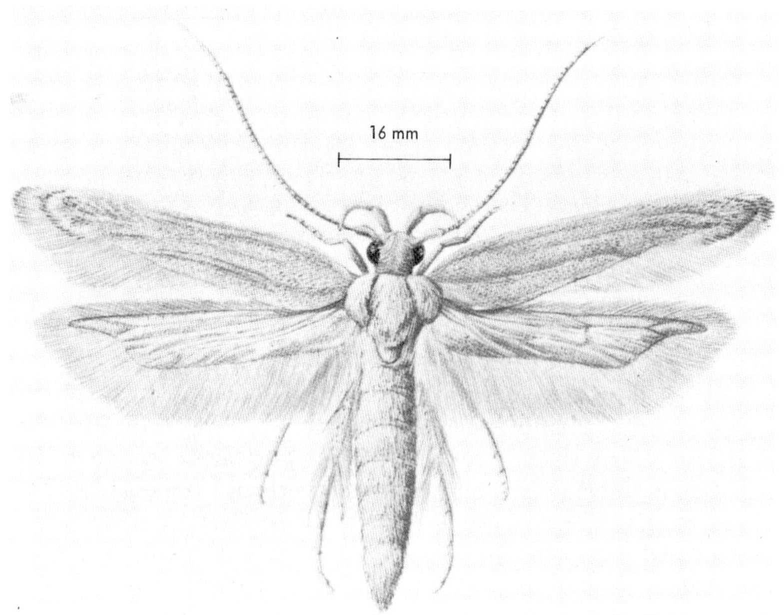

Plate 204.   Angoumois Grain Moth, *Sitotroga cerealella* (Lepidoptera, Gelechiidae).

the adults emerge from the tunnels they leave the apical half of the old pupal exuvium protruding. Two other species have been collected as adults but have not yet been identified.

Gelechiidae

Closely related to Sessiidae is a series of small families of physically small moths but containing a number of species of considerable economic importance and ecological interest. Together they constitute the super-family Tinaeoidea. The Gelechiidae includes the almost completely cosmopolitan Potato Tuber Moth *(Phthorimaea operculella)*, so far not recorded locally, the completely cosmopolitan Angoumois Grain Moth *(Sitotroga cerealella)* (Plate 204), a serious pest of stored foodstuffs, and the Pink Bollworm *(Pectinophora gossypiella)*, a major cotton pest in China and can be expected to occur in Hong Kong on Hibiscus, Okra or other Malvaceae.

Gracillariidae

The Gracillariidae are minuscule silvery-white moths some 2 mm in body length. The larvae mine in the leaves of various plants, and species of some importance mine in the leaves of cotton and Macadamia in many parts of the world. Locally there is the very common Citrus Leaf Miner *(Phyllocnistis citrella)*. This attacks all species of *Citrus* and other species in the Rutaceae, and is often most common on Grapefruit and Pumilo where it seems to prefer the larger leaves. The tunnel mines are extensive, and appear white in colour because of the air underneath the upper epidermis (Plate 205). The mines are identificable as caterpillar mines because of the central line of faecal pellets.

The other main group of leaf miners are the fly maggots of the Agromyzidae but the mining maggots lie on their sides while feeding and their faecal pellets are deposited along the edges of the tunnel and scarcely visible; the tunnel appearing to be clear as a result.

Pupation of the Citrus Leaf Miner takes place in a tiny silken coccoon tucked under a very small part of the leaf edge which is pulled over dorsally for protection; as many as 12 different tunnels and pupae have been recorded on a large Grapefruit leaf. The caterpillars prefer young leaves, and sometimes in very young leaves they cause considerable leaf distortion (Plate 206).

A similar group of tiny leaf miners is found in the Lyonetiidae, where the Coffee Leaf Miners (*Leucoptera* spp.) are important, but these have not been recorded locally. Another species very much like *Phyllocnistis* in appearance mines the leaves of *Ficus microcarpa,* typically starting as

a tunnel mine and ending as a blotch mine (Plate 207). One curious
species of Gracillariidae has just been reared from large round leaf-galls
on *Glochidion hongkongensis* and is identified as *Caloptilia* species.

Yponomeutidae

Yponomeutidae are the ermine moths, small size, often white in
colour and with spotted wings. Several unidentified local species can be
seen. Recently the cosmopolitan Diamond-back Moth *(Plutella
xylostella)* was taxonomically removed from the family Plutellidae and
placed in this family. The caterpillars are small. They feed on leaves of

Plate 205. Citrus Leaf Miner, *Phyllocnistis citrella* (Lepidoptera, Gracillariidae),
mines in leaf of Grapefruit.

(Above) Place 206. Young apical leaves of Orange rolled and distorted by Citrus Leaf Miner attack.

(Left) Plate 207. Leaf Miner damage (Lepidoptera, Gracillariidae) in leaf of *Ficus microcarpa*.

brassicas and are serious vegetable pests in the New Territories. They make small 'windows' in *Brassica* leaves, feeding on the upper surface and leaving the lower epidermis intact (Plate 208), but that thin layer soon breaks leaving the feeding site as a round hole in the leaf lamina. Heavy infestations are common and can result in almost complete plant defoliation. If disturbed, the small caterpillars will drop off the leaves on strands of silk from their mouth-parts.

Glyphipterygidae

The Glyphipterygidae are grass moths, small and brown (or metallic) in colour, generally quite numerous, and the caterpillars feed on grasses and sedges. Close examination of local grassland habitats should reveal species of this group.

Tinaeidae

The Tinaeidae is the largest group and is best known for a few cosmopolitan household species. These include the Common Clothes Moth *(Tineola bisselliella)*, the Case-bearing Clothes Moth *(Tinea pellionella)* and the Tapestry Moth *(Trichophaga tapetzella)*. As a group the small white caterpillars feed on wool, skins, furs, and a wide range of animal material in homes and godowns. In many buildings in Hong Kong the Casebearing Clothes Moth is at times common and the small 'bagworms' can be seen moving up walls. To confuse matters, the Common Clothes Moth is regarded by some taxonomists as belonging to the family Phycitidae. *Epipyrops,* the small brown moth whose caterpillar parasitizes *Pyrops candelaria* (p. 000) in the New Territories belongs to the Epipyropidae, and the other members of this tiny family parasitize other fulgoroids and leafhoppers (Cicadellidae).

Oecophoridae

The Oecophoridae contains the European Parsnip Moth and two common species of House Moth, *Hofmannophila pseudospretalla,* which is the ubiquitous Brown House Moth, and *Endrosis sarcitrella,* the White-shouldered House Moth. They both eat wool (as do the larvae of clothes moths) but they also eat vegetable matter.

Cossidae

The Cossidae include goat and leopard moths, and are sometimes called carpenter moths. Several large specimens of adult goat moths have been collected locally but nothing is known about them. The Red Coffee Borer *(Zeuzera coffeae)* is very widespread in the Colony, and is a

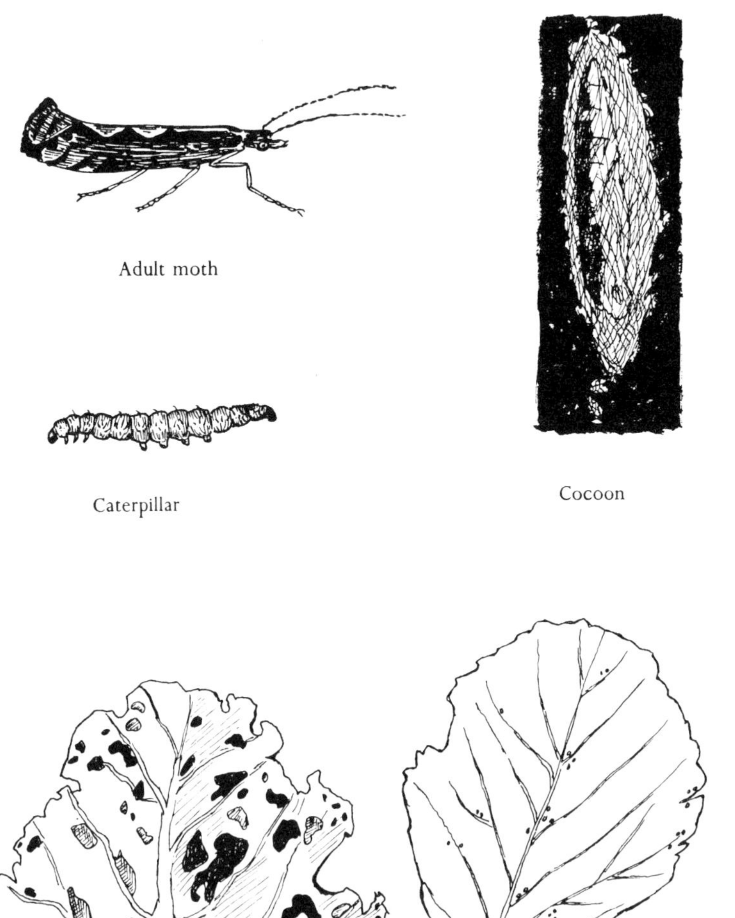

Adult moth

Caterpillar

Cocoon

Damaged leaf

Eggs on leaf

Plate 208. Diamond-back Moth, *Plutella xylostella* (Lepidoptera, Yponomeutidae).

Plate 209.   Adult Red Coffee Borer, *Zeuzera coffeae* (Lepidoptera, Cossidae); body length 23 mm.

regular minor crop pest on coffee, figs, tea and cocoa in many parts of the warmer world. The adult is almost identical to the European Leopard Moth and is a medium-sized silvery-white moth with dark spots on the wings and body (Plate 209); it sometimes flies to light at night. The larvae are red-coloured, hard-bodied, and bore in the branches and trunks of many local trees. The life cycle takes a year to complete. Adults are found in the spring. As with the Sesiidae, the emerging young adult moth leaves the old pupal exuvium half protruding from the retreat hole; these are very common on Acacia trees in the spring.

Psychidae

The Psychidae are known vernacularly as 'bag-worms', for the caterpillars and the wingless females typically construct cases of silk and pieces of plant material in which they live (Plate 210). The young caterpillars enlarge the case as they grow and when feeding they grasp the host plant leaves with their thoracic legs which protrude from the front of the case. At pupation the case is firmly fixed to a twig for safety. The females emerge from the pupal case but remain in the 'bag', and

Plate 210. Close-up of frass tube and damage by larvae of Wood Borer Moth (*Indarbela* sp.; Metarbelidae).

they are wingless and physically very reduced, often being legless and without antennae or mouth-parts for feeding. The adult males are normal moths and can usually be recognized by their long abdomen. Their wings have so few scales as to be almost transparent, especially near the apex. They are swift fliers and move quickly searching for the apterous females which remain in their 'bags'. The female is fertilized by the male through the opening at the end of the bag; shortly afterwards she lays her eggs inside the bag and then she dies. The young caterpillars hatch and immediately produce silken threads in which they hang down from the bag; often some are blown by the wind on to other plants where they start eating the leaves and making their own cases.

Some bag-worm caterpillars only feed at night, and it has been shown that for some African species the bag prevents the caterpillar from drying up in the hot sunshine. The case also appears to be protective as it is well camouflaged, and the caterpillar only extends the head and first few body segments when feeding.

There are several quite common local species of Psychidae to be found on False Acacia *(Acacia confusa)*, *Casuarina,* Chinese Arbor-vitae *(Thuja orientalis)* (Plate 210), and also on some grasses (Plate 211) and some mangrove plants (Plate 212). In some cases the 'bag' is made only of silk without the foliage pieces incorporated (Plate 212 & 213). Several

Plate 211. *Acacia* trunk showing several frass tubes of Wood Borer Moth larvae.

Plate 212. A *Gordonia* bush, 3 m high, killed by boring larvae of Wood Borer Moth.

Plate 213. Bag-worm case of *Clania* sp. (Lepidoptera, Psychidae) on a branch of *Thuja orientalis*.

species have been identified; the one on False Acacia is *Clania* sp. and the one with a smaller case is labelled *Clania crameri*. The Grass Bagworm is *Amictoides* sp. A common urban species is shown in Plate 216 *(Hyalarcta)*; it is polyphagous and feeds readily on leaves of orchids and Bamboo Palm.

Metarbelidae

A closely related small group of moths found in the Ethiopian and Oriental regions is the Metarbelidae. The adults are nocturnal and fly to lights at night, and the larvae are common wood-borers. *Indarbela* sp. nr *disciplaga* is common locally and looks rather like a smaller, darker *Zeuzera*. The larvae bore in the branches and trunks of many local trees, leaving a characteristic frass-tube on the outside (Plate 214 and 215). They bore in Chinese Banyan *(Ficus microcarpa)*, False Acacia *(Acacia confusa)* and many other trees; they sometimes kill small trees (Plate 216). The larvae bore a deep hole in the wood, other at a fork, as a retreat, and they feed at night on the bark and outer layers of sapwood under the frass tube as a shelter. If disturbed they withdraw to their retreat.

Limacodidae

The Limacodidae (= Cochliidae) are a strange group. The adults resemble each other quite closely, but there are two very distinct types of caterpillar. The first type of larva is called a 'stinging caterpillar', usually quite brightly coloured with a yellow or pale green background and contrasting stripes and spots, and with a series of fleshy protuberances (scoli) bearing short, sharp spines which are said to be of an urticating nature (i.e. 'stinging') (Plate 217). The adult moth of *Parasa lepida* (= *Latoia lepida*) is a distinctive green and brown colour (Plate 218).

Plate 214. Grass Bagworm, *Amictoides* sp. (Lepidoptera, Psychidae) on grasses in New Territories.

Plate 215. Bag-worm case (Lepidoptera, Psychidae) on mangrove plant (*Hyalarcta* sp.)

(*Above*) Plate 216. Urban bagworm (*Hyalarcta* sp.) feeding on leaves of Bamboo Palm.

(*Left*) Plate 217. Caterpillar of *Parasa lepida* (Lepidoptera, Limacodidae), a 'stinging' caterpillar, with urticating bristles; length 30 mm.

(*Below*) Plate 218. Adult moth of *Parasa lepida* (Lepidoptera, Limacodidae), with its characteristic green and brown wings; wingspan 38 mm.

The other larval form is called a 'jellygrub' or 'slug caterpillar', with a short, thick, fleshy, glabrous body, and a small retractile head (Plate 219). The thoracic legs are minute and the body segmentation indistinct. There are no abdominal prolegs, but there is a series of secondary suckerdiscs used for attachment to the leaves on which they feed. At pupation, a short, stout pupa is formed inside an almost completely spherical brown, hard, silken coccoon (Plate 220). This group is quite well-represented in tropical regions, and a number of species are common minor pests on coffee, cocoa, capsicums and oil palm in Africa and S.E. Asia; several species (such as *Thosea sinensis*) can be found occasionally in Hong Kong (Plate 221), often on tea at *Eurya chinensis* (Theaceae).

Plate 219. Slug caterpillar of *Thosea sinensis* (Lepidoptera, Limacodidae); body length 22 mm.

Plate 220. Spherical brown pupal coccoon of *Thosea sinensis* stuck to a leaf.

Plate 221. Brown drab adult moth of *Thosea sinensis* (Lepidoptera, Limacocidae), wingspan 38 mm.

Zygaenidae

Locally there is a number of day-flying moths with quite bright stripy coloration, and they look rather like butterflies. One of the commonest local species is the black and yellow *Cyclosia papilionaris* (Plate 222). These belong to a family called Zygaenidae. The red-brown burnets and green foresters are characteristic of the Palaearctic region but do not occur here. The larvae are typically short and stout with bodily protuberances (verrucae) from which arise short hairs (Plate 223). They live exposed on leaves and typically pupate in a tough membraneous coccoon

Plate 222. Adult moth of *Cyclosia papilionaris,* a day-flying moth of the family Zygaenidae; wingspan 70 mm.

spun over part of the leaf, often folding the leaf over in the process (Plate 223). The mobile pupa forces up part of the edge of the coccoon membrane prior to the emergence of the adult. As a group these moths have not been studied at all in Hong Kong.

Tortricidae

The small Tortricidae are a difficult group taxonomically and various important species have been shunted from family to family within the Tortricoidea. However for present purposes it is simpler to regard them all as one family. Together with the Tinaeoidae, they are referred to as Microlepidoptera for obvious reasons. The adult moths are called tortricids and the caterpillars are often called bud-worms because of their habit of boring in buds on trees and flowers. Many species have a similar distinctive adult silhouette when at rest, with pronounced 'shoulders', and most have distinct forewing 'tips' (Plate 224). The caterpillars are always small, usually dark, and if disturbed they wriggle violently backwards and will drop from the plant on a silken thread. They either bore in buds of trees and flowers, or sometimes bore in shoots (e.g. Coffee Tip Borer), or bore into fruits, or else they roll up leaves with their silken threads. They pupate in the same locality in which they feed.

There are many important crop pests in this group and they are

Plate 223. Caterpillar and pupal coccoon of *Cyclosia papilionaris* (Lepidoptera, Zygaenidae); caterpillar body length 30 mm.

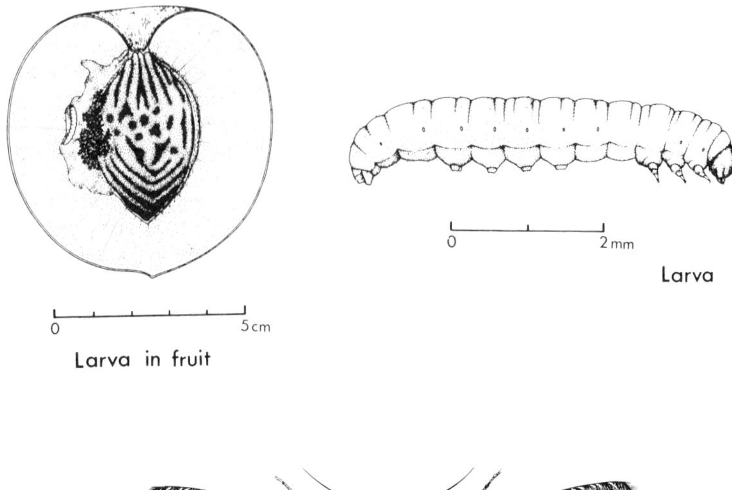

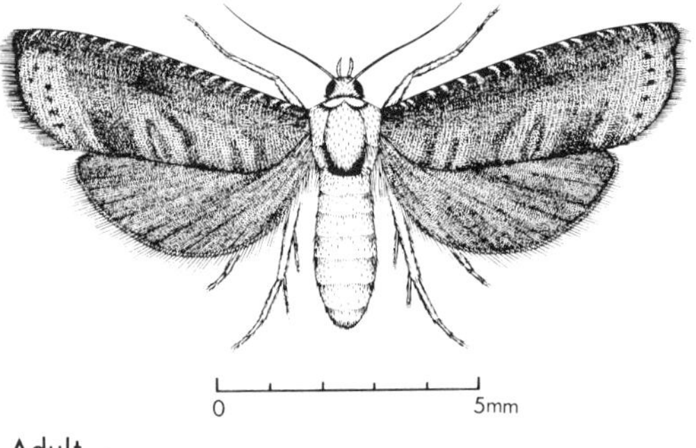

Plate 224. Oriental Fruit Moth, *Cydia molesta* (Lepidoptera, Tortricidae).

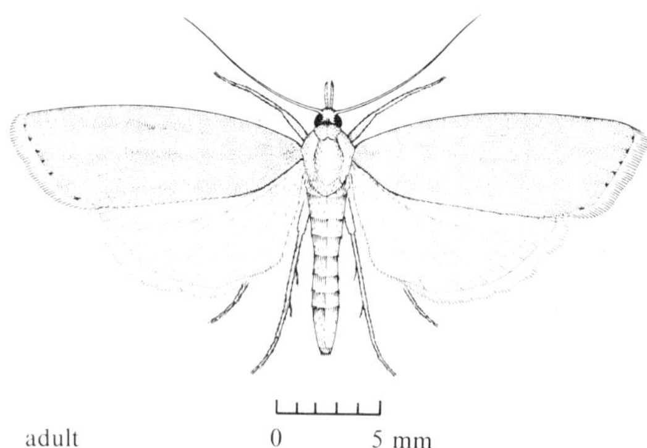

Plate 225. Rice Stalk Borer, adult moth, *Chilo supressalis* (Lepidoptera, Pyralidae).

particularly serious on fruit trees, strawberry, tea, coffee, and various flowers with large buds (e.g. rose and carnation). Some species prefer to bore into the fruit rather than the buds. The most serious pest of apple is probably the Holarctic Codling Moth *(Cydia pomonella)*, which occurs in northern China but not in Hong Kong. The closely related Oriental Fruit Moth *(Cydia molesta)* (Plate 224) occurs here in peaches and plums. *Cydia pulverula* bores into the buds of *Ficus microcarpa,* and other species bore into *Hibiscus* buds, *Citrus,* lychee, guava, and avocado *Rhyaciona cristata,* a very pretty small brown/red/white species, has a larva that bores in the shoots of the local *Pinus massoniana.*

Super-family **Pyralidoidea**

Crambidae

The first family within the Pyralidoidea is the Crambidae, or grass moths. They are tiny moths with narrow, elongate wings, whose caterpillars bore into grass stems. It is almost certain that these grass moths occur in areas of grassland in the New Territories but none has been collected and identified as such. They are a group of ecological interest but their small size and non-economic nature result in their not having been studied locally.

Pyralidae

The Pyralidae is another group of moths which in the past has had rather elastic limits taxonomically. It is a very large family of small moths with rather delicate bodies, and many are important agricultural crop pests, causing several different types of damage. The adults are nocturnal; many fly to lights and are caught in light traps.

A large number are stem-borers in the larger grasses and cereal crops, including the Rice Stalk Borer *(Chilo suppressalis)* (Plate 225), the Yellow Paddy Stem Borer *(Tryporyza incertulas)* (Plate 226), and the Asian Corn Borer *(Ostrinia furnacalis)* (Plate 227). The eggs are usually laid on the leaf-sheath or leaves, and the first instar caterpillars typically feed under the sheath on the young leaves where by biting through a young folded leaf they produce a series of 'windows' which enlarge as the leaf grows and expands. After some days the young caterpillars penetrate the stem and bore up and down the hollow pith region in cereals and grasses. In sugarcane the stem is solid and without a pith, and vertical movement is more limited here owing to the solid nature of the stem; in this host a pyralid caterpillar usually bores out only one (or part of one) short internode. Plates 228–230 show the typical damage

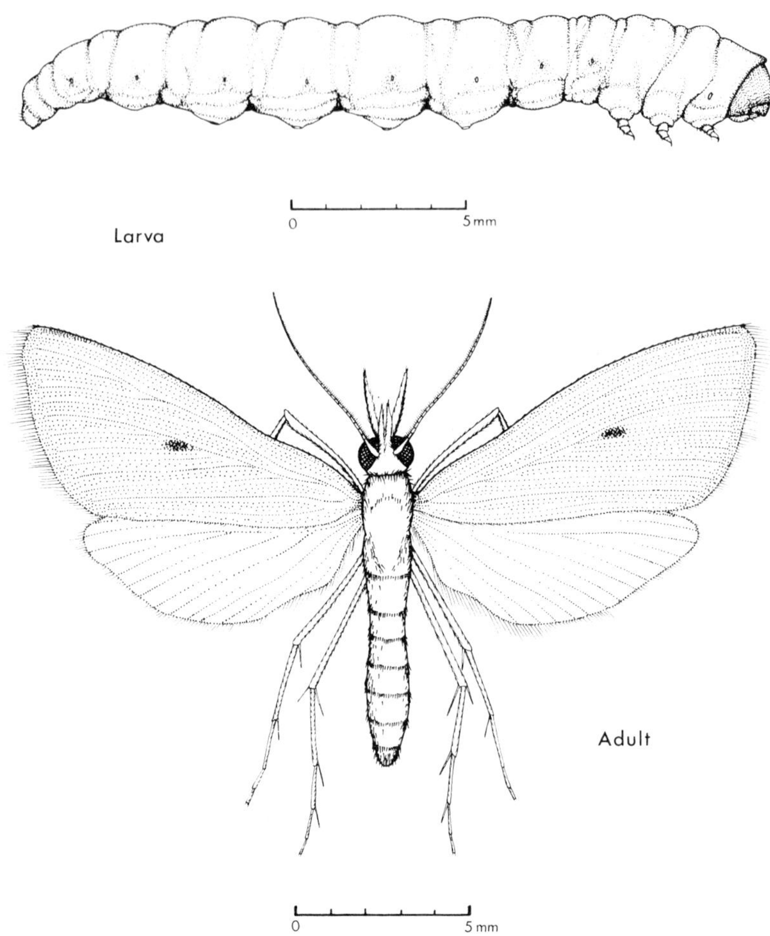

Plate 226. Yellow Paddy Stem Borer, *Tryporyza incertulas* (Lepidoptera, Pyralidae); caterpillar and adult moth.

done to sugarcane locally by *Chilo sacchariphagus,* including both the eaten-out stem internode and the emergence hole for the adult moth (pupation usually takes place in the plant tissues), and 'windowing' of the leaves by the young caterpillars.

Other local Pyralidae include the small brown Rice Leaf-Roller *(Cnaphalocrocis medinalis)* (Plates 231 & 232) whose caterpillars make longitudinal leaf rolls on wild grasses and rice plants in which they live and pupate, and the Rice Caseworm *(Nymphula depunctalis),* a semi-aquatic caterpillar which lives on the lower leaves of paddy rice and

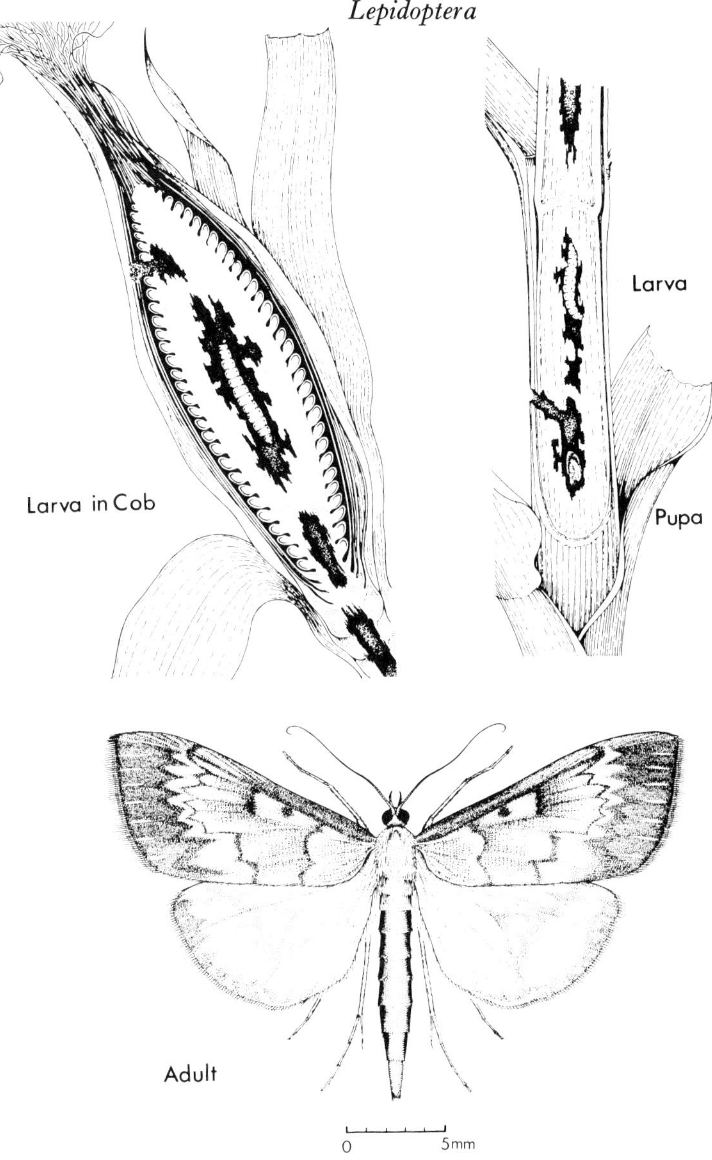

Plate 227.   Asian Corn Borer, *Ostrinia furnacalis* (Lepidoptera Pyralidae); showing larvae and pupa *in situ,* and an adult moth.

constructs a small case or tube of leaf material in which it lives (Plate 233). *Maruca testulalis* (Plate 234) bores in the pods of various Leguminosae. The Cotton Leaf Roller *(Sylepta derogata)* which is quite a serious pest of Cotton in Africa and tropical Asia, is common locally on

(Left) Plate 228. Sugarcane stem showing the emergence hole of the stem borer moth *Chilo sacchariphagus* (Lepidoptera, Pyralidae).

Plate 229. Sugarcane stem cut to show larval damage.

one of its alternative hosts of *Hibiscus* bushes (Plates 235–237). Also *S. iopasalis* is found on sweet potato leaves. The distinctive black and white species is *Palpita indica*, the local Pumpkin Leaf Roller (Plate 238) and *Glyphodes bivitralis* is a pest of legume crops in the New Territories (Plate 239).

At rest many Pyralidae typically sit with their wings partly open and with their antennae deflected backwards over their body (see Plate 232). Several of these important crop pest species are amongst the most common local species of Pyralidae as represented by the moths that fly to lights at night or are caught in light traps. On this basis the most

Plate 230.   Damage to sugarcane foliage by young caterpillars of *Chilo sacchariphagus* (Lepidoptera, Pyralidae).

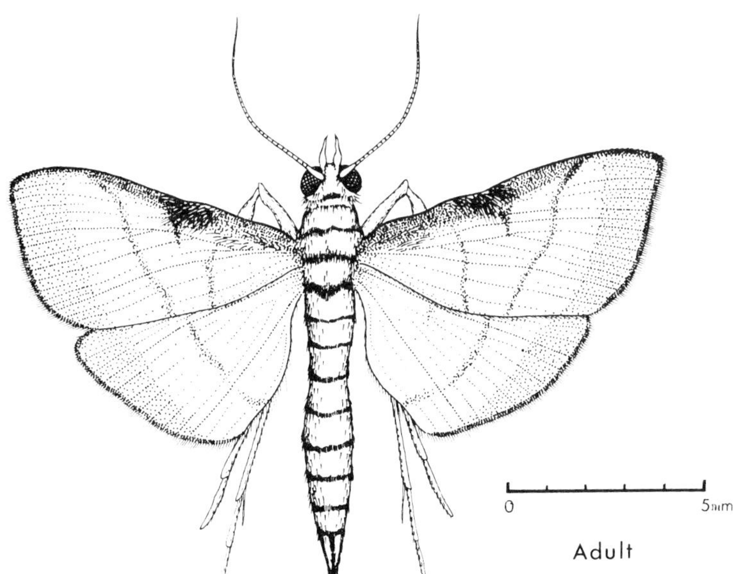

Plate 231.   Rice Leaf-roller Moth, *Cnaphalocrocis medinalis* (Lepidoptera, Pyralidae).

Plate 232. Adult Rice Leaf-roller Moth in typical resting position with deflected antennae.

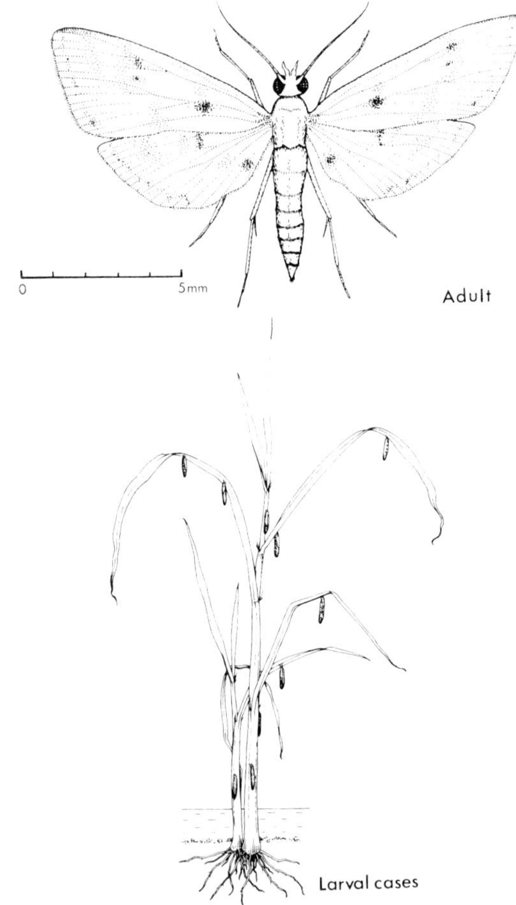

Plate 233. Rice Caseworm, *Nymphula depunctalis* (Lepidoptera, Pyralidae); adult, and larval cases attached to a growing rice plant.

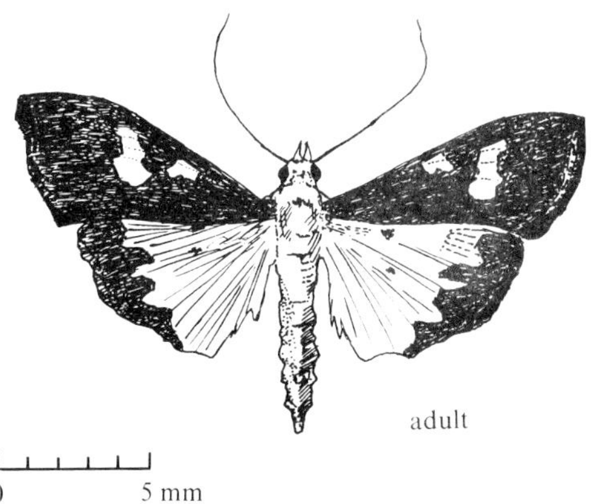

Plate 234.   The Mung Bean Moth, *Maruca testulalis* (Lepidoptera, Pyralidae).

Plate 235.   Leaves of *Hibiscus* rolled by Cotton Leaf-roller caterpillars, *Sylepta derogata* (Lepidoptera, Pyralidae).

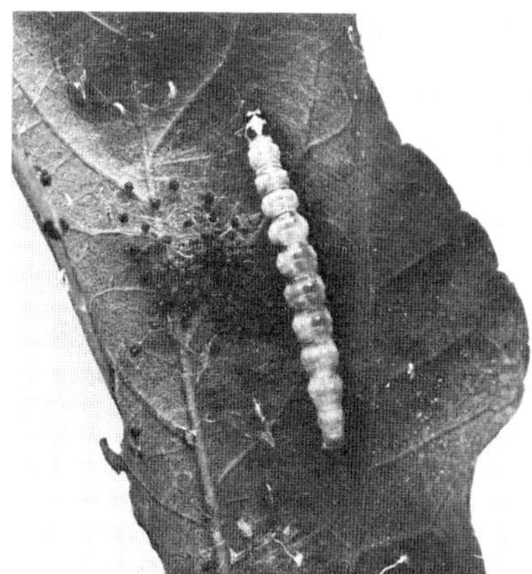

Plate 236.   *Hibiscus* leaf unrolled to show caterpillar inside.

Plate 237.   Adult moth of Cotton Leaf-roller, *Sylepta derogata* (Lepidoptera, Pyralidae); wing-span 25 mm.

Plate 238. Adult moth of Pumpkin Leaf-roller, *Palpita indica* (Lepidoptera, Pyralidae); wingspan 25 mm.

Plate 239. Common local moth, *Glyphodes bivitralis* (Lepidoptera, Pyralidae); a pest of pulse crops; wingspan 25 mm.

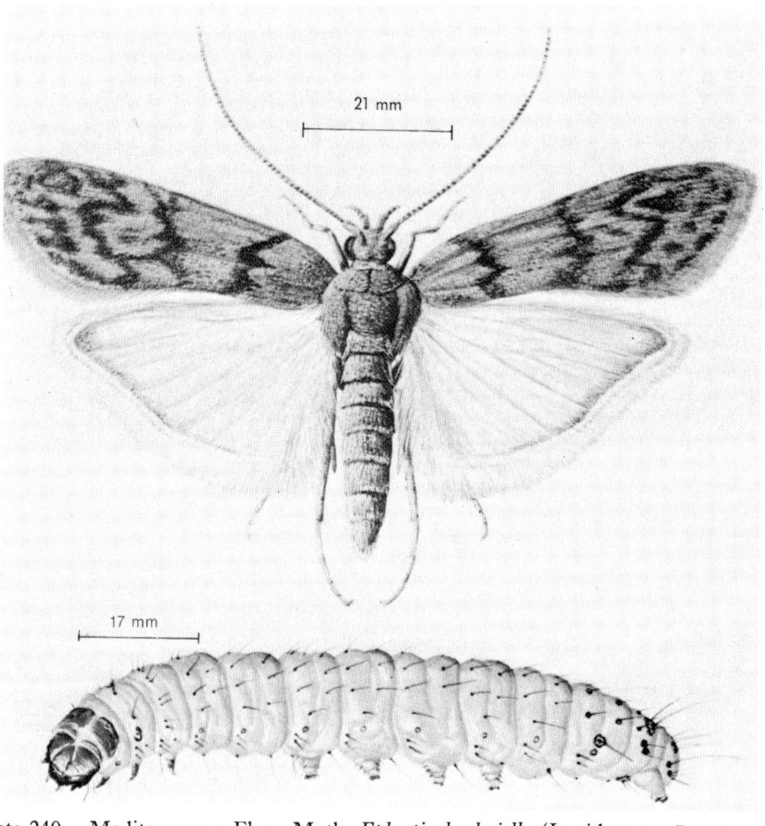

Plate 240. Mediterranean Flour Moth, *Ephestia kuehniella* (Lepidoptera, Pyralidae).

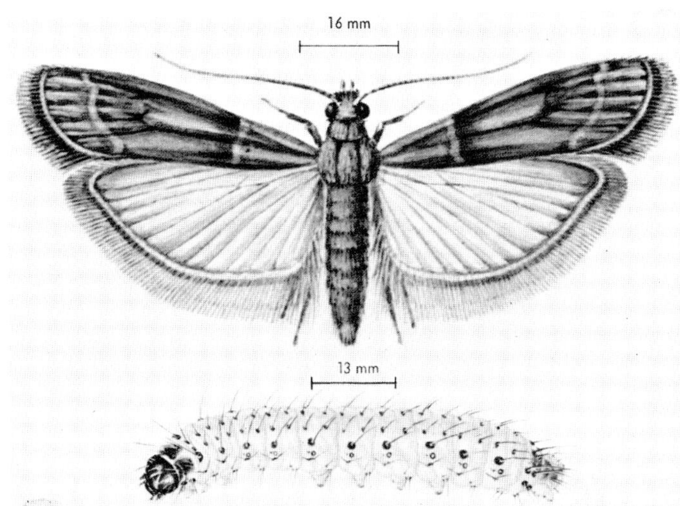

Plate 241. Tropical Warehouse Moth, *Ephestia cautella* (Lepidoptera, Pyralidae).

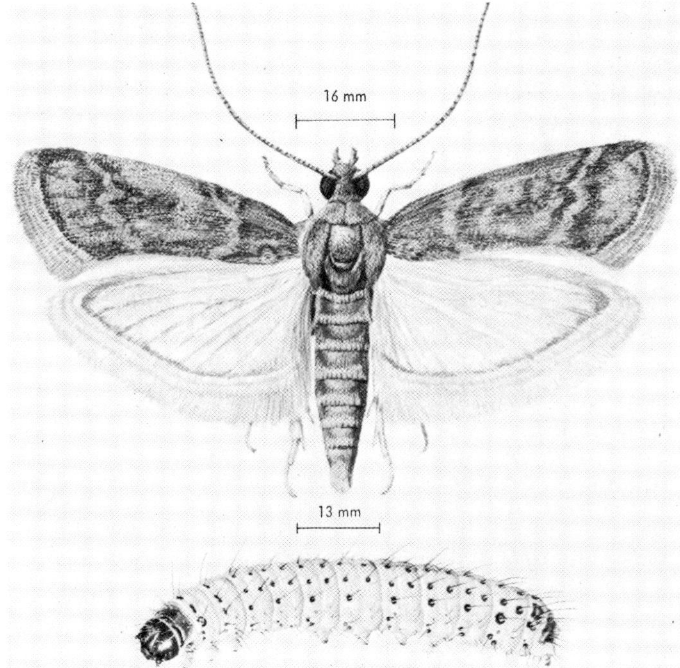

Plate 242. Warehouse (Stored Tobacco) Moth, *Ephestia elutella* (Lepidoptera, Pyralidae).

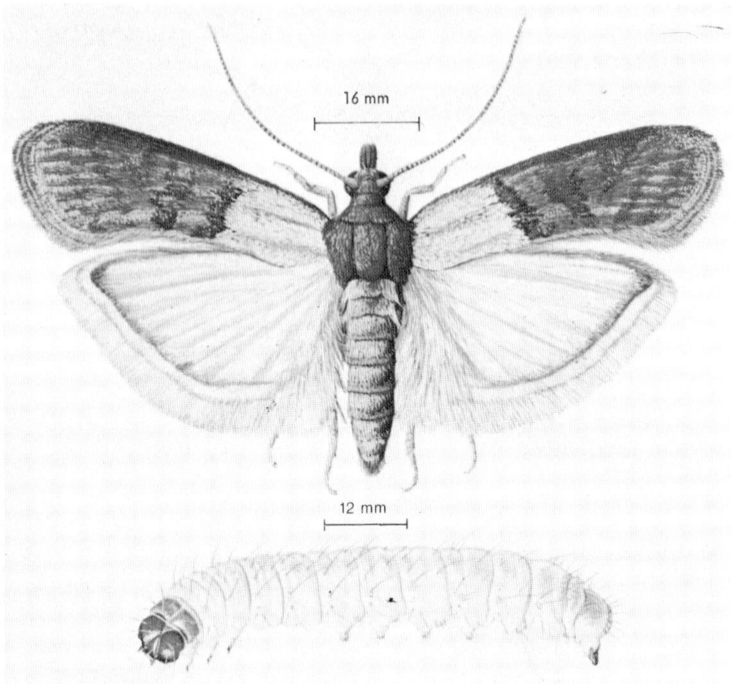

Plate 243.   Indian Meal Moth, *Plodia interpunctella* (Lepidoptera, Pyralidae).

abundant local species are *Cnaphalocrocis, Maruca, Sylepta, Palpita* and two small dark unidentified ones. Important stored products pests are *Ephestia kuehniella* (Mediterranean Flour Moth) (Plate 240), *Ephestia cautella* (Tropical Warehouse Moth) (Plate 241), *Ephestia eleutella* (Warehouse Moth) (Plate 242), and *Plodia interpunctella* (Indian Meal Moth) (Plate 243).

Pterophoridae

Still within the Pyralidoidea is the Pterophoridae or plume moths, characterized by the adult moths having the wings deeply fissured. The forewing is typically cleft into two divisions and the hind-wing into three. There are two species which are locally common in parts of the New Territories, *Aciptitia* sp. and *Ochyrotica* sp. The larvae of the former bore in the stems (vines) of sweet potato plants but the latter is known as the Sweet Potato Leaf Roller.

Super-family **Bombycocidea**

The super-family, Bombycoidea, including the families Lasiocampidea, Saturniidae and Bombycidae, contains the most striking moths in the world.

Lasiocampidae

The Lasiocampidae are called lappets and eggars, and are large brown, hairy moths. In some cases their larvae live gregardiously in large web tents (e.g. European Lackey Moth) in the foliage of bushes and small trees. Several local species have been collected including *Dendrolemus punctatus* on *Pinus massoniana* and *Gastropacha quercifolia* on Peach foliage, and the spectacularly coloured caterpillar of *Trabala vishnou* (Rose Myrtle Lappet Moth) is common on leaves of *Rhodomyrtus* and sometimes *Melastoma* (Plates 337 & 338). The adults are sexually dimorphic in that the female moth is large and yellow in colour (Plate 339), but the male is smaller and green.

Saturniidae

The Saturniidae are the beautiful and spectacular emperor moths, or, as they are sometimes called, giant silkworm moths. The local Atlas Moth *(Attacus atlas)* is probably the largest and certainly one of the most beautiful moths in the world (Plate 244), with a wingspan of up to 25 cm or more, and the females usually being appreciably larger. The caterpillars feed on the leaves of camphor, tallow *(Sapium discolor)*, *Hibiscus,* and the Duck's Foot Tree *(Schefflera octophylla)*. The large

Plate 244.   Newly emerged male Atlas Moth, *Attacus atlas* (Lepidoptera, Saturniidae); wingspan 210 mm.

Plate 245.   Caterpillar of Atlas Moth, *Attacus atlas* (Lepidoptera, Saturniidae) on twig of Camphor tree; body length 90 mm.

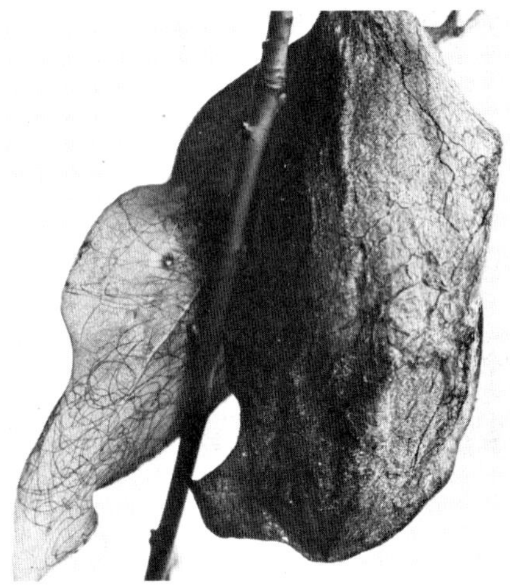

Plate 246.   Coccoon   of   Atlas Moth   wrapped   in   leaves   of camphor tree.

caterpillar is white with a waxy dust (Plate 245), up to 10 cm in length, and with a series of fleshy, spiny tubercules along the body. The pupal soccoon of spun silk is held between several adjacent leaves or in a folded single leaf (Plate 246). So far as is known the Atlas Moth here has only one generation per year, over-wintering as a pupa inside the coccoon; some pupae remain thus from mid-summer right through to the following spring in March or April.

A similar moth, though smaller, but more common locally is the Lesser Atlas Moth (Samia cynthia) (Plate 247), about half the size of its larger relative. The caterpillar is white with a series of black spots

Plate 247. Lesser Atlas Moth, *Samia cynthia* (Lepidoptera, Saturniidae); adult male; wingspan 110 mm.

Plate 248. Caterpillar of Lesser Atlas Moth, on leaf of *Michelia alba;* body length 60 mm.

and small white tubercules on each body segment (Plate 248). It has been recorded feeding on lantana and the white jade orchid tree *(Michelia alba)*.

The other spectacular local emperor moth is the Moon Moth *(Arctias selene)* (Plates 249–252) with its pale greenish-white wings and long

Plate 249.   Adult male Moon Moth, *Arctias selene,* in typical daytime resting position on tree trunk.

Plate 250.   Pinned adult Moon Moth, *Arctias selene* (Lepidoptera, Saturniidae); wingspan 140 mm.

(*Above*) Plate 251. Caterpillar of Moon Moth, *Arctias selene* (Lepidoptera, Saturniidae); body length 70 mm.

(*Left*) Plate 252. Coccoon of Moon Moth, *Arctias selene,* shrouded in leaves.

trailing tails to the hind wings. At rest the moth sits as shown on the tree trunk with the leading edges of the forewings at right-angles to the body. The caterpillars feed on the leaves of camphor and tallow trees, and there are probably several generations (up to four?) per year locally. In this species the large fat green caterpillar has small body tubercules bearing bristles.

Another local species, quite common in most parts of the Colony and especially so in the New Territories, but smaller than the previous species, is the blue/grey Giant Silkworm Moth *(Eriogyna pyretorum)* with its distinctive 'eye-spots' on each wing (Plate 253). The long slender caterpillar is up to 80 mm long and is pale blue/green in colour

Plate 253. Adult female Giant Silkworm Moth, *Eriogyna pyretorum* (Lepidoptera, Saturniidae); wingspan 110 mm.

(Above) Plate 254. Giant Silkworm, or Giant Silkworm Moth, *Eriogyna pyretorum* (Lepidoptera, Saturniidae) on the foliage of camphor tree; body length 75 mm.

(Left) Plate 255. Two pupal coccoons of Giant Silkworms.

with blue prolegs and conspicous body bristles (Plate 254). It feeds on camphor leaves, amongst which it spins its pupal coccoon (Plate 255). There is probably only one generation each year, although it is possible that there could be a second generation in the summer sometimes.

A less common (unidentified as yet) attractive emperor moth is the Golden Emperor whose caterpillars are reported to feed on *Saurauia tristyla*. This lovely moth is only 7–8 cm in wingspan with four brown eye-spots on yellow wings. Emperor moth pupae collected in the wild and taken back for rearing often yield large fat maggots or flies instead of moths. These are Tachinidae which are naturally common larval parasites of Lepidoptera.

Bombycidae

Closely related to the Saturniidae is the Bombycidae which contains the famous Japanese Silkworm *(Bombyx mori)* which feeds on mulberry leaves. The silkworms (Plate 256) can be bought at stalls locally and it

(Left) Plate 256. Japanese Silkworms, larvae of *Bombyx mori* (Lepidoptera, Bombycidae) feeding on leaves of Mulberry *(Morus alba)*.

(Below) Plate 257. Pupal coccoons of Japanese Silkworms.

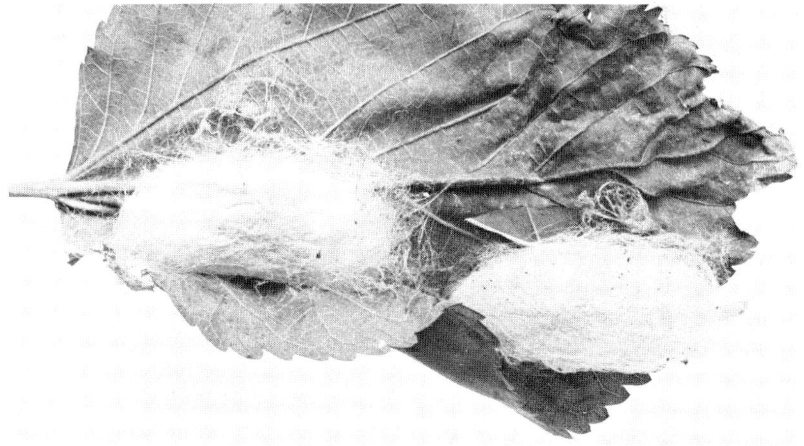

is possible that the species also occurs naturally in the New Territories. The host plant is mulberry, which, however, is not common locally. Pupation takes place within silken coccoons (Plate 257) from which commercial silk is obtained. The adult moth (Plate 258) is a small cream-coloured moth with banded wings. Small white caterpillars can be found on the leaves of *Ficus microcarpa*. They skeletonize areas of the leaf (Plate 259) which subsequently turn brown, but as they grow they change in colour to brown and become twig-like in shape with a curious small hooked 'tail' at the end of the abdomen (Plate 260). They pupate inside a folded leaf (Plate 261) and give rise to the small brown adult moth which has now been identified as *Trichola ficicola* (Plate 262).

Plate 258. Adult Japanese Silkmoth, *Bombyx mori* (Lepidoptera, Bombycidae); wingspan 25 mm.

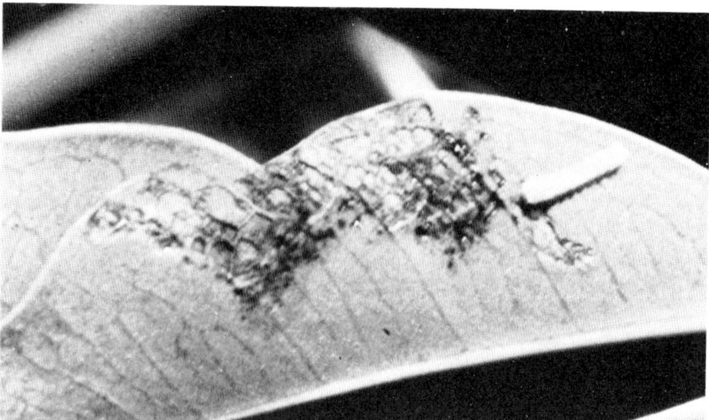

Plate 259. Banyan Silkmoth, *Trichola ficicola* (Lepidoptera, Bombycidae); young larvae skeletonizing a leaf of *Ficus microcarpa*.

(Above) Plate 260. Mature Larva of Banyan Silkmoth.

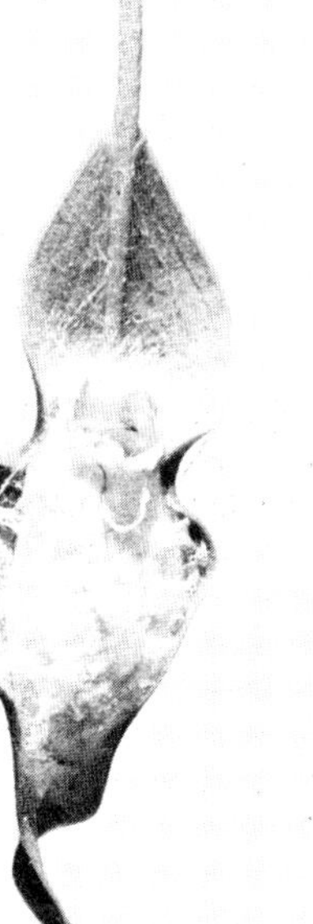

(Left) Plate 261. Pupal coccoon of Banyan Silkmoth.

(Below) Plate 262. Adult Banyan Silkmoth; wingspan 20 mm.

Super-family **Papilionoidea**

Butterflies

The next groups in taxonomic sequence are the butterflies, which constitute six major families and a few minor ones. Since the butterflies of Hong Kong have been well covered in the books by Marsh (1968) and Johnston & Johnston (1980), they will not be discussed in great detail here.

As a group they are characterized by slender antennae with an abrupt club, and at rest the wings are held vertically above the body. One small complication in studying butterflies is that they sometimes show seasonally different body forms, so that in general the dry season form (DSF) is smaller and less distinctly marked than the wet season form (WSF). In most temperate countries the local butterflies have been extensively studied and all species have suitable common names. Unfortunately, there are very few accepted common names for the butterflies occurring locally, and accepted Malayan and Indian common English names (Hill, Johnston and Bascombe, 1978) will be used here.

Nymphalidae

The first family is the Nymphalidae, which are called the four-footed butterflies because the forelegs are reduced and small and are held pressed to the thorax; at a quick glance these butterflies appear to possess only two pairs of legs. This family is a little awkward taxonomically in that it contains at least five very distinct sub-families that are in some books regarded as separate families.

Sub-family **Danainae**

The Danainae are called monarchs, or milkweed butterflies, referring to the fact that the larvae of the many species feed on plants in the milkweed family, the Asclepiadaceae. There are 12 recorded species locally. They are all large butterflies, leathery and tough, and quite difficult to kill by pinching the thorax. The gregarious and brightly coloured caterpillars generally feed on poisonous plants (mostly in the Ascelepiadaceae) and the alkaloids they eat result in their bodies being both distasteful and quite poisonous to predators. The toxic substances persist in the insect body right through metamorphosis into the adult. So in general, neither the caterpillars nor the adult Danaine butterflies are preyed upon much by birds and other predators, and both stages display themselves quite openly. The adults fly slowly so that their conspicous coloration is evident.

The caterpillars typically bear a number of long, fleshy filaments protruding from their backs. *Euploea midamus* has striking yellow caterpillars with black-tipped fleshy filaments (Plate 263 & 264), and they turn into a most beautiful golden chrysalis which hangs free from the leaf or twig (Plate 265). The adult butterfly has a bright blue sheen to the wings that is optical in origin and not the result of pigments, and hence is only visible if viewed at the correct angle (Plate 266). The males of many species have a pair of protusible 'hair-pencils' at the apex of

Plate 263. Caterpillar of *Euploea midamus* (Lep., Nymphalidae, Danainae); length 45 mm.

Plate 264. Caterpillar of *Euploea midamus* just about ready to pupate.

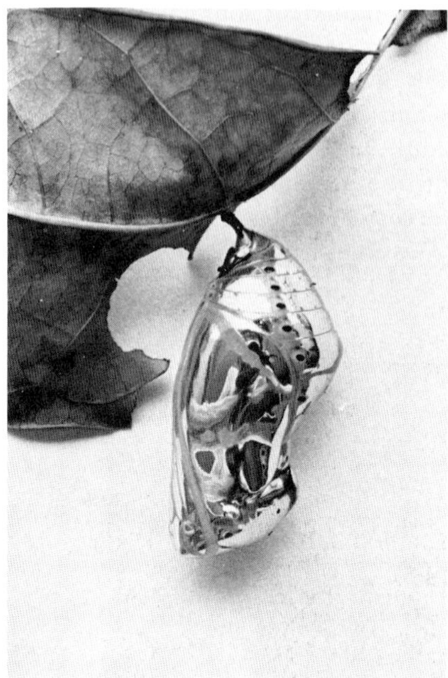

Plate 265. Chrysalis of *Euploea midamus*.

Plate 266. Adult butterfly of *Euploes midamus* (Lep., Nymphalidae, Danainae); wingspan 80 mm.

Plate 267.   Adult *Danaus genutia* (Lep., Nymphalidae, Danainae); wingspan 70 mm.

the abdomen, which are thought to be for dispensing scent with which they attract females. There are three main types of Danainae in Hong Kong. First is the typical orange and black species with white spots (*Danais* species) (Plate 267); the second is the bluish-grey with black spots and streaks (*Danais* and *Scleopsis* species); and finally is the dark-brown, having white or blue spots with a bright purple—blue sheen (*Euploea* species) (Plate 266). As will be mentioned later, mimicry is practiced by local butterflies where edible species closely resemble poisonous species of Danainae.

Sub-family **Satyrinae**

The many Browns (14 local species) seen locally represent the Satyrinae. As a group they tend to live in the shade of woodlands where their cryptic coloration blends with the background of dead leaves, and the dead-leaf resemblance is greatest when the wings are closed. Many have eye-spots on the top of the forewings and on the undersurface of the hind-wings. In a few cases the wings are shaped to represent a dead leaf when at rest. The larvae tend to be elongate, thin, and brown coloured with longitudinal lines, and they have a peculiar angled (horned) head. As a group they tend to feed on Gramineae. They usually have a short, stout, hanging chrysalis.

Sub-family **Amathusiinae**

In the Amathusiinae is the brown species *Faunis eumeus* (Plates 268 & 269) which resembles the Satyrinae both in appearance in behaviour.

Sub-families **Morphinae & Acracinae**

The Morphinae are represented by the beautiful irridescent blue *Morpho* species in South America but they do not appear locally. Neither are there any Acracinae here although *Acraea acerata* is the very common Sweet Potato Butterfly of tropical Africa.

Plate 268.   Adult *Faunis eumeus* (Lep., Nymphalidae, Amathusiinae); wingspan 74 mm.

Plate 269.   Adult *Faunis eumeus* sitting on a leaf in characteristic resting position.

Sub-family **Nymphalinae**

The Nymphalinae are well-represented locally (26 species) as would be expected since this is the largest sub-family. The cosmopolitan Painted Lady *(Vannessa cardui)* and the Indian species of Red Admiral *(Vannessa indica)* are found in small numbers on the Peak, Tai Mo

(Above) Plate 270. Adult *Charaxes polyxena* (Lep., Nymphalidae, Nymphalinae); wingspan 70 mm.

(Left) Plate 271. Pupae of *Charaxes polyxena* hanging free form leaf of *Ficus elastica*.

Shan, and some other hilltops. Species of *Precis, Neptis,* and several species of *Charaxes* (Plate 270) are quite common in most parts of the Colony. These butterflies will come to baits quite readily; rotting fruit or animal faeces are both successful baits. This does facilitate catching specimens for they all fly fast and with great dexterity; they are difficult to net when flying free. Most larvae are cylindical and bear long fleshy tubercules (scoli). The pupae are characteristic in that they hang free and bear many spines and tubercules (Plate 271).

Riodinidae

The Riodinidae is an obscure small family of quite small reddish/ brown butterflies. The two local species are quite common and noticeable in several parts of Hong Kong.

Lycanidae

The Lycanidae are the Blues, Coppers, and Hairstreaks. They are small in size, and have metallic blue or brown upper wing surfaces but dark, sombre undersurfaces, often with eye-spots or fine linear markings and sometimes with delicate 'tails' from the hind-wings. There is often striking sexual dimorphism with regard to coloration. The larvae are characteristically onisciform (resembling woodlice), with a broad, flattened body whose projecting sides obscure the legs and tapering to each extremity. Some species are smooth, others are hairy, and often they are green in colour. The pupa is small and stout; it is typically held in place with a median silken band. Several local species are interesting in that they are predatory upon aphids, and the blue *Chilades legus* can be found on Citrus shoots associated with Citrus Aphid upon which it is apparently predatory.

Some species are attended by ants and it appears that when the caterpillar's body is stroked by the antennae of the ants the caterpillar yields drops of secretion from a special dorsal gland situated on the seventh abdominal segment. A few species carry the association with ants even further, and some are predatory in ants' nests. The Large Blue in Europe is actually carried down into the nest by a worker ant, where it feeds on the young ant larvae until it pupates deep inside the ant nest. There is a large number of small Blues in Hong Kong (37 species) but they are difficult to identify and to the casual observer many species look very similar to each other.

Pieridae

The Pieridae contains the Whites and the Yellows, the most spectacular of which is the large Orange-tip *(Hebomia glaucippe)* (Plate 272), which is common in most parts of Hong Kong for almost the whole period of the warmer weather (April to November inclusive). It has a rather strange caterpillar, to be found feeding on leaves of *Capparis cantonensis* (Plate 273). The chrysalis is typical of the Pieridae, being upright with a single apical spine; it is anchored at the anal end by the cremaster, and supported by a girdle of silk around the thorax (Plate 274). In this group the caterpillars do not have fleshy filaments or horns. The Small White Butterfly *(Pieris canidia)* (Plates 275–278) is abundant locally and is a serious pest of *Brassica* crops and other Cruciferous plants. In the illustration it is seen on one of its alternative hosts, Nasturtium. The first butterflies to emerge in the spring are *Delias pasithoe* (Plate 279) which can be found in large numbers feeding upon the flowers of Chinese Privet, Lantana, and *Bauhinnia*. This and other closely related species are striking in that their best colorations are on the undersurface of the wings and are only really visible at rest. They do emerge at odd times throughout the summer and autumn, but are most noticeable in the spring because they are virtually the only butterflies found flying in March. The small yellow and black *Eurema* species (Plate 280) are common, and also are a number of different species of the Whites. There are 17 recorded species of Pieridae in Hong Kong.

Plate 272. Adult Orange-tip Butterfly, *Hebomia glaucippe* (Lepidoptera, Pieridae); wingspan 75 mm.

(Above)   Plate 273.   Caterpillar of Orange-tip Butterfly, *Hebomia glaucippe* (Lepidoptera, Pieridae); body length 55 mm.

(Left)   Plate 274.   Chrysalis of Orange-tip Butterfly.

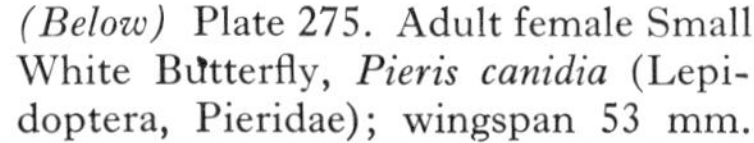

(Below)   Plate 275.   Adult female Small White Butterfly, *Pieris canidia* (Lepidoptera, Pieridae); wingspan 53 mm.

(*Above*)   Plate 276.   Caterpillar of Small White Butterfly, *Pieris canidia* (Lepidoptera, Pieridae); body length 30 mm.

(*Left*) Plate 277. Chrysalis of Small White Butterfly on a Nasturtium leaf.

(*Below*)   Plate 278.   Adult Small White Butterfly newly emerged from the chrysalis.

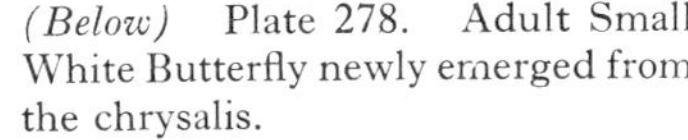

Plate 279. Adult *Delias pasithoae*; wingspan 65 mm.

Plate 280. Adult Common Grass Yellow Butterfly, *Eurema hecabe* (Lepidoptera, Pieridae); wingspan 44 mm.

Papilionidae

Swallowtails constitute the family Papilionidae although not all species actually have trailing tails from the hind-wings. This is probably the best represented family butterflies in Hong Kong although there are only

16 species here. The beautiful Birdwing, *Troides helena,* can be seen sometimes in a few locations in the New Territories. A number of Swallowtails are rather rare and they only occur in a few localities where the food plant for the larvae is to be found. The majority of caterpillars are quite specific in their choice of food plant and so in general the distribution of the species of butterfly is very closely linked to the distribution of the food plant. The more polyphagous the caterpillar, the more widely occurring will be the butterfly, or alternatively if the host plant is widespread so will be the butterfly. This is why the most abundant butterflies in Hong Kong are probably of the species of *Papilio* which feed on *Citrus* and other Rutaceous plants.

The *Citrus* leaf-eating species of *Papilio* are interesting in that the young caterpillars look remarkably like bird droppings (Plate 281). They are rather inactive and move little and this aids the deception. As they grow and moult they suddenly change form and colour completely, becoming large, smooth, fat and green with various tangential stripes across the flanks (Plate 282 & 283). It is said that the colour of the dorsal marks and diagonal stripes enable specific identifications to be made. Six species of *Papilio* caterpillars may be found feeding on *Citrus* leaves locally but the most common are *Papilio demoleus* (Orange Dog, or Lemon Butterfly) (Plate 284), *P. polytes* (Plate 285), and *P. memnon* (Plate 286). The mature caterpillars possess eye-spots behind the head and if alarmed they will rear-up and take a curious snake-like pose (see Plate 283). They also have a strange red horned structure called an osmeterium which is extruded from behind the head and emits a strong smell. At pupation time the mature larvae spins a pad of silk on which

Plate 281.   Young caterpillar of *Papilio* on a Kumquat leaf; body length 14 mm.

(Above) Plate 282. Mature caterpillar of *Papilio*; body length 42 mm.

(Left) Plate 283. Mature caterpillar of *Papilio* showing the distinctive eyespot on the thorax, and the hunched snake-like appearance frequently adopted.

(Below) Plate 284. Lemon Butterfly, *Papilio demoleus* (Lepidoptera, Papilionidae); wingspan 90 mm.

Plate 285.   Common *Citrus* butterfly, *Papilio polytes*; male; wingspan 70 mm.

Plate 286.   Common *Citrus* butterfly, *Papilio memnon*; female; wingspan 110 mm.

the posterior end rests. In a vertical position on a twig or wall it spins
a girdle of silk strands from the support surface around its thorax, and
it then pupates. The chrysalis form varies, the typical shape is elongate
and pointed (Plate 287) with two cephalic spines (c.f. one in Pieridae),
but sometimes it is cryptic and twig-like (see Plate 293).

Plate 287. Chrysalis of *Papilio* species on *Citrus* twig; length 34 mm.

Plate 288. Swordtail Butterfly, *Graphium antiphates* (Lepidoptera, Papilionidae); wingspan 63 mm.

Plates 288–290 illustrate three other local species of Swallowtail butterflies. As will be mentioned later many local species of Papilionidae are polymorphic, and some are mimic species of distasteful or poisonous Danainae. A notable example is the Common Mime *Chilasa clytia*. The larvae feed on leaves of the widespread *Litsea glutinosa*. When they are small, they resemble black and white bird droppings, but later they are very conspicous in black with yellow patches and red tubercules (Plate 291). After pupation (Plate 292) their advertisement ceases and the chrysalis looks remarkably like a broken twig (Plate 293). The adult butterfly (Plate 294) is similar in appearance in both sexes but the form *clytia* is brown and appears to be a mimic of *Euploea core* (Danainae), and the form *dissimilis* is striped blue/grey and mimics *Danaus similis*. In order for the mimic to be successful it must be recognized first in order to be avoided by predators. So although the Swallowtails in general fly fast, high, and with a swerving motion, *Chilasa* has adopted the slow leisurely flight typical of Danainae.

Plate 289. *Graphium sarpedon*; wingspan 70 mm.

Plate 290. Adult *Lamproptera curius* (Lepidoptera, Papilionidae); wingspan 35 mm.

Plate 291. Caterpillar of *Chilasa clytia* (Lepidoptera, Papilionidae) with its distinctive bold coloration, on a leaf of *Litsea glutinosa*.

Plate 292. Caterpillar of *Chilasa clytia* about to pupate, in typical swallowtail position.

Plate 293. Pupa of *Chilasa clytia* in cryptic brown twig-like chrysalis.

Plate 294. Adult butterfly of *Chilasa clytia (form dissimilis)* (Lepidoptera, Papilionidae); wingspan 80 mm.

Super-family **Hesperoidea**

Hesperiidae

On the basis of species, the most well-represented group of butterflies in Hong Kong is the Hesperiidae, or the Skippers, with 45 species here; but because of the superficial resemblance between a number of the small brown species their real abundance is not obvious. The Skippers are placed in a separate super-family Hesperoidea, and are generally recognized as butterflies although they bear some resemblance to moths. At rest Skippers sit with their wings half spread (Plate 295) as opposed to the vertical position practiced by other butterflies, and the antennal club has a definite hooked shape. As a group they are of interest ecologically in that the larvae are almost entirely confined to Gramineae as hosts. They are so-named because of their rapid, erratic, darting flight.

Several species are agricultural pests of some consequence on various cereal crops; this includes locally the small, brightly coloured orange and brown Rice Skipper *(Parnara guttata)* (Plate 296) which is a widespread rice pest and can also be found locally on a number of grasses. Several species feed on leaves of bamboos; locally there is the striking Banana Skipper *(Erionota torus)*, which with a wingspan of 40–50 mm must be one of the largest skippers ever known (Plate 297). The spherical and ribbed eggs are laid in small groups on the banana leaf (Plate 298) and

Plate 295.   Skipper Butterfly at rest on rice leaf.

Plate 296. Rice Skipper Butterfly *Parnara guttata* (Lepidoptera Hesperiidae); wingspan 32 mm.

Plate 297. Adult Banana Skipper, *Erionota torus* (Lepidoptera Hesperiidae), showing the typical hooked antennae; wingspan 65 mm.

the larvae move to the edge of the leaf where they bite into the lamina and make a tiny roll of leaf tissue around their body (Plate 299); as the larvae grow the leaf rolls increase in size until by the time of pupation the roll may be as large as 15 cm long and 3–4 cm broad (Plate 300). The caterpillar is typical of the family in that it has a peculiarly humpbacked body with a small black head capsule and a distinct 'neck' region. The

(Above) Plate 298. Eggs of Banana Skipper, *Erionota torus* (Lepidoptera, Hesperiidae); on a Banana leaf.

(Right) Plate 299. Young caterpillars of Banana Skipper starting their leaf rolls.

Plate 300. Large leaf rolls containing pupae of Banana Skipper; rolls about 120 mm long and 35 mm broad.

soft white body is covered with a white waxy dust (Plate 301) and segmentation is indistinct. The pupa is elongate, soft, and thin-skinned but is protected by the roll of leaf material. The rolls of leaf tissue are maintained in that shape by fastenings of silk. There are certainly two generations of Banana Skipper each year in Hong Kong, but it is possible that there are more; in Malaya the developmental cycle takes only about 40–50 days. Distributions of this species in Hong Kong is very irregular; in 1974 a one acre banana plot at Tai Lung Farm was almost completely defoliated by Skipper caterpillars in November, but in adjacent areas

Plate 301.   Mature caterpillar of Banana Skipper removed from its leaf roll; body length 45 mm.

near Fanling and Tai Po no larval cases could be seen on the small clumps of banana plants. In the same year in the spring there were localized infestations of Banana Skipper in the Stanley area on Hong Kong Island where it is not common.

The interesting local phenomenon of butterfly polymorphism is associated with mimicry. This appears to be classical Batesian mimicry where an edible species of animal very closely resembles another species which is quite poisonous or at least distasteful. As already mentioned, caterpillars of the Danainae and also the Swallowtail *Atrophaneura aristolochiae* contain poisonous alkeloids derived from their host food plants; the poisons accumulate in the insect body and persist through metamorphosis into the adult butterfly. As shown in Table 1, these butterflies are mimicked by various species of edible Papilionidae and Nymphalinae. For the purposes of biological advertisement the poisonous species of butterflies are distinctively coloured, usually with bold contrasting patterns of red and black, or yellow and black, etc. Venomous insects (and other animals), as well as distasteful ones, also have this type of conspicous advertising coloration (e.g. wasps, hornets, some assassin bugs). The object is to teach predators to avoid species with these warning colour patterns. The mimic thus achieves a level of immunity from predation because of its resemblance to a distasteful species, provided however that its numbers remain relatively low. Thus the mimetic form of butterfly has become through evolution a rare polymorph.

Butterflies are unusual in that most of the colour pattern of the wings is controlled by a single gene on one of the pairs of chromosomes, so that production of different forms, or morphs as they are sometimes called, is genetically not a very complicated affair. However, this polymorphic mimicry must have evolved over a period of thousands of years for any selective survival advantage must have been slight initially and very gradual. The inheritance of the tailed condition, for example, is *Papilio memnon* females, appears to be more complex and under the control of two distinct pairs of genes, one of which is linked with the gene controlling the pattern mimicry while the other is not.

The best local examples of polymorphic species are probably *Papilio polytes, Papilio memnon,* and *Chilasa clytia.* The *P. polytes* females occur in three forms in Ceylon, but in only two here. The female form *mandane* resembles the male, except that the white band on the hind-wing is broader, while the rarer form *polytes* is a reasonably good mimic of the Swallowtail *Atrophaneura aristolochiae,* except that it has a black and not a red tinted body. The caterpillars of *A. aristolochiae* in Hong Kong feed on the leaves of the creeping poisonous vine *Aristolochia tagala.* In *Papilio memnon* the female form *agenor* is like the male in being tail-less, but it has large white patches on the hind-wings. The female form *alcanor* has distinctive black tails and smaller white patches on the hind-wings, with two dark red patches at the bases of the forewings; it is thought that it might be mimicking the smaller *A. aristolochiae.* The third form of *memnon* occurs in Malaysia. *Chilasa clytia* occurs in Hong Kong in two forms, both of which mimic different species of Danainae (see Table 1). The female of each form resembles the male; both brown *(clytia)* and striped blue/grey forms *(dissimilis)* freely interbreed, and the offsprings consist of a proportion of the brown form, the remainder being striped.

TABLE 1: **HONG KONG BUTTERFLY MIMICS**

FAMILY	MODEL	FAMILY	MIMIC
Papilionidae	*Atrophaneura aristolochiae*	Papilionidae	*Papilio polytes*—(♀) *polytes*
Nymphalidae (Danainae)	*Euploea core*	Papilionidae	*Chilasa clytia*—(♂ & ♀) *clytia*
	Danaus similis		*Chilasa clytia*—(♂ & ♀) *dissimilis*
	Danaus chrysippus ⎱ *Danaus genutia* ⎰	Nymphalidae (Nymphalinae)	*Argyreus hyperbius* (♀)
	Danaus chrysippus		*Hypolimnas misippus* (♀)
	Euploea midamus		*Hypolimnas bolina* (♀)
	Euploea core		*Hypolimnas antilope*
	Danaus spp. (Blue/grey species)		*Hestina assimilis* (♂ & ♀)

After Marsh (1968)

The difference in behaviour and flight between the distasteful Danainae and the edible Papilionidae and Nymphalinae is quite striking, for they fly slowly and often glide, which is to be expected since the success of their avoidance by predators clearly depends upon their being easily recognized. This difference is also shown by the mimics because their success as mimics also depends upon easy recognition.

It ought to be mentioned that the story of polymorphism (and mimicry) is based on observation of morphological differences and careful breeding and cross-breeding experiments. The mimicry aspect involves a considerable amount of supposition as detailed studies on butterfly predation have not usually been carried out, but the story does fit with the observable facts available. Details of experiments on polymorphic butterfly inheritance using species obtained from Hong Kong are available in the article by Clarke, Sheppard & Thornton (1968), and also in other articles in the series.

Super-family **Geometroidea**

The super-family Geometroidea contains the Drepanidae, Geometridae, Uraniidae, Epiplemidae and Sphingidae.

Drepanidae

This is a small family mostly occurring in the Indo-Malayan region. The adult moths often have slightly projecting wing-tips (i.e. falcate) and are called hook-tips. The caterpillars are thin, sometimes humped in appearance, without claspers on the thirteenth segment, and with the anal extremity prolonged into a raised projection. They are called 'tailed caterpillars'. Occasional specimens of both moths and caterpillars have been seen locally, but they have not been identified.

Geometridae

The Geometridae proper are the 'loopers', so called because the elongate caterpillar has only one pair of prolegs on the sixth abdominal segment and the terminal claspers on the tenth (last) segment. They walk by drawing up the posterior segments close to those of the thorax, and the body thus forming a loop, the body is then extended forwards in the desired direction and the looping repeated. In a few species rudimentary prolegs may appear on the fifth segment. The vast majority of the larvae are brown or green and quite twig-like in appearance; they feed nocturnally and spend the day resting across a space between twigs.

In Hong Kong a large brown looper is often found on Citrus and

(Left) Plate 302. Large Brown Looper caterpillar (Geometridae) on Kumquat bush; body length 48 mm.

(Below) Plate 303. Adult moth of *Calospilos* sp. (Lepidoptera, Geometridae), resting on a wall; wingspan 42 mm.

Kumquat bushes where considerable defoliation can result. It is thought that this larvae (Plate 302) may belong to the adult moth *Calospilos* sp. (Plate 303); owing to their habit of usually pupating in the soil, this species has not yet been reared through. The adults are nocturnal, and have rather thin bodies and large rounded wings; they often sit on walls or leaves with the wings held firmly pressed to the substratum. A common local species (*Abraxas* sp.) (Plate 304) resembles the European Magpie Moth and also appears to be polymorphic with regard to wing coloration and pattern. Quite a large number of different adult Geometrids have been collected in very small numbers, and small looper caterpillars are the nest-provisioning prey of the Common Potter Wasp (*Eumenes pyriformis*) locally.

An unusual and interesting local species is *Problepsis* sp (Plate 305),

Plate 304.   Magpie Moth, *Abraxas* sp. (Lepidoptera, Geometridae); wingspan 48 mm.

Plate 305.   Eye-moth *Problesis* sp. (Lepidoptera, Geometridae); wingspan 25 mm.

known in S.E. Asia as the Eye-moth; it is attracted to animal fluids and will visit mammals (including man) in order to suck tears (lachrymal fluid) from the eyes. Some closely related species will suck blood from wounds in the skin of various large mammals, but this has not been recorded locally.

Uraniidae

Occasionally in Hong Kong during the summer can be seen a large greyish moth, quite butterfly-like in appearance, some 10–13 cm in wingspan, with pale bands down the wings and with two distinctive 'tails' projecting from the hind-wing. This is a species of *Nyctaleamon* in the family Uraniidae, about which nothing is known, but typically the larvae have the full complement of legs and pupate in loose coccoons.

Epiplemidae

The Epiplemidae is a small group of little importance, but does contain the African Coffee Leaf Skeletonizer. The adults typically rest during the day with the forewings spread but slightly rolled up in a peculiar manner and the hind-wings applied laterally to the sides of the body. Possible members of this family have been seen in Hong Kong.

Sphingidae

The next group is the interesting Sphingidae or the hawk moths. The adults are easily recognized by their large, tapering bodies, streamlined shape with the outer wing margins very oblique and with their antennae thickened in the middle region coming to a pointed terminal hook. They all possess a long coiled proboscis which in the larger species measures up to 15 cm or more when fully extended. With a couple of exceptions they are all nocturnal, fast-flying moths which are attracted to lights and so are caught in light-traps; also they often can be found on buildings in the morning. At rest they sit with the wings pointing backwards, making an arrow-head shape (Plate 306), but they are typically pinned and mounted with the wings held as though in flight (Plate 307). They feed typically from flowers with a long clayx, such as fire-cracker vine (*Pyrostegia ignea*), *Convolvulus*, *Ipomoea*, spider lily (*Hymenocallis americanum*) and other lilies, where the nectar is unavailable to the shorter 'tongued' species. The caterpillars are collectively known in the U.S.A. as 'hornworms' because dorsally on the eighth abdominal segment there is almost always a curved horn-like process projecting posteriorly (Plate 308). The largest and most interesting local hawk moth is the Death's Head Hawk Moth (*Acherontia atropos*) (Plates 305 & 306)

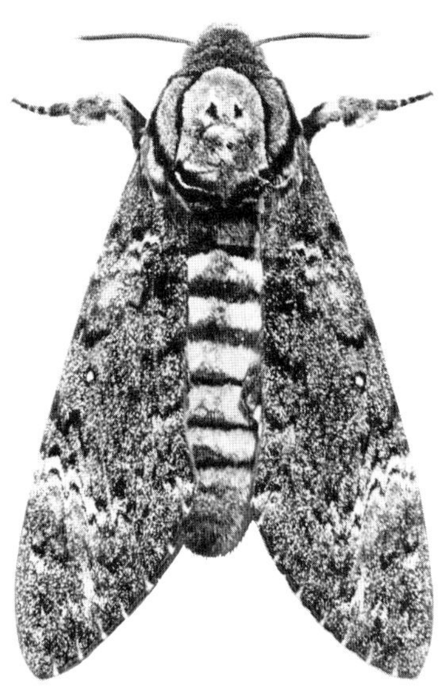

Plate 306. Death's Head Hawk Moth, *Acherontia atropos* (Lepidoptera Sphingidae) in natural resting position.

Plate 307. Mounted Death's Head Hawk Moth showing distinctive yellow hindwings; wingspan 100 mm.

which is found throughout Europe, Africa and Asia as a minor pest on most Solanaceous crops (i.e. potato, brinjal, tomato, tobacco, peppers). Pupation of all Sphingidae takes place in the soil, and in the pupa the long proboscis is evident as a large coiled projection under the head externally (Plate 309), but in some species the coiling is not distinct (Plate 310). The commonest hawk moth here is undoubtedly *Agrius convolvuli* called either the Convolvulus or Sweet Potato Hawk Moth.

(Above) Plate 308. Caterpillar of *Panacra mydon* feeding on a leaf of *Alocasia odora*; body length 50 mm.

(Left) Plate 309. Pupa of hawk moth.

(Below) Plate 310. Pupa of *Panacra mydon*; length 35 mm.

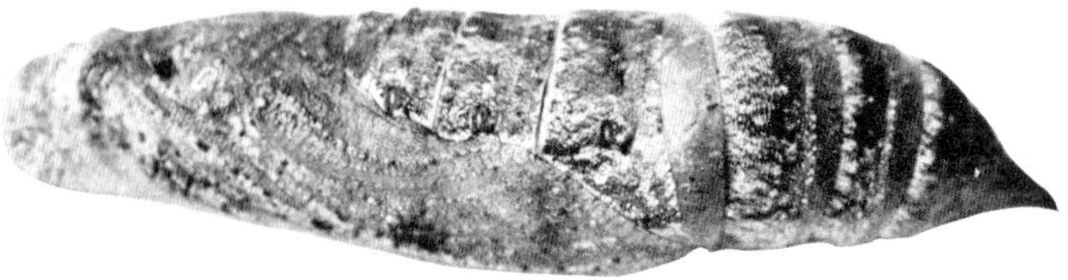

Plate 311. Sweet Potato Hawk Moth, *Agrius convolvuli*; wingspan 96 mm.

Plate 312. Silver-striped Hawk Moth, *Hyles lineata*; wingspan 68 mm.

The pink and black-banded abdomen with a median dorsal grey stripe and the grey wings are quite characteristic (Plate 311). The caterpillars feed on the leaves of Convolvulaceae. Several other hawk moths can be found regularly here, including the small Silver-striped Hawk *(Hyles lineata)* (Plate 312), the Pine Hawk *(Hyloicus pinastri)* (Plate 313), Oleander Hawk *(Daphnis neu)*, the Coffee Hawk Moth *(Cephonodes*

Plate 313.   Pine Hawk Moth, *Hyloicus pinastri*; wingspan 85 mm.

Plate 314.   Coffee Hawk Moth, *Cephonodes hylas* (Lepidoptera Sphingidae); wingspan
65 mm.

hylas) (Plate 314) with its hyaline wings, the common small *Macroglossa saga,* and the two diurnal species—the Humming-bird Hawk *(Macroglossa belis)* and the Narrow-bordered Bee Hawk *(Hemeris tityus)*.

One of the few cases where the larvae have been found is *Panacra mydon* (Plate 315); the large green and brown caterpillars were located on the underside of the huge spade-shaped leaves of *Alocasia odora*

Plate 315. Adult Hawk Moth, *Panacra mydon* (Lepidoptera, Sphingidae); wingspan 46 mm.

(Plate 308). In view of the abundance of Oleander bushes in most of the Colony it is rather surprising that records of the Oleander Hawk are so sparse. A few other records of unidentified hawk moths have been made locally, but these mostly appear to be scarce species.

Super-family **Noctuoidea**

The final group of Lepidoptera in the systematic sequence is the superfamily Noctuoidea, containing several families of importance: Arctiidae, Notodontidae, Amatidae, Noctuidae and Lymantriidae.

Arctiidae

The Arctiidae are the day-flying tiger moths with their characteristic bright coloration. The distinctive yellow or red/black colour combination in these moths is thought to be warning colours showing that they are distasteful to predators. Some experiments have been carried out in Europe with mice as predators and it has been demonstrated that mice do find some European tiger moths distasteful. The two commonest local species are *Dysphania militaris* (Plate 316) with its distinctive yellow/black coloration, and the brown/black species *Obeidia tigrata* (Plate 317), and there are quite a few other smaller unidentified species to be found here.

Plate 316.   Common Tiger Moth, *Dysphania militaris* (Lepidoptera, Arctiidae); wing-span 74 mm.

Plate 317.   Brown Tiger Moth, *Obeidia tigrata*; wingspan 60 mm.

Usually the caterpillars are fat bodied and very hairy; in Europe they are called 'woolly bears', but the larvae of the local species have not been recognized yet. The well known European Cinnabar Moth occurs in this group. It is renowned as the destructor of ragwort *(Senecio jacobaea)* and the bright yellow and black-banded caterpillars were deliberately introduced into New Zealand to attempt control of this pernicious weed.

Notodontidae

The Notodontidae are called prominents and several local species can be seen, but nothing is known about them.

Amatidae

The Amatidae is used to be called Syntomidae, and there are two common local species of *Syntomis* to be found in small numbers through most of the warmer part of the year, up to and including December. *S. polymita* (Plate 318) has the whole body banded yellow, but *S. sperbius* (Plate 319) is black, having a single median abdominal yellow band with smaller bands across the anterior and posterior edges of the thorax. In both cases the narrow wings are black with hyaline (clear) patches. They

Plate 318.   Adult Wasp Moth, *Syntomis polymita* (Lepidoptera, Amatidae).

Plate 319.   Adult Wasp Moth, *Syntomis sperbius* (Lepidoptera, Amatidae); wingspan 36 mm.

are sometimes called wasp moths but should not be confused with clearwing moths (Sesiidae), although apparently they do show affinities with the day flying moths (Zygaenidae). Several other species have now been collected but not yet identified.

Noctuidae

The Noctuidae (= Agrotidae) is probably the largest family within the Lepidoptera; it contains over 6,000 species, and many are important agricultural crop pests. As the name suggests, the moths are, with a few exceptions, nocturnal. They are stout-bodied moths of moderate size, most being 20–30 mm in body length. Amongst the important agricultural pests, certain groups are collectively named according to the habits of the caterpillars, for example, armyworms (*Spodoptera, Mythimna*

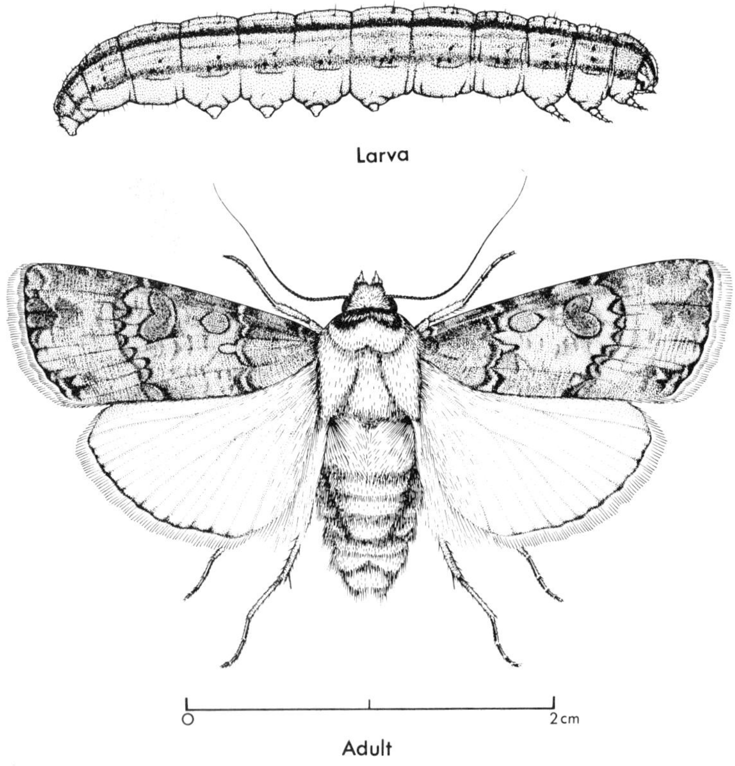

Plate 320.   Black (Greasy) Cutworm and Moth, *Agrotis ipsilon* (Lepidoptera, Noctuidae).

spp. Plates 332 & 333), cutworms (*Agrotis, Noctua, Euxoa* spp.), leafworms (*Spodoptera, Plusia* spp.), bollworms, stalk-borers (*Sesamia* spp.), etc. These categories are of only limited use entomologically for several widespread species do more than one type of damage and in some cases the differently aged caterpillars fall into different categories.

The local species of Noctuidae whose identities are known are all important crop pests, and typically they are often the most abundant local species. Amongst the local species are the Black Cutworm *(Agrotis ipsilon)* (Plate 320), Common Cutworm or Turnip Moth *(Agrotis segetum)* (Plate 321), American Bollworm (*Heliothis armigera*) (Plate 322), Purple Stem Borer *(Sesamia inferens)* (Plate 323), Paddy Armyworm *(Spodoptera mauritia)* (Plate 324), and Rice Cutworm or Fall Armyworm

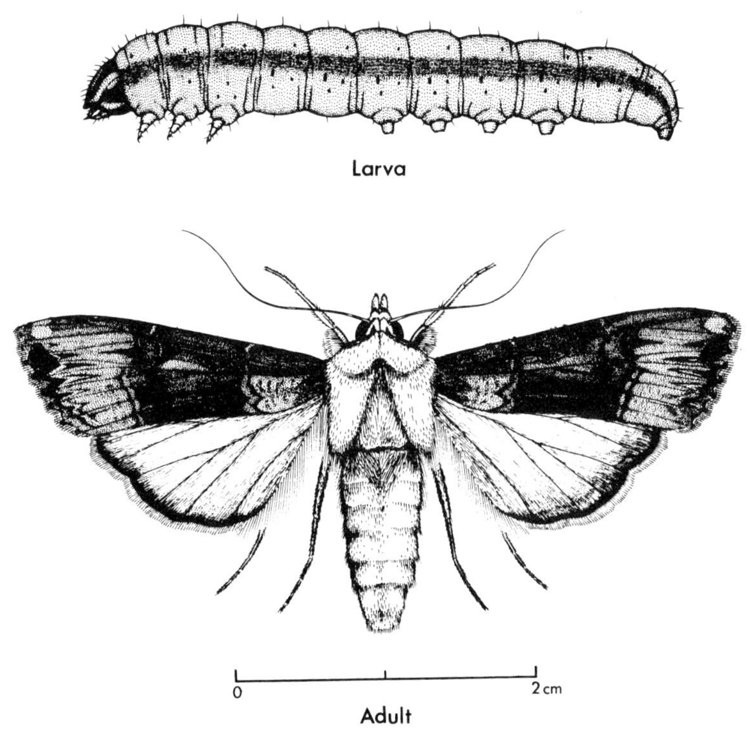

Plate 321. Common Cutworm and Turnip Moth, *Agrotis segetum* (Lepidoptera, Noctuidae).

 Insects of Hong Kong

Plate 322. Adult American Bollworm, *Heliothis armigera* (Lepidoptera, Noctuidae); wingspan 35 mm.

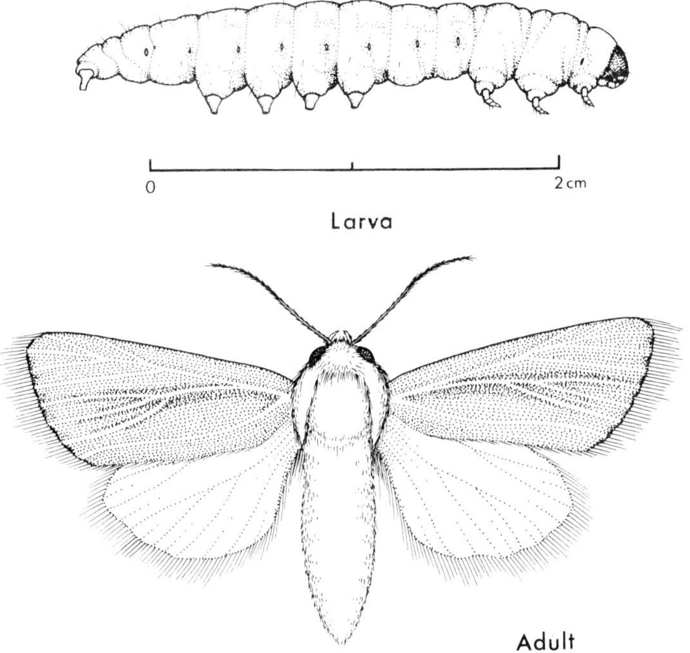

Plate 323. Purple Stem Borer caterpillar and Moth, *Sesamia inferens* (Lepidoptera, Noctuidae).

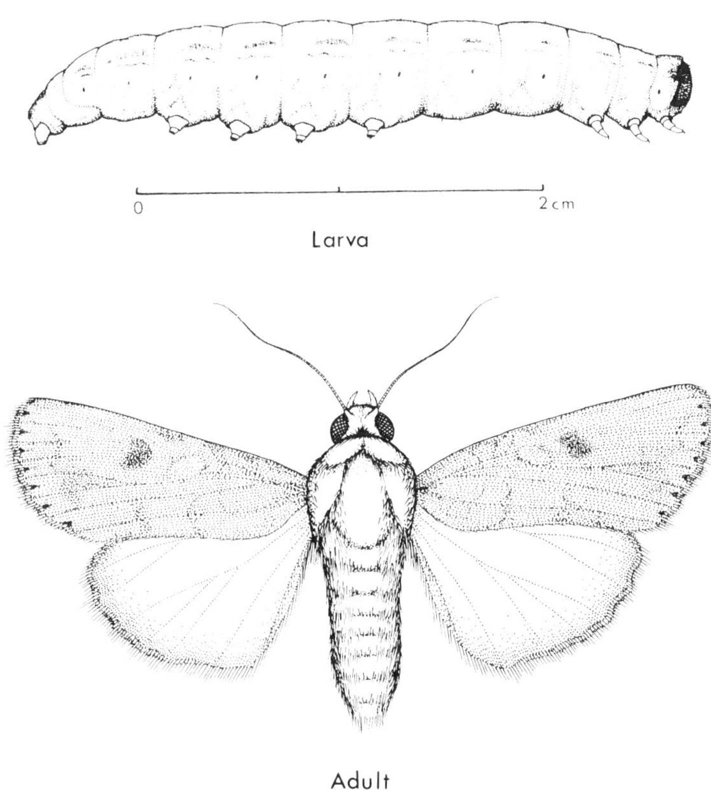

Plate 324. Paddy Armyworm and adult moth, *Spodoptera mauritia* (Lepidoptera, Noctuidae).

(Spodoptera litura) (Plate 325 & 326). The commonest species are probably the polyphagous and widespread species *Spodoptera litura* and *Agrotis ipsilon,* closely followed by *Mythimna* spp. and *Heliothis armigera.*

A group of interesting local species are to be found in the sub-family Ophiderinae, and are fruit-piercing moths. *Othreis fullonica* (Plate 327) is a large moth that at rest looks leaf-like, the forewings are brown and shaped rather like a leaf and the hind-wings are yellow with dark markings; it has a short, stout, toothed proboscis with which it pierces thick-skinned ripening fruits (*Citrus, Litchi,* etc.). The slightly smaller *Ophiusa tirhaca* has a thinner and weaker proboscis and can only pierce soft-skinned fruits such as peach (Plate 328). In Malaya it has been recently discovered that the noctuid *Calpe eustrigata* is a blood-sucker

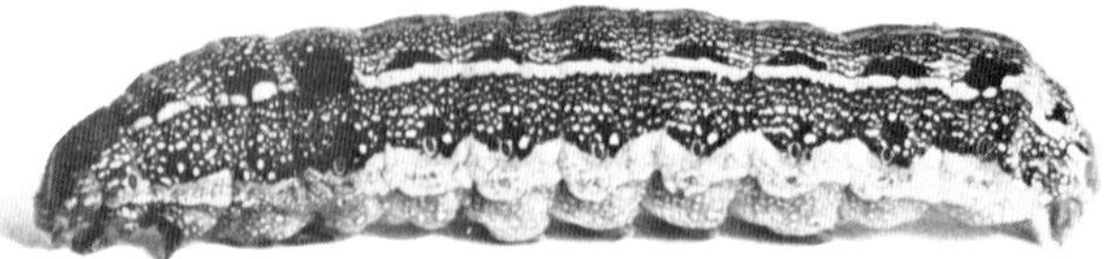

Plate 325.   Caterpillar of *Spodoptera litura* (Lepidoptera, Noctuidae), called the Rice Cutworm or Fall Armyworm; body length 40 mm.

Plate 326.   Adult *Spodoptera litura* (Lepidoptera, Noctuidae); wingspan 35 mm.

Plate 327.   Adult Fruit-piercing Moth, *Othreis fullonica* (Lepidoptera, Noctuidae); wingspan 100 mm.

feeding upon large mammals such as tapir, rhinoceros, buffalo, elephant, and deer. The blood is obtained by direct piercing through the skin, and it is thought that this habit has evolved subsequent to the skin-piercing of fruits. This phenomenon of blood-sucking by a moth has not been recorded locally. *Ophiusa* further resembles *Othreis* is being leaf-like at rest.

Plate 328. Adult Fruit-piercing Moth, *Ophiusa tirhaca* (Lepidoptera, Noctuidae); wingspan 65 mm.

Plate 329. Adult *Trichoplusia chalcites* (Lepidoptera, Noctuidae); wingspan 35 mm.

Plusiinae

One distinctive group within the Noctuidae is the Plusiinae where the caterpillars have only two, or sometimes a small third, pair of abdominal prolegs. They progress in a somewhat looping manner which earns them the common collective name of semi-loopers. These are in general

Plate 330.   Cotton Semi-looper caterpillar on *Hibiscus*; body length 30 mm.

Plate 331.   Adult moth of Cotton Semi-looper, *Anomis flava* (Lepidoptera, Noctuidae, Plusiinae); wingspan 28 mm.

leaf-eaters on a wide range of vegetables, flowers and herbaceous plants. They include the well-known European Silver-Y Moth *(Plusia gamma)*, also the local *Plusia chalcites* (Plate 329) and the Cabbage Semi-looper *(Trichoplusia brassicae)*, both of which are serious pests of *Brassica*

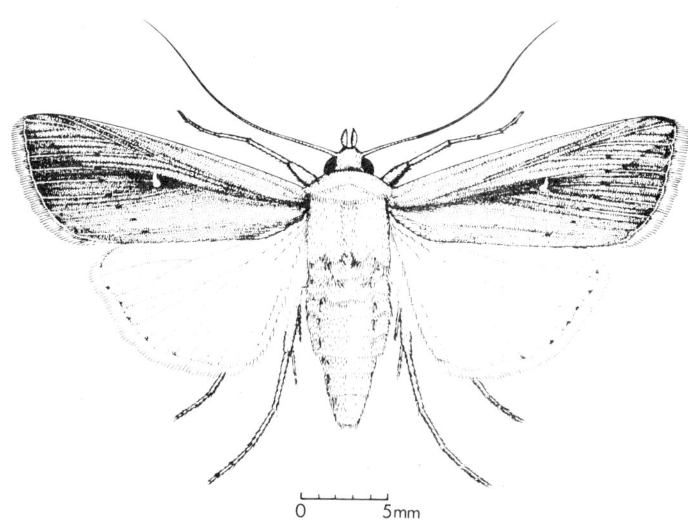

Plate 332.   Rice Armyworm Moth, *Mythimna loreyi* (Lepidoptera, Noctuidae).

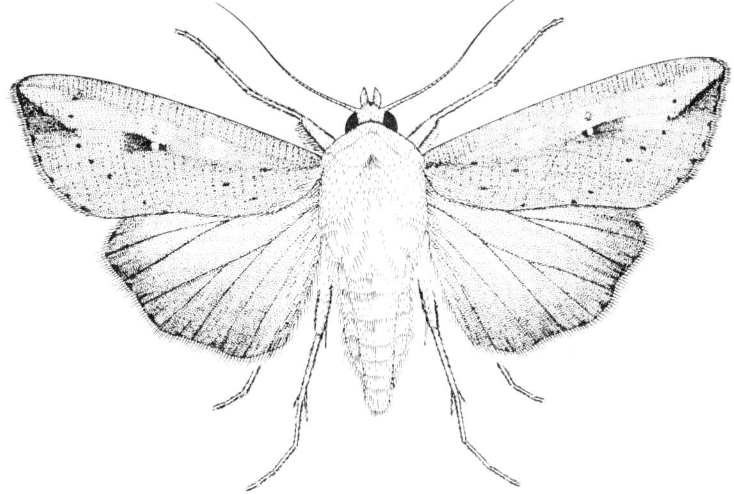

Plate 333.   Rice Armyworm Moth, *Mythima separate* (Lepidoptera, Noctuidae).

crops and lettuce in the New Territories. Another local species is the Cotton Semi-looper *(Anomis flava)* (Plates 330 & 331) found defoliating *Hibiscus* plants (*Hibiscus* and cotton are both in the same family, the Malvaceae).

Lymantriidae

The final family of Lepidoptera is the Lymantriidae where the caterpillars are distinctively hairy. Some species are just very hairy, but more typically they have tufts of long hair at the front and rear extremities and coloured tufts (often yellow) on the thoracic segments; this earns them the common name of tussock moths. There is pronounced sexual dimorphism shown in this group and various European species have wingless females (e.g. *Orgyia antiqua*, the Vapourer). Species of *Euproctis* in Europe are well-known for their urticating bristles which can cause serious skin irritation. The females often have a terminal tuft of hairs on the abdomen which are removed and used to cover the egg mass for protection. Species of *Euproctis* have been recorded locally on *Bauhinia*, and *Orgyia* on *Citrus* and peach. The commonest local tussock moth is *Perina nuda* and is found on the leaves of *Ficus microcarpa*. The tufted caterpillars (Plate 334) and the pupae under their thin film of webbing (Plate 335) can be found several times a year. The adult female is larger and yellow/brown in colour whereas the adult male has the larger part of the wings hyaline (Plate 336).

Plate 334. Caterpillar of Fig Tussock Moth, *Perina nuda* (Lepidoptera, Lymantriidae); body length 32 mm.

Plate 335. Pupa of Fig Tussock Moth on leaf of *Ficus microcarpa*.

Plate 336. Adult female *(above)* and male *(below)* Fig Tussock Moth, *Perina nuda* (Lepidoptera, Lymantriidae); wingspan 35–40 mm.

Two species of *Lymantria (L. dissoluta* and *L. nebulosa)* have been collected in pheromone traps at Tai Lung Farm, New Territories, but up to date no specimens of the notorious Gypsy Moth *(Lymantria dispar)* have been recorded locally.

Plate 337.   Caterpillar of Rose Myrtle Lappet Moth, *Trabala vishnou* (Lepidoptera, Lasiocampidae); body length 60 mm.

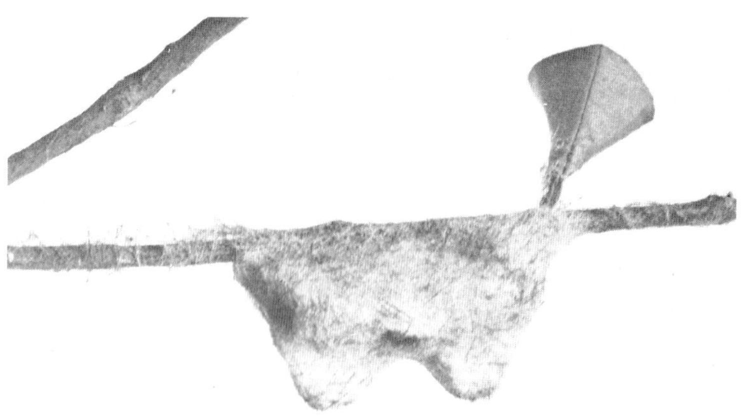

Plate 338.   Pupal coccoon of Rose Myrtle Lappet Moth on a twig.

Plate 339. Adult female Rose Myrtle Lappet Moth, *Trabala vishnou* (Lepidoptera, Lasiocampidae); wingspan 60 mm.

ORDER **TRICHOPTERA**
(Caddis-flies)

These are small to medium-sized insects, rather delicate in appearance and looking like small moths. They are from 2–40 mm in body length, but most are rather small. They are in fact regarded as being closely related to moths. The four membraneous wings are hairy and sometimes also scaly, and the adult has long tapering antennae and long hairy legs with conspicuous spurs on the hind-legs (Plate 340). The wings are held roof-like over the back at rest, like many moths. The adult mouth-parts are essentially of the biting and chewing type, but are poorly developed and weak. Most adults only imbibe liquids and some species elsewhere survive for up to one month as adults. They are weak fliers and are seldom found far from freshwater streams. During the day they typically rest under the leaves on overhanging trees; at night adults are attracted to lights and many be traped in light traps at some distance from water, but generally adults are seldom found. Caddis-flies undergo complete metamorphosis and have characteristic aquatic larvae.

The caddis larvae are rather caterpillar-like with well developed thoracic legs, head capsule and mouth-parts, and usually with terminal

Plate 340.   Adult local caddis-fly (Trichoptera), *Stenopsyche* sp. body length 15 mm.

claspers or hooks. Elsewhere caddis may sometimes be found in ponds and lakes; they are typical freshwater stream species, and are mostly to be found in the upper reaches of streams, in riffles or torrent stretches, where the water is fast-flowing, cool and highly oxygenated. The larvae either respire through the general body surface or they have abdominal gills. The gills are either abdominal tracheal gills held at the sides or under the abdomen in tufts, or are small anal blood gills.

The striking feature of caddis larvae is that most of them build elaborate cases or structures in which they live. Most of the cases are transportable and when feeding or moving the caddis protrudes its head and thorax out of the case and walks dragging the case behind. The terminal hooks or claspers hold the case firmly as it is dragged along by the thoracic legs. The shape and nature of the case is specific to each species of caddis, usually to each genus, but frequently it appears that the case-type is not a family characteristic, which makes field identification of caddis even more difficult. Not all the cases are portable, some are constructed like small fish-traps and stuck on to rocks (F. Hydropsychidae), others are in the form of a flattened silk web with no obvious entrance and also stuck on to rocks in torrent regions (F. Hydropsychidae).

Eggs are usually laid in clusters in mucilage, often in the water but sometimes on overhanging vegetation; each female may lay from 300–1,000 eggs. The whole life cycle usually takes a year. The eggs hatch in a few days and the pupal stage takes 2–3 weeks; the adult lives for 3–4 weeks and the remainder of the time is spent as the larva in the water. Some larvae do not live in cases of any sort but are free-ranging and pre-dacious; they hunt for other insects in the stream gravel, and these groups are generally considered as being primitive. However, all caddis can produce silk and all caddis pupate inside a silken coccoon usually fixed firmly to some solid substrate. The pupa is unusual in that it is always mobile and possesses the well-developed larval jaws. It cuts its way out of the coccoon and then swims to the surface of the water where it crawls out onto a rock; here it moults and the adult emerges.

Trichoptera are a very difficult group of insects to study casually for several reasons. Firstly the adult taxonomy uses a series of rather esoteric characters which are difficult for the non-expert to appreciate. Secondly, there appears to be only a little ecological correlation between the members of the two dozen different families. In some of the larger families case structure and larval habits vary tremendously. Finally it is very difficult to determine the diet of many species, especially the net-spinning ones. Some authorities say they are herbivorous and feed on diatoms, desmids, and pieces of leaf fragments caught in the nets, others say they are carnivorous, and again it is reported that they are omnivorous and eat almost anything organic that they catch in the nets. Quite possible the latter case is more likely, with different species showing different preferences within a general omnivorous habit.

Identification of Trichoptera is very much a case for the expert, and experts on this group are very scarce. Thus to date only the two species of *Macronema* as adults have been authoritatively identified to species (British Museum (Nat. Hist)), however we were fortunate in having been able to enlist the support of the late Prof. H. H. Ross, the world authority on Trichoptera, and he had identified many local specimens to the level of genus.

The larvae of caddis-flies may be described as being of two body types. Campodiform means it has the body rather compressed (but not always slender) with the head protruding in line with the body, and these are active forms, usually predacious and they seldom construct transportable cases. They seldom have tracheal gills and respiration is either cutaneous or sometimes by a tuft of anal blood gills. The eruciform type look more like sedentary caterpillars with a fatter body and abdominal tracheal gills; the head is held at a marked angle to the body and they live in portable cases.

In Hong Kong to date at least 20 species of caddis have been collected from streams in the New Territories, and they appear to belong to 10 different families. A few species are very common, but on the other hand some have only been found once or twice. For Trichoptera the most prolific streams are those that flow through areas of forest. This is thought to be partly because the larvae require leaves and twigs for both food and for making cases. Furthermore the weak-flying adults are protected from wind and can find refuge in the foliage of the overhanging trees. It should perhaps be pointed out that it is only recently (spring, 1977) that some of the local species were reared. Prior to this a few adults were collected by chance but they could not be associated with any of the caddis larvae that could always be found in some of the New Territories streams. It is hoped that in the near future we shall have a series of both adults and larvae of the common species identified correctly (see Dudgeon, 1981). At present the species recorded here have been identified to genus and allocated to the following families which have been arranged in sequence according to the relative abundance of the species listed.

Hydropsychidae (Net-spinning Caddis)

This widespread and abundant family is the commonest group found in Hong Kong. Two species of *Macronema* have been identified to species, they are the very common *M. latum* which is a dark blackish caddis with small white mottling on the wings, and about 17 mm in length. It has occasionally been seen in Tai Po Kau Forest Reserve alongside the stream when the trees have been festooned with hundreds of newly emerged adults. *M. fastosum* is slightly smaller, and is yellow-brown in colour with black wingtips and a black bar across the closed wings; the body length is about 16 mm (including the folded wings) and it is quite uncommon.

The larvae live on boulders in the torrent part of the stream and they weave a stiff silken net in the shape of a river fish-trap with a stone-covered retreat at the end to one side in which they sit (Plate 341). The whole structure is firmly attached to the rock surface at the edges. The retreat (case) is open posteriorly to allow water through, and act as an escape route for the caddis if necessary. The larvae are campodiform (Plate 342) with small anal blood-gills and are thought to be either omnivorous or carnivorous feeders; they eat the small organisms caught in the web. They are very abundant on the rocks in the swifter flowing

Plate 341. Rock with caddis net *in situ* (*Macronema latum*).

Plate 342.   Larva of *Macronema* (Trichoptera, Hydropsychidae) and fish-trap type of web removed from rock surface; length of larva 15 mm.

regions of many streams in the New Territories. Although the two species of *Macronema* are distinct as adults, the larvae have not yet been positively separated, and presumably most of the larvae seen are the more common *M. latum.*

In addition there are two species of *Hydropsyche* which make softer nets on the surface of flat rocks, and a species of *Cheumatopsyche* which also makes a simple, soft net on rather flat rocks slightly away from the actual torrent regions.

Calamoceratidae

In streams with extensive leaf-litter can be found a species of *Ganonema*; it lives in a flat case made out of two pieces of leaf which if cuts with its mandibles. In the Tai Po Kau stream the favourite leat seems to be *Liquidamber formosana* (Plate 343). As the larva grows, so at intervals it has to cut a new and larger case. The adult is a small orange-brown caddish-fly with pale brown mottled forewings and plain dark hindwings; the body is about 10 mm long.

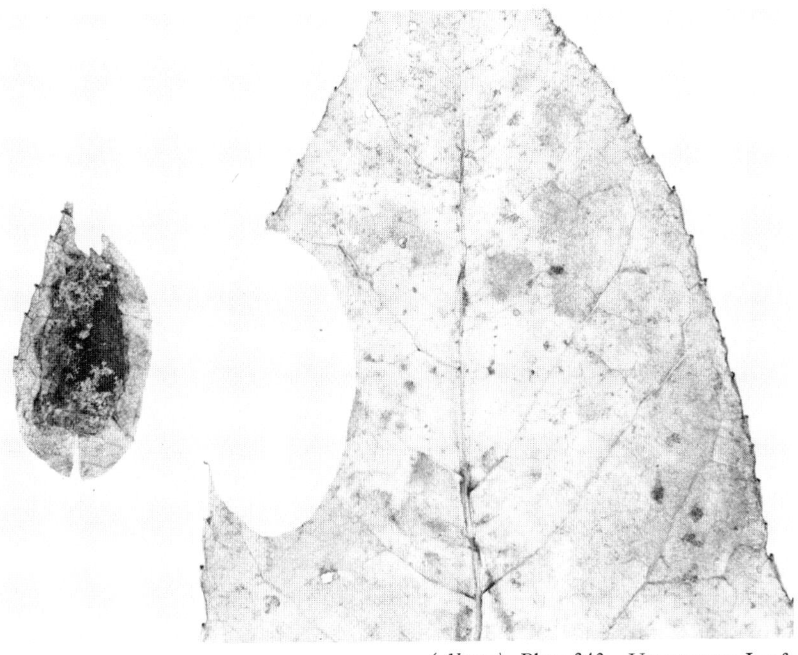

(*Above*) Plate 343. Uncommon Leaf-case Caddis, with the leaf from which the case was cut; length of case about 30 mm.; the species is *Ganonema* sp. (Calamoceratidae).

 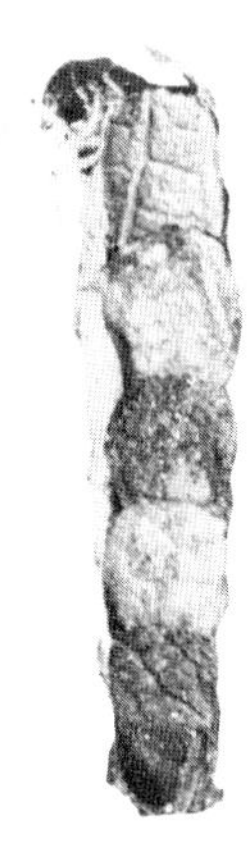

(*Left*) Plate 344. Caddis case made of small pieces of leaf (*Lepidostoma* sp.); case length about 15 mm.

Leptoceridae

In a few streams can be found caddis that live in a case made of small leaf fragments which is square in section and about 10 mm long (Plate 344); at times this species is quite abundant, and it has been identified as a species of *Lepidostoma*. The larvae are peculiar as the smallest cases are constructed of tiny sand grains (which is typical of the family elsewhere), but as the larvae grow they suddenly make the square case out of cut leaf fragments. One or two cases have been collected where the posterior half is made of sand grains and the anterior part of leaf fragments.

Also in this family are a few other less common species. *Mystacides* makes a case of wrapped-over leaf fragments that are fairly large in size, and these cases can be found in a number of streams locally. *Triplectides* has a larvae which bores out a short twig, but these hollow-twig cases are not very common.

Stenopsychidae

Stenopsyche is probably the largest caddis to occur in Hong Kong. Fully grown larvae measure 45 mm in length and 5–6 mm broad across the abdomen, and the adult (Plate 340) has a body length of 15 mm, not including the wings. The species is common under stones in swiftly flowing water, where it appears to be free-ranging hunting in the gravel for small invertebrates. At the time of pupation it constructs a large silk and stone shelter stuck firmly on to the side of a stone, or between two adjacent stones, and the pupa lies within a silken coccoon. The larva is quite ferocious and appears to be carnivorous in its feeding habits, but it is quite possible that it really is omnivorous and eats most organic material it encounters on the rock. The pupal cases are usually found low down on the side of boulders or actually underneath where they are presumably less susceptible to fluctuations in the water level in the streams.

Odontoceridae

This group typically contains small species, with adults generally 10–12 mm in body length. The larvae are eruciform, and the species provisionally called *Odontocerum* is common locally in stream riffles where they make a strong cylindrical but slightly curved case of large sand grains stuck together (Plate 345). The sand cases may persist long after the larvae have gone, and the cases may be collected by sieving the gravel from the stream. At pupation the sand case is typically stuck on to the side of a rock in the stream, and both ends are sealed with silk. The larvae are fiercely predacious and if kept in a container with other insects they will invariably kill and usually eat the other insects.

A second species which it has not yet been possible to place to genus apparently belongs to this family. It builds a case covered with small twigs of different sizes and can be found occasionally in some of the streams.

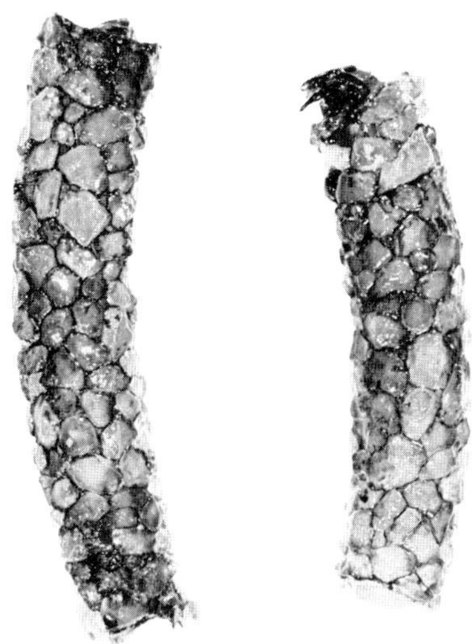

Plate 345.   Caddis-fly cases made of large sand grains (*Odontocerum* sp.); case length 15 mm.

The last few families are represented locally each by a single species, and none of which are common although some are fairly abundant:

Helicopsychidae (Snail-case Caddis)

A few specimens of the larva of a species of *Helicopsyche* with its quite characteristic case in the shape of a snail-shell have been found in two different streams in the New Territories.

Molannidae

Molanna has a characteristic case which is in a shield-shape and made of sand grains stuck together with a central cylindrical tube in which the larva lives. Such cases have been found on several occasions.

Philopotamidae (Finger-net Caddis)

Elsewhere this group is known collectively as Finger-net Caddis after their small finger-like tubular nets. The nets in which they catch their food are made of silk and attached to stones. The species which occurs here is small, carnivorous and apparently is free-living in habits; it is to be found on the sides of and underneath boulders in the higher reaches of local streams. It has been identified as a species of *Chimarra*.

Sericostomatidae

The larvae of *Silo* can sometimes be found locally. They live in a case made of small stones surrounding a larger one, and the case is stuck on to the side of a rock. Only two specimens have actually been seen, and only one was collected.

ORDER **DIPTERA**
(True Flies)

As a group the Diptera are rather inconspicuous and apart from the bloodsucking species and the houseflies, they are likely to escape casual notice. They are moderate to small in size and all have only a single pair of wings, the hind-wings being modified into balancing organs called halteres. The mouth-parts are modified for sucking purposes and for imbibing liquid food, and sometimes also for piercing. The larvae are worm-like and legless, either terrestial, aquatic or parasitic, but the actual appearance varies somewhat with the group concerned. The pupa in the higher forms develops inside a smooth rounded puparium, usually without a coccoon.

Sub-order **Nematocera**

The first Sub-order is Nematocera, where the flies are usually small with long antennae, often feathery, and the larvae are not too reduced and have a distinct head capsule. The pupa is sometimes active.

Tipulidae

The Tipulidae are known as either crane flies or 'daddy-longlegs', and these are large flies, rather thin, and with long, delicate legs. Several species are common locally, the first being very much like the European *Tipula*—large and brown with clear wings; it is called *Holorusia*. Another attractive orange and brown species with dark-spotted wings and rather short legs is *Ctenophora flavibasis* (Plate 346). Several other species do occur here, including a completely black-bodied (including wings) species now identified as *Hexatoma* species, the larvae of which are known as 'leather-jackets' and are damaging to lawns and high quality turf; however they are very seldom seen here. Some species have aquatic larvae, and the long tapering whitish larvae can be found frequently in freshwater, in the substrate, or in the soil of paddy fields.

Psychodidae

In the Psychodidae are two very distinct groups. The minute sand flies (*Phlebotomus* spp.) are well-known vectors of several tropical diseases including Kala-azar and other forms of leishmaniasis which fortunately do not occur in Hong Kong. Their counterparts are the small, stout-bodied, hairy, dark coloured moth flies (*Psychoda* spp.) to be found in cool, dark, damp corners of buildings, bathrooms, sewers, etc.

Plate 346.   Adult ♂ Crane Fly, *Ctenophora flavibasis* (Diptera, Tipulidae); body length 16 mm.

These are quite harmless flies, and their larvae have once been found in wet soil at the edge of a paddy field.

Culicidae

All mosquitoes are classified into one family, the Culcidae. This is one of the best known, and most disliked, group of flies in Hong Kong. The feeding bites are irritating and may be quite painful in susceptible individuals, and the nocturnal aerial manouvers of a hungry female mosquito are most disturbing, for the high pitched sound of the beating wings is clearly audible at close quarters. In many countries mosquitoes spread serious diseases such as malaria, dengue, yellow fever, and elephantiasis, and for this reason the Malaria Bureau of the Urban Services Department has premises all over the Colony to check for local populations of mosquitoes. Mosquito control is largely effected by treatment of water bodies with oil films to kill the larvae and prevent breeding. Unfortunately, the oiling of streams results in complete ecological impoverishment as it destroys both plant and animal life in the water. However, most of the permanent streams in the New Territories are not oiled and so some of the local freshwater fauna does survive. As a result of the Malaria Bureau's activities Hong Kong is now regarded as malaria-free. The rare cases that occur in Hong Kong from time to time are almost invariably either chronic cases or in persons who have been abroad recently to a malaria area.

The mosquitoes which transmit malaria are species of *Anopheles* which only occur in the New Territories. The causal organism of malaria is a microscopic protozoan (single-celled animal) *Plasmodium* occuring as several different species which live and multiply in the blood stream and liver of man. If taken into the stomach of a feeding mosquito, the *Plasmodium* parasites continue to grow, develop and multiply in the insect. Eventually, the infective parasites reach the mosquito's salivary glands and wait there for the next feeding period. When the mosquito next feeds, it injects after making the feeding puncture some saliva to stop the blood from clotting and in the saliva are some of the *Plasmodium* parasites. The parasites then multiply in the man and in the process of so doing they produce the clinical symptoms of malaria. Thus malaria can only be transmitted from an infected patient by a feeding *Anopheles* mosquito which then goes on to feed on another man after a few days. If the parasite could be eliminated from the human population then it would not matter if the vector mosquitoes were present or not. However, elimination of the parasite from man is difficult, for this is a chronic disease and the parasites can remain dormant in the liver for months or even years between successive attacks. Tourist travel is now so common-place to many parts of S.E. Asia, and there is constant re-infection of the local population taking place. The two most obvious methods of combating malaria are by the drug prophyllaxis which kill the *Plasmodium* parasite in the blood, and by preventing the mosquite from breeding. This latter action includes draining of unwanted bodies of water and oiling streams, etc. The incidence of indigenous malaria which was very high when the Colony was established more than one hundred years ago is now non-existent for all practical purposes.

Some 52 species of mosquito are found in Hong Kong. They all have their own preferences with regard to breeding places, resting places, feeding, etc. The males seldom feed on blood, and usually pierce plants for sap, but the females almost invariably require a blood-meal for successful development of their eggs. The mosquito bites by inserting four blade-like stylets into the skin and penetrating down until a blood capillarly is pierced. Two other stylets are modified to form a double tube, which is inserted behind the cutting edges. One tube is used for pumping down saliva into the feeding wound to prevent the blood from clotting (agglutinating), and the other tube is the feeding canal up which the blood (or sap) is sucked into the pharynx and stomach of the insect.

Some species of mosquito prefer to feed upon human blood (i.e. anthropophilic), some on birds, and others on mammals, reptiles, amphibia, fish, etc. Each female mosquito may lay several batches of eggs; each batch takes four or five days to mature after the blood meal, and each batch must be preceded by a separate blood meal. Eggs are laid on or near the surface of water, either singly (as in *Anopheles* or *Aëdes*) or in batches of up to 300 or more *(Culex)* in collective rafts. The aquatic larvae hang on to the water surface by special bristles, and feed on microscopic plants and animals and organic debris, using their water-filtering mouth bristles. When at rest and feeding, the larvae of *Anopheles* species float horizontally just under the water surface, whereas the Culicines float in a head-down position.

A few species have carnivorous larvae which feed on other mosquito larvae. The large black/blue species in Hong Kong is *Toxorhynchitis splendens* and its larvae feeds upon the larvae of *Aëdes albopictus* in cans, broken bottles and other small domestic water bodies as well as holes in trees. The transparent larvae ('glass-worms') of *Chaoborus* (Ghost Midge) are mosquitoes and not chironomids. They are pelagic and predatory, seizing their prey (usually other mosquito larvae) in their prehensile antennae. They are curious in that their respiration is cutaneous; they are to be found in Plover Cover and other reservoirs in small numbers. The pupae of mosquitoes are also aquatic and active but are more rounded with two small respiratory 'horns'.

The two most important species of *Anopheles* which can transmit malaria in the New Territories are *Anopheles minimus* and *A. jeyporiensis*; both are domestic, crepuscular species, also feed on pigs and cattle and breed in ditches and rice paddies. At least another four species can transmit malaria, but these are rarer; *A hyreanus* breeding in stagnant water with vegetation; *A. fluviatilis* breeding in hill streams; *A. maculatus* breeding in hill streams, ditches and seepages; and *A. tessalatus*. In the Culicinae occur the two most annoying domestic mosquitoes in Hong Kong, the large brown *Culex fatigans* which breeds in polluted water and ditches up to an altitude of about 700 m, and the smaller black species *Aëdes albopictus* (Plate 347) which has a white line along the thorax, and black and white banded legs. This species flies quietly and bites aggressively.

Aëdes albopictus is highly urban and breeds in discarded receptacles containing rain water and in water trapped in holes in trees and in leaf-axils. In these minute water bodies is also to be found the large predatory larvae of the blue/black mosquito *Toxorhynchitis splendens*.

Aëdes togoi and *Culex sitiens* breed in rain-filled rock pools along the

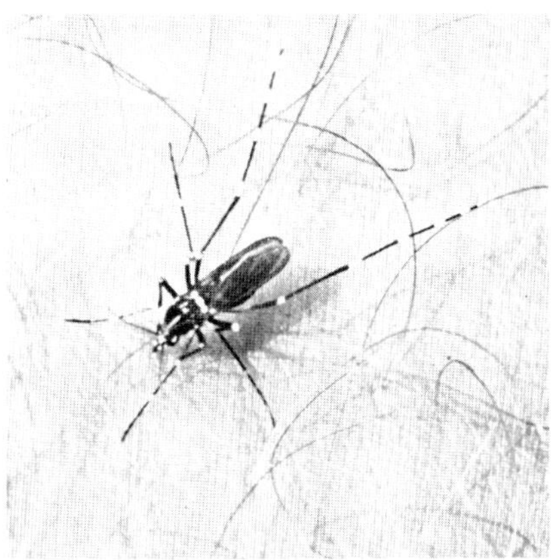

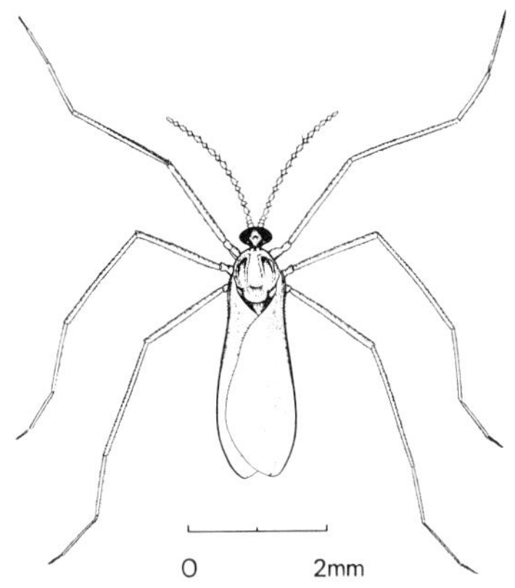

Plate 347. Adult female House Mosquito, *Aëdes albopictus* (Diptera, Culicidae), feeding on the leg of the senior author.

Plate 348. Rice Stem Gall Midge, *Pachydiplosis oryzae* (Diptera, Cecidomyiidae).

seashore above the high tide level. The larvae only develop in water with a small salt content. *Mansonia uniformis* breeds in swamps and the larvae respire by piercing the stems of water plants to gain access to the internal air spaces within the plant body. At least five local Culicine mosquitoes can transmit the minute parasitic nematode worm *Wucheria bancrofti*, which causes filariasis in man. The worms multiply in the blood and lymph systems of man and in severe cases the mass of worms blocks various lymphatic ducts causing extreme swelling of limbs or genital organs known as elephantiasis. The normal incidence of filariasis in Hong Kong is quite low, and the extreme form is not usually found.

Cecidomyiidae

Cecidomyiidae are the Gall Midges, a very widespread group of tiny flies which attack almost all known types of plants throughout the world. Most species are phytophagous and live inside plants; the saliva from the feeding larvae produces a growth reaction on the part of the plant and the resulting distortion and growth of the plant is referred to as a gall. In some cases the gall is slight and not too conspicuous (e.g. rice stem galls, violet leafcurling, etc.) but in other cases large globular growths are produced. *Pachydiplosis oryzae* (Plate 348) is the Rice Stem Gall Midge, which occurs in the New Territories. Very conspicuous twig galls are made by *Asphondylia* c.f. *morindae* on *Aporusa chinensis* (Plate 349); other local conspicuous galls include swollen stem galls on *Bischofia trifoliata* (Plate 350) and spherical leaf galls on *Glochidion*

(Left) Plate 349. Stem galls on *Aporusa chinensis* made by larvae of *Asphondylia morindae*.

(Below) Plate 350. Twig galls on *Bischofia trifoliata* made by an unidentified gall midge.

Plate 351.   Leaf galls on *Glochidion hongkongense* made by an unidentified gall midge.

hongkongense (Plate 351). In general the presence of Cecidomyiidae is indicated by the widespread plant galls they produce. The adults are seldom seen, and neither is it easy to rear them from their galls. Sometimes elongate, finger-like galls are produced on leaves of some plants, and other times the tiny maggot may be found in leaf-curls or in the shoots and flowers of a number of different plants. Leguminoseae and Compositae are favoured hosts of many different gall-midges. Not all species are phytophagous however, for some are predatory and feed on scale insects (Coccoidea) and even other flies and midges in their larval stages; a few species are saprophagous and feed on decaying vegetable matter.

Sciaridae

Sciaridae resemble midges as adults but their tiny white larvae usually have a distinctive black head capsule and typically are saprophagous in soil or feed on fungi. However, a few species are gall makers, as shown by the leaf galls on *Glochidion wrightii* by *Bradysia* species (Plate 352).

Mycetophilidae

Fungus gnats (Mycetophilidae) are small midges with long antennae. The small, smooth, white maggots have a black head capsule and are to be found in soil and litter feeding on decaying vegetable matter and fungi. Certain species are very damaging to cultivated mushrooms and are introduced into the mushroom cellars in the manure used in the beds. Sometimes the entire mushroom may be completely eaten out by these small, black-headed maggots. As a group they may be encountered

Plate 352. Leaf galls on *Glochidion wrightii* produced by larvae of *Bradysia* sp. (Diptera, Sciaridae).

in ecological studies of soil and litter where the organic content is high.

Simuliidae

Simulium species are known as black flies. They are small, stoutly-built, black, blood-sucking flies to be found in the vicinity of running water. They are a worldwide group and of some importance as pests. In Africa *S. damnosum* (and other species in Central America) transmits the fliarial worm *Onchocerca,* causing filariasis or onchocerciasis, and in humans the worm usually settles in the eye and causes 'River Blindness'. In many parts of the world they are important because of their sheer numbers and they cause extreme discomfort to man and domestic animals. Several species are important in North America, biting man, fowls, and domestic animals. Some specimens of *Simulium* larvae have been collected from the streams of Plover Cove Reservoir and Tai Po Kau and Lam Tsuen Valley, and in some places they are quiet common. The larvae live in swiftly running water and are attached to submerged rocks or vegetation, although in other countries they are sometimes attached to the gill chambers of freshwater crabs (*Potomon* spp.), or to the body of mayfly nymphs in Africa. Pupation takes place in a small conical coccoon in the same locality as the larvae.

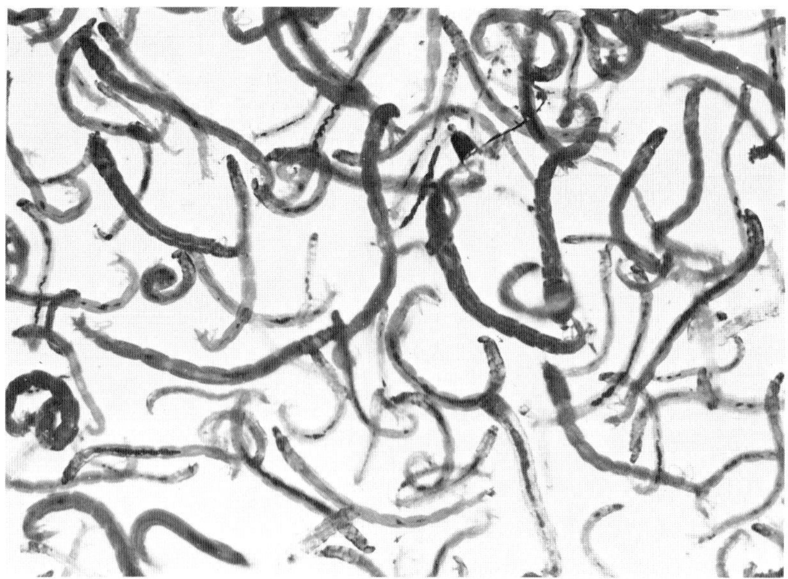

Plate 353.   Bloodworms, larvae of *Chironomous* sp. (Diptera Chironomidae).

Chironomidae

Chironomidae are lake midges—delicate midges looking rather like mosquitoes. The males have densely plumose (feathery) antennae and their wings are without scales. The larvae live in freshwater, more commonly in lakes and ponds rather than in flowing water but different species prefer different habitats. Certain species have the red respiratory pigment haemoglobin in their blood and these bright red larvae are known as 'blood-worms' (Plate 353). These usually burrow in the mud, or make small mud-tubes, whereas the few surface-haunting species are usually green in colour. Enormous quantities of 'blood-worm' are collected from ponds in the New Territories and are sold in Hong Kong as food for tropical fish. Each larva has two pairs of elongate blood gills on the eleventh body segment, and two pairs of papilla-like anal gills. A few species are pelagic and not confined to living in the mud on the lake bottom. In many lakes and reservoirs the population of chironomids can be so large at certain times of the year that the swarms of adults constitute a serious inconvenience to the nearby inhabitants. In parts of the tropics, for example Lake Victoria in East Africa, the swarming midges are sometimes collected in nets and their bodies compressed into blocks of 'midge butter' to be used for culinary purposes.

It is of interest to note that when Plover Cove Reservoir was first established there was a considerable lake midge problem at times during the summer. After the Wild Carp (*Hemiculter* spp.) were introduced the biological production of the reservoir evened out somewhat and now only small swarms of adult midges are to be seen over the reservoir surface. Not all of the larvae are detritus-feeders and a few are carnivorous; one group (Clunioninae) contains the European *Clunio* which has its larvae living on the seashore amongst algae (and the females are apterous).

Ceratopogonidae

The final family in this group is the Ceratopogonidae or biting midges. They are not common in Hong Kong but in wooded parts of Hong Kong Island there is *Lasiohelea stimulans* which causes some nuisance as a blood-sucker. They are tiny flies, with an irritating piercing bite and feed on the blood of man and various animals; some feed on other insects. In Africa they are popularly called 'no-see-ums', alluding to their tiny size. The larvae are either aquatic or living in decaying organic matter. *Culicoides circumscriptus* feeds on various vertebrates in forested areas locally, and *C. anophelis* apparently feeds on local mosquitoes. Another species of *Culicoides* is abundant in certain mangrove areas (e.g. Three Fathom Cove, N.T.).

Sub-order Brachycera

The sub-order Brachycera includes 14 families, some of which are rather obscure, characterized by their wing venation and short, protruding (porrect) palps. The antennae are variable but usually three-segmented and shorter than the head. The larvae have a small retractile head with vertical biting mouth-parts, and the pupae have a spiky appearance. They are very much intermediate between the other two well-defined sub-orders.

Stratiomyiidae

In the family Stratiomyiidae there is a very common local species, *Stratiomyia* sp. (Plate 354), with distinctive short, erect antennae, and dark brown body. Nothing is known of its biology.

Plate 354.   Adult Strationmyidae, *Stratiomyia* sp.?; body length 15 mm.

Tabanidae

Three other families are locally well represented enough to be worth mentioning. The Tabanidae are called horse flies or clegs. They are large, robust flies with large eyes. The wings are either clear or spotted, and the mouth-parts piercing for sucking blood. They feed readily from both man and domestic animals in a most insidious manner. They fly straight on to the body of the host and while sitting and inserting the proboscis they are so intent that they can be swatted with no difficulty;

(Left) Plate 355. *Tabanus* feeding on the knee of the senior author.

(Below) Plate 356. Common Brown Horsefly, *Tabanus* sp. (Diptera Tabanidae); body length 15 mm.

they can even be picked off the body by hand (Plate 355). The local large brown species is *Tabanus* sp. (Plate 356), and an equally large grey-bodied horse fly is not yet identified. The very widespread genus *Chrysops* is slightly smaller and has large black spots on the wings with distinctive large, shiny, green eyes. It has been seen locally but has not been collected yet. The larvae are semi-aquatic and live in moist habitats and prey on small worms, Crustacea and other insect larvae. In other areas some larvae are truly aquatic.

Asilidae

A distinctive group of predatory insects is seen in the Asilidae (robber flies). These are large flies with an elongate body (presumably for balance in flight) and large, stout, bristly legs ending in large paired hooks. They are predatory on other insects which are caught in flight and the short, stout, ventrally projecting proboscis is used for sucking the body fluids of the prey. They are extremely voracious and regularly catch dragonflies, bees, wasps, beetles, butterflies (including the distasteful Danainae), as well as many other flying insects. Several species are regularly seen in

(Left)   Plate 357.   Adult Robber Fly, *Promachus indigenous* (Diptera, Asilidae), in typical resting position; body length 20 mm.

Robber   Fly,   *Philodious   javanicus*

(Below)   Plate 358.   Mounted adult Robber   Fly,   *Philodious   javanicus* (Diptera, Asilidae); body length 20 mm.

Hong Kong in small numbers, including *Promachus indigenous* (Plate 357), and *Philodicus javanicus* (Plate 358) is usually found hunting on sandy beaches. The larvae usually live in soil and litter and are predacious or scavenging. They are cylindrical in shape with a small dark, pointed head, and the pupae are characteristically spined.

Bombylinidae

The Bombylinidae are called bee flies. They are medium-sized, stout-bodied, and often densely hairy, with dark wing patches, long slender legs, and a distinctive long proboscis. A common local species is the small *Anthrax* sp. A larger darker species, *Ligyra tantalus* (Plate 359), is often to be seen feeding on flowers. At rest the wings are usually held outspread. They feed in a manner similar to bees on the nectar of flowers, and they hover in a very bee-like manner over the flowers as they prepare to feed. Several other species can be seen occasionally in Hong Kong but they are not easy to catch, and as yet only the two above have been identified positively.

Sub-order **Cyclorrhapha**

The last sub-order is the Cyclorrhapha in which the flies have very short antennae fitting into sockets on the face and so are scarcely visible. The antennae are three-segmented and bear a sub-terminal arista (bristle). This is generally regarded as one of the most evolutionarily

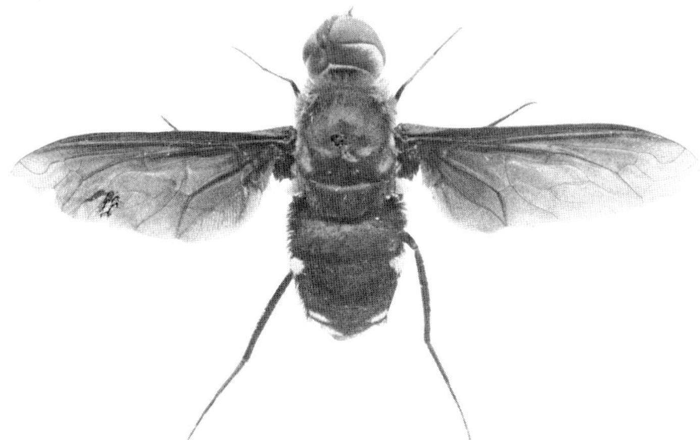

Plate 359.   Adult Bee Fly, *Ligyra tantalus* (Diptera, Bombylinidae); length 12 mm.

advanced group of insects. The larvae are maggots with a vestigial head and they have paired black mouth-hooks with which they feed. Pupation takes place inside a smooth rounded puparium.

Section **Aschiza**

Syrphidae

The Syrphidae are hover flies. They are small, brightly coloured flies, often banded yellow on a black or blue background, which gives them a resemblance to wasps. Some are stout-bodied and bristly and look rather like bumble bees. The adults are nectar feeders and to be found hovering over flower heads. Many species are common locally but only a few have been identified. The larvae are interesting in that some are aquatic, like the peculiar rat-tailed maggot of some *Eristalis* species, with its long terminal respiratory siphon. This has been found at Wu Kwai Sha in the New Territories. Other larvae are predacious on aphids and sometimes on scale insects and other bugs. The predatory species are of considerable importance agriculturally for they devour large numbers of aphids on many crops and host plants. Being blind and legless maggots, they do not in the least resemble predatory insects. The female fly only lays her eggs on plants which are already infested by aphids or other plant bugs, so that the maggots will develop in a site with a dense prey population and their limited sensory receptors will be sufficient for close quarter predation. They pierce a bug with their mouth-hooks and suck out the body juices. When feeding they often take aphid nymphs at a rate of about one per minute and usually they destroy several dozen each day.

Two of the commonest genera of Syrphidae locally are *Eristalis* spp. and *Syrphus* spp. recorded preying on Turnip Aphid and Citrus Aphid. The adults of many species are of some importance as pollinators of a wide range of plants. Another common species is *Eristalinus* sp. and is also of interest as the prey of the wasp *Ectemnius* sp. nr. *fossorius* which makes nest tunnels in old dead tree stumps in Tai Po Kau forest.

Section **Schizophora**

Tephritidae

The Tephritidae (= Trypetidae) are the fruit flies, a distinctive group of cosmopolitan distribution, usually with conspicuously marbled (mottled) wings and a flattened horny ovipositor in the female. At rest the wings are held laterally and often are waved as the fly walks. The larvae are phytophagous and well known pests of soft fruits, but some

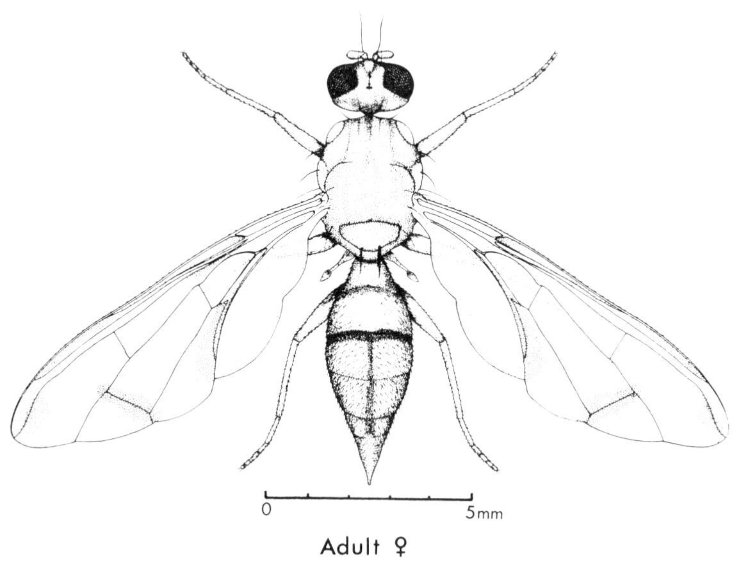

Plate 360.   Adult Melon Fly, *Dacus cucurbitae* (Diptera, Tephretidae).

species are confined only to the flower-heads of Compositae; some are leaf and stem-miners (e.g. the European Celery Fly), and a few species make galls on various parts of plants.

The colourful species of *Ceratitis* with their marbled wings and green eyes are rather scarce in Hong Kong, being seen only occasionally. The common local species is the larger, more drab, brown species, the Melon Fly *(Dacus cucurbitae)* (Plate 360), its hyaline wings have a brown leading edge, and it is found in several different cucurbit fruits. And as yet unidentified species of *Dacus* causes extensive infestations in ripening guava fruits locally and in parts of the New Territories, local guava crops may be totally infested, with each fruit containing a dozen or so large white maggots feeding on the white flesh. In fruit fly infestations the female oviposits through a puncture in the skin of the fruit while it is quite young and far from ripe; the maggots develop and increase in size as the fruit enlarges and ripens, and by the time the fruit is ripe the maggots are mature.

Agromyzidae

Agromyzidae are leaf miners and common locally on a wide range of plants. As previously mentioned (p. 223) their leaf mines would not be confused with caterpillar leaf mines because in the former the larval faecal pellets are deposited at the sides of the tunnel and so are scarcely visible whereas the caterpillars deposit their faecal pellets along the middle of the tunnel where they are very distinct. Some species of Agromyzidae make distinct tunnels (Plate 361), whereas others make one large cavity which is called a blotch mine. Some species start their larval existence in a small narrow tunnel mine which, as they develop suddenly and abruptly, is converted into a large blotch mine.

As a group they tend to be difficult to rear and so several species are locally common but as yet unidentified. As a group they make leaf mines, stem and twig galls, and root mines. One genus is endoparasitic inside the Cottony Cushion Scale *(Icerya purchasi)*. Several groups are to be found in the flower-heads of Compositae. The Bean Fly *(Ophiomyia phaseoli)* is found in the New Territories, and the Pea Leaf Miner *(Phytomyza horticola)* occurs in leaves of peas, chrysanthemum and various *Brassica* spp. (Plates 362 & 363). Other species of *Phytomyza* and *Liriomyza* occur in the leaves of local creepers, Bougainvillea (Plate 364), and other plants. An interesting and new species is common on *Ficus microcarpa* where it is distinctive in making elongate galls in the leaves (Plate 365). Adults have now been successfully reared in the

(Below) Plate 361. Leaves mined by larval Agromyzidae.

(Right) Plate 362. *Brassica* leaf mined by the widespread Pea Leaf Miner, *Phytomyza horticola* (Diptera, Agromyzidae).

(*Above*) Plate 363. *Brassica* leaf, ventral view, showing leaf miner pupae *in situ*.

(*Left*) Plate 364. *Bougainvillea* leaf mined by local Agromyzidae (Diptera).

laboratory, and they have been named *Ophiomyia fici sp. nov.* (Spencer & Hill, 1977).

Ephydridae

Other fly leaf miners are found in the family Ephydridae which

Plate 365. Leaf galls of *Ophio-myia fici sp. nov.* (Diptera, Agromyzidae), on leaves of *Ficus microcarpa*.

contains mainly flies that live in rotting seaweed on the beaches. *Hydrellia griseola* (Rice Leaf Miner) larvae mine leaves of rice in Southeast Asia and in wheat in Europe. Other species mine leaves of various pondweeds (*Potamogeton* spp.). *Notiphila* is found in paddyfields in the New Territories, and a species of *Scatella* mines the leaves of Chinese cabbage.

Drosophilidae

The Drosophilidae are small red-eyed vinegar flies (sometimes called small fruit flies) containing of course the ubiquitous *Drosophila* used for teaching genetics. These small flies essentially only attack fruits that are overripe and odoriferous. Being of small size they are able to have a generation quite quickly, as distinct from the fruit flies proper the Tephritidae.

The final major groups in the Acalyptrate Diptera are the Gasterophilidae (some bot flies) which do not appear to occur locally, the Chloropidae and Opomyzidae (grass flies).

Chloropidae and Opomyzidae

These two groups have phytophagous maggots which bore in the shoots of young grasses and cereals (Gramineae) producing typically a cereal 'dead-heart'. When the apical shoot is severed, it becomes dead and brown and can be pulled out of the plant easily; but sometimes a large swollen and distorted shoot results. These groups are of interest ecologically in that they are confined to Gramineae and quite cosmopolitan in distribution. A species of *Chlorops* has been found making large shoot galls in *Cynodon dactylis* (Plate 366) in the Pokfulam area. Also in

Plate 366.   Shoot gall of *Chlorops* sp. (Diptera, Chloropidae); *Cynodon dactylis* grass.

the Chloropidae are species of *Siphunculina* which are thought to transmit conjunctivitis and other eye diseases in the Orient. They have abdominal spines which appear to make slits in the conjunctiva and aid the entry of pathogens carried on their bodies.

Oestridae

The Calyptrate Diptera are probably the best known flies generally. The Oestridae (warble or bot flies) is a small family of stout, bristly flies looking like bees and seldom seen as adults. The larvae are endoparasitic in cattle, living either under the skin along the back (warbles, *Hypoderma* spp.) or in the nostrils of sheep (Sheep Nostril Fly, *Oestrus ovis*). It is not known whether warble flies are present in the New Territories but it is possible that they might be found in the local cattle. There is a Human Warble Fly *(Dermatobia hominis)* whose females seize mosquitoes (usually *Psorophora* spp.) and attach their eggs to the mosquito. When the mosquito alights on to a man to feed the body warmth causes rapid egg hatching and the larvae drop on to the skin and start boring. Fortunately this fly is confined to Central and South America.

Calliphoridae

The Calliphoridae includes a large number of species, many of which are associated with man. The larvae may be flesh feeders (*Sarcophaga*

Plate 367.   Adult Flesh Fly, *Sarcophaga* sp. (Diptera, Calliphoridae), resting on leaf of Morning Glory.

Plate 368.   Common local Blue-bottle, *Chrysomya megacephala* (Diptera, Calliphoridae); body length 8 mm.

spp.), saprophagous (*Calliphora* spp.), or in some cases parasitic on various arthropods. *Sarcophaga* is the Flesh Fly, a large fly with grey and black striped thorax and marbled abdomen (Plate 367), and occurs as many different species throughout the world. *Sarcophaga* is larviparous and each female deposits some 40–80 larvae in their first instar. Most species live in decaying flesh, but some parasitize caterpillars, grasshoppers, beetles, earthworms, snails, and other animals. *Calliphora* spp. are the ubiquitous blue-bottles or blow flies, with metallic blue bodies and general reduction of bristles (especially on the upper surface of the abdomen), again to be found in decaying flesh and carrion. The species of *Lucilia* are again cosmopolitan and include the green-bottles. In general *Lucilia* is metallic blue/green in colour with more body bristles than Calliphora. Other local species such as *Chrysomya megacephala* are quite bluish (Plate 368). These three genera are quite common in Hong Kong. A number of species have been identified and are in the collection of the Pest Infestation Control Section of Urban Services Department. The adults are frequently seen on flowers and are important pollinators. In this family are also found the Screw-worm Fly *(Callitroga macellaria)* of New World which is notorious as a cattle pest, and the African Mango or Tumbu Fly *(Cordylobia anthropophaga)* whose larvae bore in the skin of man causing cutaneous myiasis. Several species of Calliphoridae have maggots which can cause different types of mammalian and human myiasis, but they are not common locally and cases of myiasis (wound or intestinal) in Hong Kong are rare.

Tachinidae

Tachinidae look very like bristly grey or brownish blow flies and they are important parasites of caterpillars, and to a lesser extent of larvae of Hymenoptera, and of adults and larvae of Coleoptera, Orthoptera and Hemiptera. The eggs are usually laid by the female fly actually on the body of the host. In some species the eggs hatch almost immediately, and the maggots bore into the body of the host. Many Lepidopterous pupae, if kept in rearing cages, will yield tachinid flies rather than the adult moths or butterflies. In parts of the New Territories the level of parasitism of emperor moth (Saturniidae) pupae is typically very high. Agriculturally the Tachinidae are very important in their role as parasites of the many caterpillar pests. There are many local species; several local species have been reared and subsequently identified. The adults are commonly found on flowers.

Muscidae

The Muscidae contains the house flies, the African tsetse flies, and the cosmopolitan Stable Fly *(Stomoxys calcitrans)* which is very common on the cattle in the New Territories. The Stable Fly has mouth-parts specially modified for piercing skin and sucking up blood. These look like house flies (Musca spp.) and they follow the local cattle in small swarms as they wander about in the brush and grassy areas. It is also called the Biting House Fly and is found sometimes in domestic premises. The other domestic species include the ubiquitous house fly *(Musca domestica)* (Plate 369) and the smaller Lesser House Fly *(Fannia canicularis)* with its peculiar spiny larvae. Both species feed as

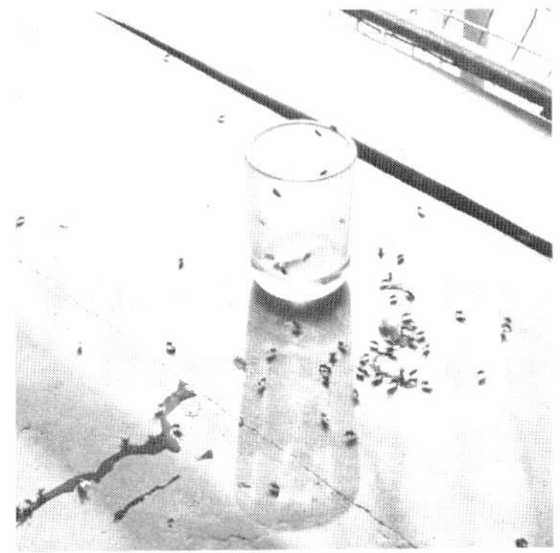

Plate 369. Common House Fly, *Musca domestica* (Diptera, Muscidae); length 6 mm.

Plate 369A. House Fly *(Musca domestica)* as a domestic nuisance in the New Territories.

adults on foodstuffs and refuse, while the maggots are to be found in rubbish, decaying foodstuffs, excrement, and most kinds of fermenting animal and plant material. In the warm climate of Hong Kong fly development is rapid and there are many generations during the year. House flies are important domestic pests for the adults transmit the casual organisms of typhoid fever and a number of other intestinal diseases by their contamination of uncovered foodstuffs (Plate 369A). *Atherigona* are shoot flies found in many parts of the world on different grasses, cereals, and other crops, and locally on rice seedings.

Anthomyiidae

The Anthomyiidae is sometimes included in the Muscidae and sometimes regarded as a separate family. It is distinguished by having the anal vein in the forewing not extending right to the wing margin. Some are called root flies or root maggots and they are important agricultural pests. The maggots live in soil and feed upon the roots or stems of a number of different vegetable and cereal crops. They are mostly of crop pest importance in temperate regions and to date only a number of synanthropic species have been collected locally. A species of what is probably *Limnophora* can be found inhabiting local pitcher plants *(Nepenthes mirabilis)* in the New Territories.

Section **Pupipara**

Hippoboscidae

The final group of flies are grouped together as the Pupipara for they are all ectoparasites of birds or mammals. One modification to their mode of life is that they have become larviparous. The eggs are retained singly in the 'uterus' of the fly and at birth a fully-formed larva is deposited and it pupates immediately, so the group can actually be said to be pupiparous rather than just larviparous as is the case in Tachinidae and some Calliphoridae. The Hippoboscidae are louse flies; they are ectoparasites, usually winged and capable of free flight, and occurring on wild birds and mammals. The species to be found locally are *Hippobosca equina* and *H. variegata,* occasionally on local cattle, but more frequently on imported Indonesian beef cattle. The Pigeon Louse Fly *(Pseudolynchia canariensis)* is common on the locally reared domestic pigeon; the shiny black spherical puparia can be found in the nests and in dark corners of the local dovecotes. On the local wild birds several species have been recorded, such as *Ornithoica trideus, Ornithophila metallica,* and *Icosta sensilis,* but this is not a group well-represented in Hong Kong.

Nycterbiidae

Nycterbiidae are wingless, very reduced ectoparasites on bats, more frequently to be found on fruit bats (Megachiroptera) than the insectivorous (Microchiroptera). The local species is *Leptocyclopodia ferrarrii* to be found on the Dog-Faced Fruit Bat.

Streblidae

The Streblidae are even further reduced ectoparasites in the fur of bats but to date none has been recorded on the local bats.

ORDER **SIPHONAPTERA**
(Fleas)

The fleas are small, compressed, wingless insects adapted for life as blood-sucking ectoparasites of birds and mammals. Generally most wild mammals and birds have their own species of fleas, which one only encounters by trapping their wild hosts. The common fleas in Hong Kong are all domestic species, more or less. The Human Flea *(Pulex irritans)* is extremely rare here, as it is now in many parts of the world, and in fact the only recent record here came from a sailor in 1951. It is the Human Flea that is used for performing in 'flea circuses' and for some years several European newspapers have carried advertisements from entertainers who wish to purchase specimens of Human Flea for 'training'. The once quite common 'flea circus' in European countries is now very much of a rarity. The common local flea is the Cat Flea *(Ctenocephalides felis)* which infests cats and dogs alike, and if deprived of its preferred hosts it will feed readily enough on man. This species is important so far as pets are concerned for it is the vector of the Small Dog Tapeworm *(Dipylidium caninum)*. The cats and dogs eat infective fleas which they catch during fur-cleaning activities and then the tapeworms develop in their intestines. The Tropical Rat Flea, or Plague Flea, is *Xenopsylla cheopis* (Plate 370) and is quite common on the local Ship Rat and other urban rats. The local populations of both rats and rat flea are carefully monitored by the Pest Infestation Control Section

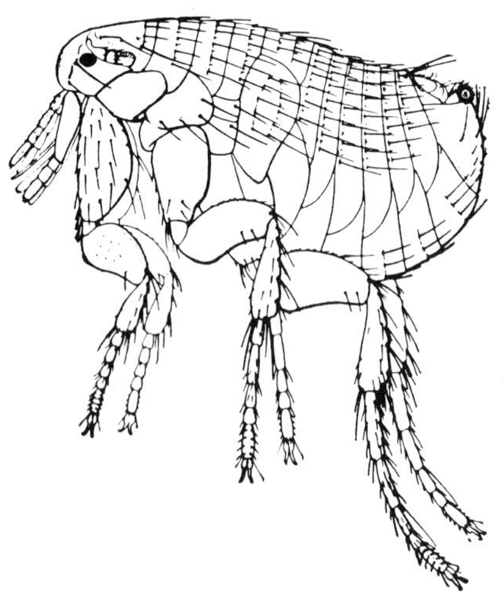

Plate 370.   Male Rat Flea, *Xeonopsylla cheopis* (Siphonptera); length 1 mm.

of the Urban Services Department, for this flea is the vector of the dreaded plague *(Pasturella pestis)*, and checks are made to ensure that the flea population does not build up to plague proportions. The last common urban flea is the Mouse Flea *(Leptopsylla signis)*.

All fleas have larvae that are non-parasitic and live in the nests and homes of their hosts feeding on dried blood and other organic debris. The pupae are also found in these crevices and typically most pupae do not hatch into adults unless stimulated by physical vibrations. This is clearly a survival mechanism for the species and helps to ensure that the young adults do not emerge into deserted nests or homes. In flats and houses, a favoured site for flea larvae and pupae is in the crevices between wooden parquet floor blocks where there is shelter and usually sufficient organic debris for food.

ORDER **HYMENOPTERA**
(Wasps, Bees, Ants, etc)

This is a large group of insects ranging in size from minute to large, and distributed well throughout the world. There are more than 50 families and over 100,000 species described. Their characteristics are the possession of two pairs of membranous wings with the hind pair smaller and joined to the fore pair by special hooks. The mouth-parts are mostly for biting and chewing but are modified in the bees for sucking nectar. The first segment of the abdomen (propodeum) is fused to the hind edge of the thorax and in many cases the second (and third) segment is constricted and narrow and forms the typical 'wasp-waist'.

Sub-order **Symphyta**

Tenthredinidae

The sub-order Symphyta contains the sawflies. These are phytophagous species where the larva is a caterpillar, but distinguished from Lepidoptera by having extra prolegs, five or more, in addition to the terminal claspers (Plate 371). They are often spotted rather than having the usual caterpillar striping, and some species feed socially, with a group of larvae around a leaf eating at the margin and each sitting with its abdomen raised in the air. This type of behaviour can be observed

Plate 371.   Gregarious sawfly larvae defoliating a plant; six pairs of abdominal prolegs are evident.

Plate 372. Adult Camphor Sawfly, *Hemichroa* sp. (Hymenoptera, Tenthredinidae); body length 7 mm.

Plate 373. Larva of Rose Sawfly, *Arge pagana* (Hymenoptera, Tenthredinidae); body length 18 mm.

in the species of *Hemichroa* (Plate 372) when they eat leaves of Camphor in the spring. A few species are pests of some importance; the large spectacular woodwasps (Siricidae) do not appear to occur here, but *Arge pagana* (Plate 373) is not uncommon locally on roses. In the New Territories *Neodiprion biremis* is of regular occurence on the rarer local pine tree *(Pinus elliotti)*. In this group there is no 'wasp-waist', and the abdomen is broadly joined to the thorax without any constriction. As a group the females have a saw-like ovipositor which is used to make holes in the plant tissue into which the eggs are placed. Generally more sawflies are found on trees and bushes than on herbaceous plants. Pupation takes place in the litter or soil in woven coccoons.

Sub-order **Apocrita**

The other sub-order is the Apocrita, in which the adults have the basal constriction of the abdomen making the characteristic 'wasp-waist'. Since the first segment of the abdomen (propodeum) is fused with the thorax, what appears to casual inspection as 'abdomen' is actually referred to as the gaster. The adults are highly specialized in their habits, and are often social, living in large communities. The female ovipositor is adapted for either piercing (Parasitica) or stinging (Aculeata) when it has well-developed poison glands. The larval habits are too diverse to generalize about, but they are all apodous and morphologically reduced.

Parasitica

The parasitic forms of Hymenoptera are lumped together in the Parasitica, with one common feature that the female ovipositor is often long and is used for piercing.

Super-family **Ichneumonoidea**

There are two very large groups, the first is the Ichneumonoidea which contains two families, Ichneumonidae and Braconidae which are separated mainly by the presence or absence of a small enclosed cell in the forewing distal to the pterostigma. Both groups are well-represented locally and of obvious biological importance in that they parasitize a large range of insects, many of which are pests, including caterpillars, beetles and bugs; one group is important as aphid parasites. In many cases the female has a long ovipositor, but in others it is shorter (Plate 374) or not protruding at all. Identification of these groups is very much a matter for the taxonomic specialist.

Super-family **Evanoidea**

Evaniidae

A small local group (Evaniidae) is represented by the small black

Plate 374.   Icheumon Wasp with distinctive short external ovipostor, *Xanthopimpla* sp. (Hymenoptera, Ichneumonidae); body length (without ovipositor) 13 mm.

Ensign Wasp (*Evania* spp.) frequently found in houses and buildings where it parasitizes the oöthecae of cockroaches. They are so named after the shape of the tiny stalked abdomen which is constantly flicked up and down as the wasp walks. The wasps are completely black in colour and have long flickering antennae.

Super-family **Cynipoidea**

Cynipidae

A well-developed Palaearctic group is the interesting Cynipoidea containing the classical gall wasps of the European entomological literature, which form spectacular galls on many trees and shrubs (i.e. oak 'apples', oak leaf galls, rose pincushion gall, etc.). They are also renowned for their biologically complex life-histories and alternation of generations. To date only one species has been found in Hong Kong; it is *Saphonecrus* sp. which inhabits globular galls in the male flowers of *Lithocarpus glabra* (Plate 375); whether it actually makes these galls is not clear, it is reported to be an inquiline usually.

Super-family **Chalcidoidea**

The other equally large group of parasitic wasps is the Chalcidoidea,

Plate 375. Galls on flowers of *Lithocarpa glabra* inhabited by Gall Wasp, *Saponecrus* sp. (Hymenoptera, Cynipidae).

but these are mostly small in size and quite inconspicuous to the untrained observer. As a group they have greatly reduced wing venation and elbowed antennae. There are 19 families commonly recognized, and some species are the smallest insects known. The majority are either parasites or hyperparasites of other insects, and are of great economic importance (even greater than the Ichneumonoidea) as a means of natural control of many species of agricultural pests. A few species are phytophagous and make galls on plants.

Agaonidae

The Agaonidae are the true fig-wasps, pollinators and symbionts which only live inside the fruit (figs or syconia) of plants of the genus *Ficus* (Moraceae). This group has been reported locally by Hill (1967a), and

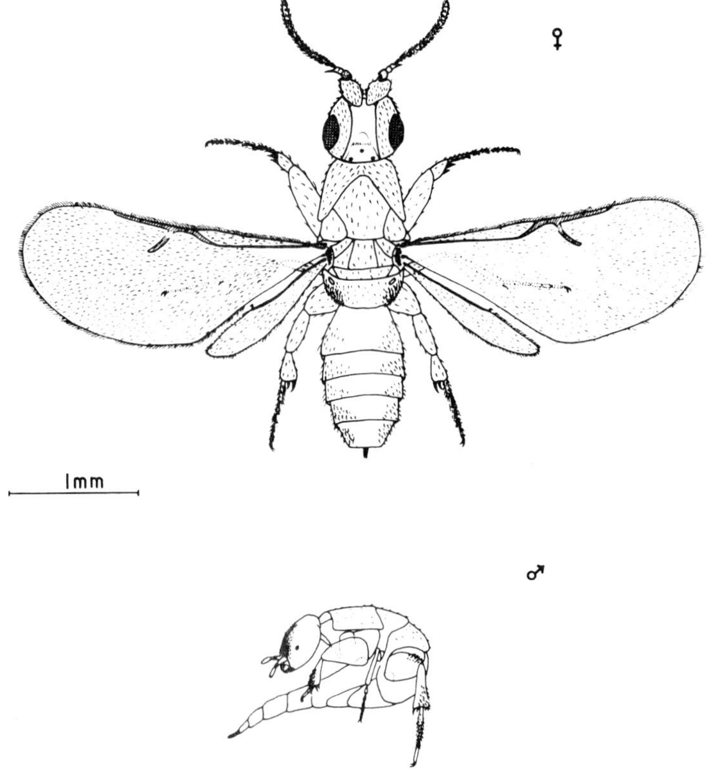

Plate 376.   Drawing of male and female *Blastophaga pumilae* Hill, 1967, the pollinator of *Ficus pumila* in Hong Kong (copyright Leiden Museum).

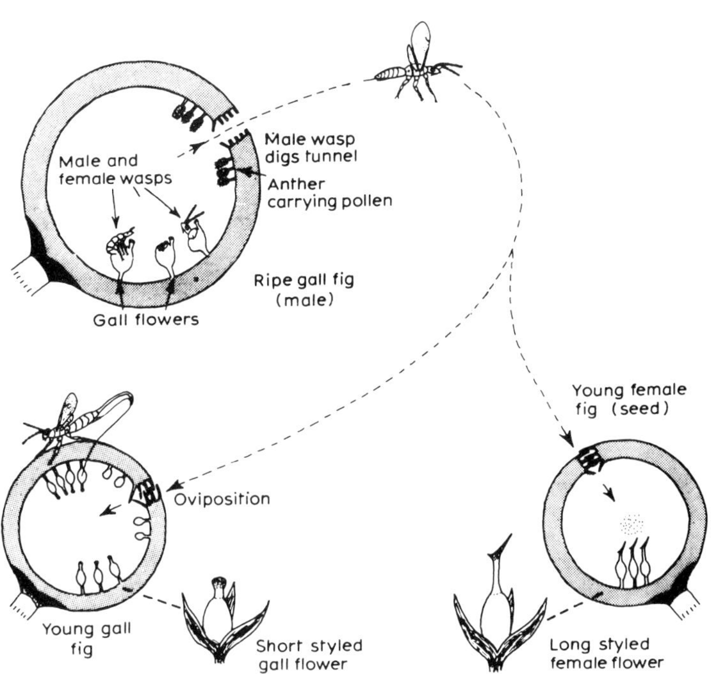

Plate 377. Diagrammatic representation of the life cycle of an Agaonid wasp in a dioecious species of fig.

will be further discussed later (p. 365). Basically each of the local indigenous species of *Ficus* has its own species of pollinating Agaonid wasp. There are a total of 15 recorded to date but two more are expected to be here. *Blastophaga* is the most common genus, but there are also three species of *Ceratosolen*. Sexual dimorphism is very pronounced (Plate 376). Fig-wasp life history is shown diagrammatically in Plate 377.

Torymidae

Torymidae is a family with considerable biological divergences. Two groups, Scyophaginae and Epichrysomallinae, are fig-wasps and thought to be either phytophagous and feeding on the developing fig ovules or else parasitic and feeding upon the developing Agaonid larvae inside the ovules (Plates 378 & 379). These groups do not penetrate to the interior of the young figs to oviposit; they sit on the outside of the young fig and using their long ovipositor they probe through the fig wall until they locate an ovule inside whereupon they lay their egg next to the fig ovule. In some cases it is thought that the female torymid does not actually

Plate 378.	Ripe fig of *Ficus superba* var. *japonica* showing opened ovary-galls from which the fig-wasps have emerged.

Plate 379.	Close-up of the scene in Plate 378 showing some wasps, the emergence hole in the ovary-galls, and three of the very large galls made by the Epichrysomallinae and the Ormyridae.

Plate 380.   Adult females of *Eukoebelea* (Hymenoptera, Torymidae) resting under *Bauhinnia* leaves adjacent to a tree of *Ficus variegata* var. *chlorocarpa*.

deposit an egg unless her chemical receptors on the ovipositor indicate that an Agaonid has already laid an egg there. Sometimes fig-wasps are found sitting on leaves near fig trees in enormous numbers, Plate 380 shows hundreds of *Eukoebelea* wasps sitting on *Bauhinnia* leaves next to a tree of *Ficus chlorocarpa* var. *variegata*.

Some other Torymidae are truly phytophagous; *Megastigmus* sp. makes galls inside the seeds in pine cones in Europe. Many others are entomophagous and are parasites of caterpillars, beetles, and other insects. An interesting local group of Torymidae is found in the genus *Podagrion* which has already been mentioned (p. 132) as parasites upon the eggs of the Mantidea. The female wasp ovipositor is long and is used to penetrate the thick foamy wall of the mantid oötheca so that the wasp eggs are laid alongside the mantid eggs. It appears that each of the local species of mantid has its own species of *Podagrion* parasitizing it, and also that each female wasp has an ovipositor of just the right length to penetrate the oötheca of its host species. The level of parasitism locally is high; mantid oöthecae collected in the wild and brought indoors very often yield *Podagrion* wasps rather than young mantids. The *Podagrion* wasps all have large swollen hind femora and curved tibiae which reflex against them as they walk.

Ormyridae

The Ormyridae are sometimes included within the Torymidae but they really are quite different. They often parasitize gall-making insects. Several species are found locally as fig-wasps, but another species of *Ormyrus* appears to make galls of its own in the leaves of *Ficus microcarpa* (Plate 381).

Plate 381. Leaf-galls made by *Ormyrus* species (Hymenoptera, Ormyridae) in leaves of *Ficus microcarpa*.

Chalcididae

Stout, curved hind femora and tibiae are characteristics of the Chalcididae. They are stout, black (and sometimes yellow) wasps with no obvious external ovipositor, and usually parasitize the pupae of Lepidoptera. Several large black species (10 mm long) can be reared from local emperor moth (Saturniidae) pupae, and some smaller ones from smaller pupae. But as is the case with most local Chalcidoidea, they have not been identified.

Eurytomidae

The Eurytomidae are biologically diverse. One group (*Sycophila* spp.) are fig-wasps and probably phytophagous, making very large ovary galls inside the local banyan figs. Other species are phytophagous gall-formers on a range of plants, and several are seed gall-makers in clovers and other legumes; some others are parasites on gall-forming Diptera and Hymenoptera. The large and widespread genus *Eudecatoma* is phytophagous and makes galls on a number of plants. Locally, some specimens of *Eudecatoma* have been reared from the stem galls on *Aporusa chinesis* (Plate 349) which yielded a number of Cecidomyiid gall midges. The nature of the relationship between the *Eudecatoma* and the Cecidomyiidae is unknown; the *Eudecatoma* may just be inquilines.

Pteromalidae

Pteromalidae is the largest chalcid family, and they mostly parasitize fly pupae, scale insects, aphids, and some wood-boring beetles. *Pteromalus puparum* is a widespread and common parasite of the pupae of various Lepidoptera, specially *Pieris* spp. One species is recorded from figs in Hong Kong, and several parasitize stored products pests, including *Anisopteromalus calandra* on *Sitophilus* weevils and *Choetospila elegans*.

Encyrtidae

The Encyrtidae live in the eggs, larvae, or pupae of various insects. The most frequent hosts are Homoptera and Lepidoptera, and most genera are specific to certain hosts. *Oöencytus* parasitizes the eggs of Heteroptera, and many different genera emerge from parasitized scale insects. There are many species locally, but few identified as yet.

Eulophidae

Eulophidae are a large group of small species; some parasitize eggs of beetles, and scarcely a group of insects is safe from these wasps, either as primary or secondary parasites (hyperparasites). Some are characteristically found on leaf-mining Lepidoptera and Agromyzidae, and many species infest galls of Diptera, Hymenoptera and Coleoptera. Two species are found in local banyan figs.

Aphelinidae

The Aphelinidae are tiny, delicate, wasps which parasitize armoured scales (Diaspididae) and aphids, and various species have been spectacularly successful as agents of biological control. A number of species are common locally, but have not yet been identified.

Trichogrammatidae

Trichogrammatidae are minute egg-parasites, the smallest being

probably the smallest insects known (0.2 mm). Lepidopterous eggs are
the most frequently used hosts and species of *Trichogramma* have been
important in many biological control programmes. *Prestwichia aquatica*
and *Hydrophylax* are ecologically of interest in that in Europe and North
America they parasitize underwater the submerged eggs of water bugs
(Notonecta, Ranatra) and beetles *(Dytiscus)*, and dragonflies, respectively. The minute chalcid swims under water using its wings as paddles.

Mymaridae

The last family to be mentioned specifically is the Mymaridae
(fairy-flies). Again they are all minute, and are egg parasites (often of
beetles). One species is aquatic. *Caraphractus* swims with its wings and
parasitizes eggs of Dytiscidae under water, but has not been recorded
locally.

Super-family Bethyloidea

The Bethyloidea are not of particular importance. They are parasitic
forms but look more like green bees. The bright green *Chrysis* with its
peculiarly hinged gaster is not uncommon in houses and flats in Hong
Kong, being found on the windows and window-sills. It is some 10–12
mm in body length. Occasional specimens of Ruby-tailed Wasps are
seen but none has been collected and identified as yet. They are parasitic
in the nests of solitary bees and various wasps.

Aculeata

The other Hymenoptera have the basic ovipositor modified into a
sting, and from the female accessory reproductive glands a large poison
gland has developed; they are collectively known as the Aculeata.

Super-family Scolioidea

These are generally regarded as being the most primitive members of
the group Aculeata. The females possess a sting, but they are mostly
parasitic, usually on beetle larvae or other Aculeata such as soil-nesting
bees and solitary wasps.

Scoliidae

A larger family worldwide and it contains many of the largest individuals of Hymenoptera. Locally it is represented by the species *Scolia
azurea,* a large black hairy wasp to be seen in parts of the New Territories
in the spring; the male is all black except for the end of the abdomen

which is red; the female is similarly coloured but she has a reddish smooth head capsule. The female is slightly larger, and measures up to 35 mm in body length. Females can be seen digging in rubbish heaps where they are presumably seeking scarab larvae on which to lay their eggs.

The species *Megacampsomeris* species has been caught flying free but nothing is known of its occurrence or biology locally.

Mutillidae

The common name for these insects is Velvet 'Ant'—the females are apterous, red or blackish in colour and with a general velvety appearance, and with an overall appearance not unlike an ant. The female insect will sting very readily if handled. The group parasitizes other Hymenoptera mostly. The two local species collected have recently been identified as *Trogaspidea oculata* and *Odontomutilla urania*.

Formicidae

The first major family is the Formicidae, which includes all the ants. These are polymorphic social insects, most forms being wingless, but they all are characterized by having one or two distinct nodes (swellings) on the petiole joining the thorax to gaster (abdomen). Ants are renowned for their ability to co-operate in carrying large items of food back to their nest (Plate 382).

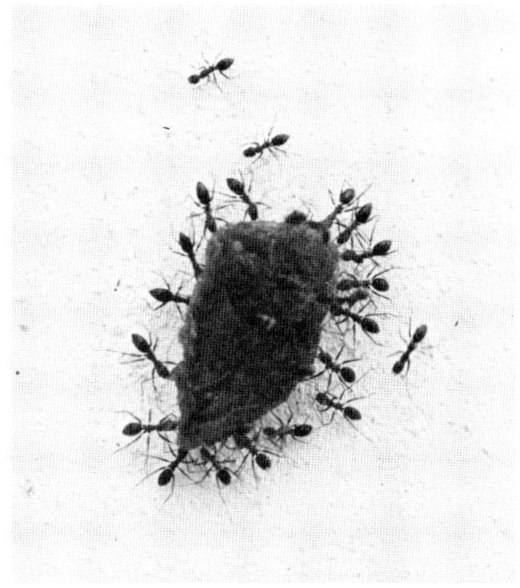

Plate 382. Example of co-operation when a group of ants carry a large piece of material collectively back to the nest.

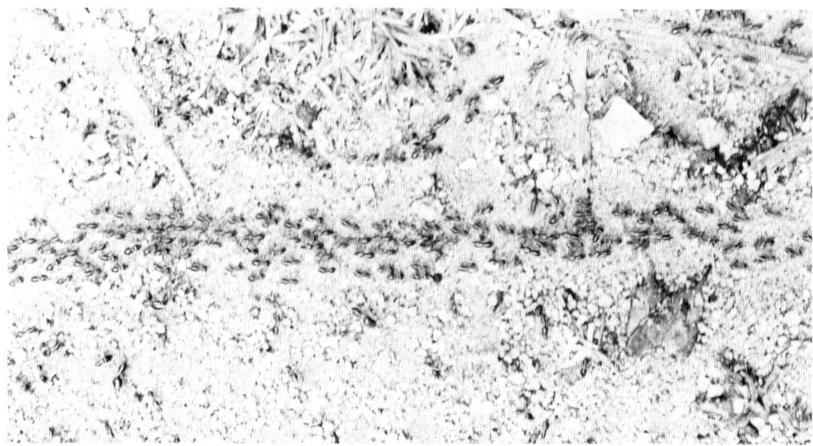

Plate 383.   Driver Ants (Formicidae, Dorylinae) on the march.

Plate 384.   Male local Driver Ant, *Dorylus* sp. (Hymenoptera, Formicidae); called in the vernacular 'Sausage-fly'; length 22 mm.

The family contains several distinct sub-families sorted somewhat according to their supposed phylogenetic history. The Dorylinae are regarded as primitive. They are predacious and nomadic, with small blind workers, and a blind wingless queen, and they include the legendary Driver Ants of Africa and S.E. Asia. On the move they are impressive because of the sheer numbers involved and the relentlessness with which they march (Plate 383), and the ferocity with which the large soldiers attack any living animal in their path. At times they stop and make a temporary bivouac, but never a proper nest. These hunting and foraging species are thought to be the most primitive ants. A very common nocturnal visitor to houses in the early part of the summer is the large blundering 'sausage-fly', which is not a fly at all but is the male *Dorylus*

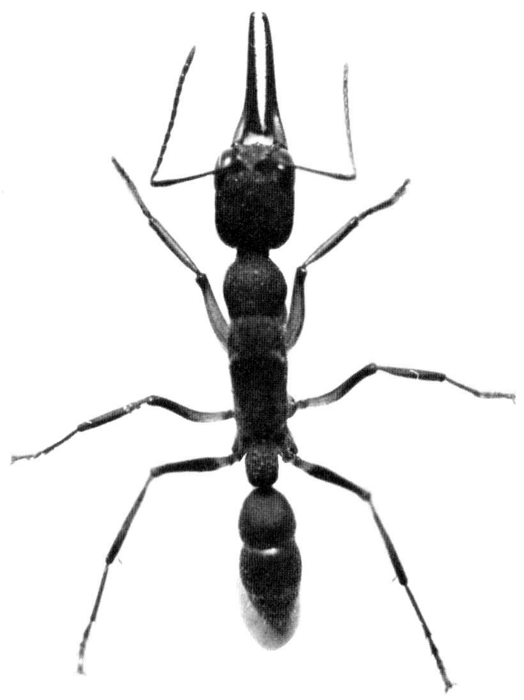

Plate 385. The Jumping Ant, *Harpegnathos venator* (Hymenoptera, Formicidae); body length 14 mm.

orientalis with its large sausage-shaped abdomen (Plate 384). But the marching hordes of the Oriental Driver Ant have not yet been reliably witnessed locally; the illustration in Plate 383 comes from Malaya.

The Ponerinae are probably the most primitive ants. They are also predacious and they make small subterranean nests with only a few dozen individuals. Polymorphism is slight between the three basic forms (workers, soldiers, and queen). This group is represented locally by the genus *Diacemma* (Winney, *in litt.*), and the uncommon *Harpegnathus venator* (The Jumping Ant) (Plate 385) which is easily recognized by its long mandibles, black body colour, and the curious manner in which it jumps for distances of 10 cm or more, using the mandibles somehow for leverage.

The Cerapachynae is a small group of little-known species which prey on other ants, and is represented locally by the genus *Cerapachys*.

In the Pseudomyrmecinae, which is a small tropical group, are the ants whose larvae are highly modified and have a peculiar feeding arrangement with the workers which involves trophallaxis and reciprocal feeding. The genus *Tetraponera* can be found in Hong Kong, usually nesting in the internodes of dead bamboo stems.

The Myrmecinae is the largest group and by far the most common; its members can be easily recognized by having two nodes on the petiole. Most species are phytophagous and complex nest-colonies are constructed. A number of serious crop pests occur in this group, such as the Harvester Ant *(Pheidole)*, and the widespread Fire Ant *(Solenopsis geminata)* (Plate 386). The genus *Crematogaster* is represented locally by

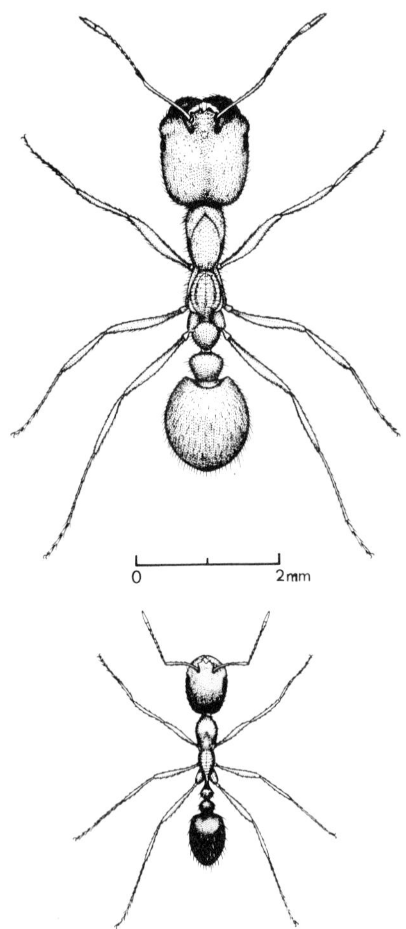

Plate 386.   Workers of Fire Ant, *Solenopsis geminata* (Hymenoptera, Formicidae).

Plate 387. Sub-nest of *Crematogaster* ants enclosing colonies of *saisselia* scale insects.

Plate 388. Winged adult female of *Pheidologiton* ant (Hymenoptera; Formicidae); body length 16 mm.

species of widespread distribution. They are mostly arboreal and build small nests in tree foliage or hollow out dead twigs. They are frequently found in association with aphids and scale insects. As already mentioned on p. 201 an extreme case was found in 1977 in the New Territories where on a tree of *Ervatamia divaricata* (Oleaceae), workers of *Crematogaster* had built small sub-nests in the forks of small branches and enclosed within each was a small colony of *Saissetia* scales (Plate 387). Apparently this mutualistic phenomenon is well known and in fact there is a species of *Saissetia* known as *S. formicarius* which is only found associated with ants.

Pharoah's Ant is *Monomorium pharaonis* and this tiny ant is reknowned for being a ubiquitous pest of domestic premises. The nest is typically hidden in an inaccessible location down in the base of the building, and the tiny workers and soldiers scavenge throughout the whole building for spilled food, crumbs, and general organic debris. It is really a tropical species and in temperate countries it tends to be restricted to places such as hospitals with effecient central heating and large kitchens and storerooms.

A very common genus in Hong Kong is *Pheidologiton* which occurs as many different species. On the evenings when the 'sausage-flies' come to house lights they are often accompanied by slightly smaller and more rounded flying ants; these are the males and females of *Pheidologiton* (Plate 388). Some Myrmecinae are called 'fungus ants' because they collect vegetable matter and cultivate fungus. This is done in the same manner as the mound-building termites (family Termitidae). The fungus grows on a honeycomb-like structure constructed of chopped plant material mixed with ant saliva and built in special subterranean chambers. It produces bromatia upon which the ants feed. In the tribe Attini are found the Central and South American genus *Atta* and its relatives, known collectively as the leaf-cutting ants.

The Formicinae is the second largest sub-family and contains a rather heterogenous assemblage of species, but in this group the collection of honey-dew has become very highly developed. The workers in some groups have a thin flexible skin which can allow tremendous extension when the crop is filled with honey-dew and nectar. Some of the foraging workers are able to engorge themselves while out in the field and then to regurgitate the honey-dew on their return to the nest. The real 'honey-ants' or 'repletes' as they are technically termed, are confined to the nest and almost immobile, being living storage containers (or honey pots) for the honey-dew brought in by the foraging workers. Chocolate-coated honey-ants are a major export item from Japan and other parts of the world.

In this group we have the abundant species of *Polyrachis dives* with its spiny thorax and petiole. They are generally referred to as the Black Tree Ant. They nest in the foliage of trees and bushes, but most nests are not larger than about 25 cm in diameter, and many are smaller.

Plate 389.   Aerial nest of Black Tree Ant in *Celtis* tree, seen in the winter when the leaves of the trees are shed.

Plate 390.   Nest of Black Tree Ant, *Polyrhachis dives* (Hymenoptera, Formicidae).

Plate 391.   Nest of Red Tree Ant, *Oecophylla smaragdina* (Hymenoptera, Formicidae).

Sometimes a single tree may bear a dozen or more small nests (Plate 389). These are best seen in the winter when the deciduous trees have shed their leaves. The nest is made of chewed pieces of plant material mixed with ant saliva (Plate 390). Occasionally nests are found in large tussocks of grass if there are no suitable woody plants available.

The genus *Camponotus* is well-represented here, and is very variable in its habits at the species level. It usually prefers to nest underground.

Oecophylla smaragdina, the large Red Tree Ant, is a bright yellowish red ant about 9 mm long with long legs and only one petiole node. It is of arboreal habits and nests in trees between a couple of sewn together leaves (Plate 391). It is quite common and is a nuisance in that it will bite with great ferocity if disturbed, and it also injects formic acid into the bite wound. It typically cultivates and guards arboreal colonies of aphids and scale insects. This ant is unusual in that the larvae are used when the leaves are sewn together to construct the nest. The two edges of the leaves are held together by a row of worker ants, then other workers appear holding a small larva each. Only the larvae secrete the silk used for sewing and as the workers pass the larvae back and forth, silk is extruded and hardens instantly, binding the two edges of the leaves together. The largest nests may be 20–30 cm in diameter and may consist of a number of leaves sewn together.

The final sub-family is small but in some respects is thought to be the most highly evolved ant group. The Dolichoderinae contains the local *Dolichoderus* with its peculiar elongate body. Also in this group are the two common species of Sugar Ants *(Iridomyrmex* and *Tapinoma)*. They are widespread throughout the warmer parts of the world and can be quite serious pests of domestic premises and food stores where their small size but large numbers makes them difficult to contend with.

It is estimated by Winney *(in litt.)* that the total number of ant species in Hong Kong is well in excess of 100.

Pompilidae

The Pompilidae are spider-hunting wasps, very agile and quick-moving when encountered in the field (Plate 392). They have particularly long hind legs, and some species can be quite large. The largest genus is the South American *Pepsis* which is some 70 mm long and preys on the giant Tarantula spiders. Mostly they are fossorial and make nests underground, but some make mud nests. In all cases they provision the nest with paralyzed spiders on which the developing wasp larvae will feed and they can often be seen dragging spiders back to the vicinity of their nests (Plate 393).

Vespidae

The true wasps are all placed in the one family Vespidae which is a

Plate 392.   Large Spider-hunting Wasp (Hymenoptera, Pompilidae), *Leptodialepis bipartitus*; body length 27 mm.

Plate 393.   Small Spider-hunting Wasp *(Tachypompilus analis)* dragging off a paralysed spider to its nest.

large group subdivided into about eleven subfamilies, six of which are solitary and the rest social. In some books the solitary potter wasps are regarded as a separate family (Eumenidae), but here they are regarded as a subfamily, the Eumeninae. There appears to be a number of different potter wasps locally, ranging in size from the large yellow and brown *Eumenes pyriformis* (Plate 394) and the somewhat longer yellow and black *Eumenes* sp. (Plate 395) down to a small black species some 12 mm long which builds two or three (or four) nests adjacently and also a

Plate 394. Common Potter Wasp, *Eumenes pyriformis* (Hymenoptera Vespidae); wingspan 36 mm.

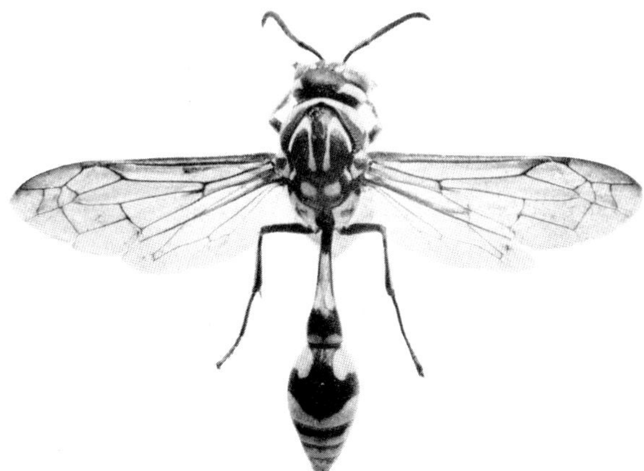

Plate 395. Local Potter Wasp, *Eumenes* sp. (Hymenoptera, Vespidae); wingspan 38 mm.

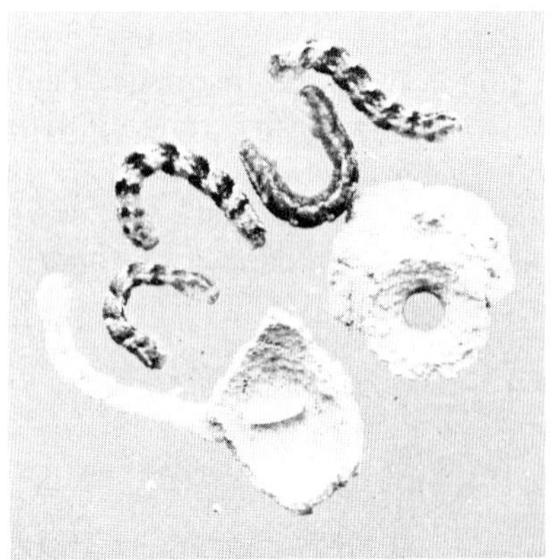

Plate 396.   Nest of Potter Wasp, *Eumenes* nest—5 paralysed looper caterpillars together with one Potter Wasp egg.

Plate 397.   Contents of Potter Wasp nest—5 paralysed looper caterpillars together with one Potter Wasp egg.

species of *Eumenes*. Different wasp species make slightly different nests although all retain the basic urn-shape. The nest is sometimes oval in shape, but the most beautiful nest is the symmetrical globular, vase-shaped nest with the wide mouth and narrow neck (Plate 396), which is reputed to have served as a model for early Indian pottery. One nest was built under observation (Plate 396) and took only 50 minutes for completion. These nests are attached to almost anything, but often on walls, twigs, wire netting, and even on curtains, and some species prefer to use a key-hole or a peg-hole in wooden bookcases. *Eumenes pyriformis* provisions its nest with up to ten or more small paralyzed looper caterpillars (Plate 397) and it suspends the single egg from the roof of the cell by a thread; the nest is then sealed and left.

The social wasps make large aerial papery nests, often in tree foliage as well as in hollow trees and in holes or corners in buildings. Typically the nest hangs free from an apical stalk. These wasps are large, stout-bodied, coloured yellow and brown or black, and they readily sting if disturbed. The commonest local wasp is probably *Vespa bicolor* (Plate 398) with its distinctive triangular black mark on the dorsum. As with the other species, it is carnivorous and predatory and feeds both itself and its larvae on other insects, although it does have a taste for nectar, honey-dew and soft fruit juices. The nest is spherical, about 20–30 cm in diameter (Plate 399) and is either in a tree or a building. Many of the larger nests contain several thousand individuals at one time. Basically

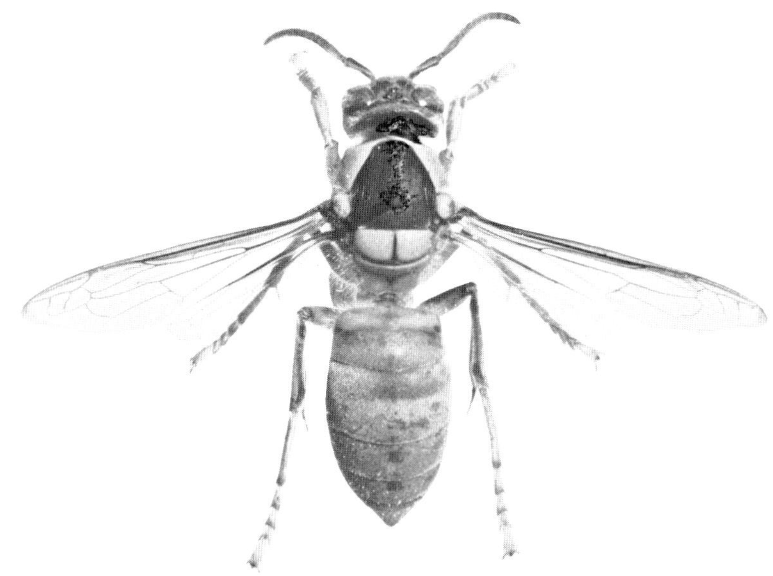

Plate 398. Common Wasp, *Vespa bicolor* (Hymenoptera, Vespidae); body length 20 mm.

Plate 399. Nest of Common Wasp, *Vespa bicolor,* in a tree, nest diameter about 350 mm.

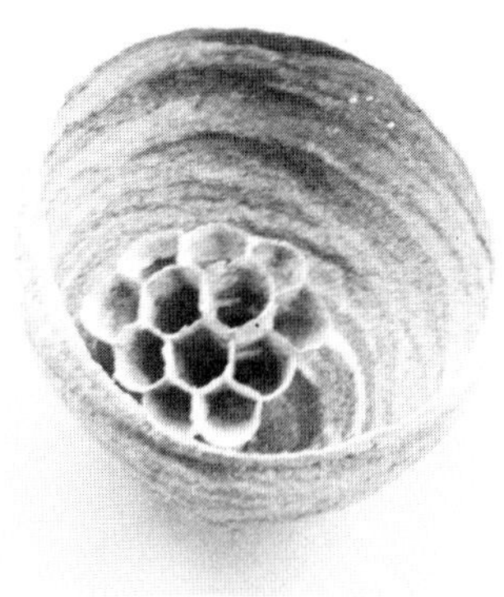

Plate 400. Initial nest made by a queen of *Vespa bicolar* in the spring after emerging from hibernation.

each nest consists of one large queen who spends most of her time just laying eggs, female workers who are developed from fertilised eggs, and male wasps who come from unfertilized eggs. Periodically, fertilised eggs develop into young queens instead of workers; these queens at maturity will swarm with a few males and several hundred workers and

Plate 401.   Common Paper Wasp, *Polistes olivaceous* (Hymenoptera, Vespidae); body length 20 mm.

Plate 402.   Large Brown Wasp, *Vespa affinis* (Hymenoptera, Vespidae); body length 30 mm.

Plate 403. Nest of Large Brown Wasp, *Vespa affinis* in Horsetail tree.

depart to start a new nest (Plate 400) and colony elsewhere.

Polymorphism is less developed in the wasps than in the ants; there is no soldier caste, and all adults are winged and capable of flight. Several subsocial species are common locally, such as *Polistes olivaceous* (Plate 401) which builds a small hanging open nest with a few dozen cells, and *Vespa affinis* (Plate 402) which are large in size and brown in body colour with an orange base to the abdomen, and typically nest in tall trees in large and elongate nest (Plate 403) measuring 20 by 50 cm. The small aggressive *Polistes* species are tropical wasps which make small open nests containing a few dozen individuals, and sometimes stuck between leaves in trees or else hanging from a slim peticle in buildings. These wasps may attack and sting with very little provocation. A number of medium to large-sized brown and yellow wasps have been identified over a period of years confusing as different species of *Vespa* and *Polistes,* but the precise limits of these two genera are not known. Social wasp colonies are annual only, even in the tropics; locally the nests are empty and the young queens are hibernating in buildings and vegetation by December, and by the end of winter the old nests will have dis-intergrated.

Sphecidae

Sphecidae includes the solitary wasps and mud-daubers. One of the most interesting recorded is the European sand-dune species *Ammophila* which provisions its nest with caterpillars. Locally there is the large, long-legged, brown *Sphex fulvohirtus* (Plate 404) which causes a nuisance with its burrows in the sandy soils of golf courses, playing fields and

Plate 404. Adult female *Sphex fulvohirtus* (Hymenoptera, Sphecidae); length 20 mm.

Plate 405. Burrowing Wasps, *Sphex fulvohirtus* (Hymenoptera, Sphecidae); outside their burrows in the sandy soil of the H.K.U. Sports Centre playing field.

other areas of short turf (Plate 405). The wasp makes a deep burrow about 10 mm in diameter, the entrance of which is marked by a small pile of sand, and the terminal nest chamber is provisioned with several paralyzed long-horned grasshoppers. The process of stinging the grasshopper to paralyze it and leave it actually alive so as to be available as

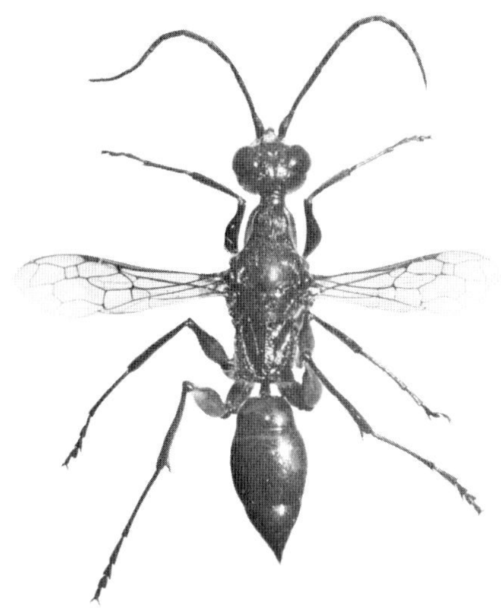

(Left) Plate 406. *Ampulex compressa* (Hymenoptera, Sphecidae), common predator of cockroaches in buildings; body length 20 mm.

(Below) Plate 407. Adult female Mud Dauber Wasp, *Sceliphron* sp. (Hymenoptera, Sphecidae); length 25 mm.

fresh food for the wasp larva has been observed in several European species and is a most complex affair. The short-winged irridescent blue wasp *Ampulex compressa* (Plate 406) to be seen in houses occasionally is predacious on cockroaches. In Malaya it is reported to pounce on a

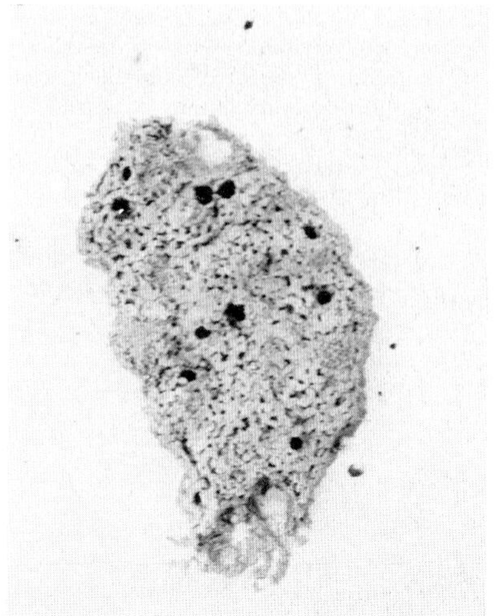 

Plate 408.   Nest of Mud Dauber Wasp,   Plate 409.   Nest of Mud Dauber Wasp
Sceliphron sp. (Hymenoptera, Sphecidae);   on a wall.
length 20 mm.

cockroach, sting it into submission, and then drag the body into a dark crevice where an egg is laid upon it.

The nest of the Mud-dauber Wasp (*Sceliphron* sp.) (Plate 407) is arranged in a series of separate cells plastered over with a mass of mud, often situated on rocks, walls or tree trunks (Plate 408). The prey taken locally are usually small spiders. Some species build a more regular shaped mud nest (Plate 409).

The prey generally taken by Sphecid wasps throughout the world includes caterpillars, spiders, Hemiptera (including cicadas), grasshoppers and cockroaches, and in some cases thrips, beetles, flies, mayflies and even Collembola.

Apidae

The social and solitary bees are placed in the Apoidea, and despite the abundance of social bees, the great majority are in fact solitary species. This group has been rather neglected in Hong Kong and little is at present known about local solitary bees. Within the social bees, the Apidae, it has already been mentioned that the Holarctic Bumble Bees (Bombinae) are absent. One attractive local bee is the stout-bodied *Anthophora* species (Anthophorinae) (Plate 410) with its distinctive blue-green horizontal banding on the abdomen. This is often seen feeding from flowers and it is presumed to nest in the ground.

Plate 410. Adult Blue-banded Bee, *Anthrophora* sp. (Hymenoptera, Apidae); body length 14 mm.

The ubiquitous Honey Bee *(Apis mellifera)* or Hive Bee, is reared commercially in different parts of Hong Kong Island and in some quantity in the Sheung Shui region of the New Territories. Wild nests are seldom seen.

The Xylocopinae includes bees which nest in wood or plant stems, and there are at least two or three local species. These large hirsute bees look like dark bumble bees but the abdomen is bare dorsally. The large, black *Xylocopa irridipennis* is the Bamboo Carpenter Bee (Plate 411) which makes holes in the internodes of certain larger species of bamboo, but probably only in dead stems, and it makes a nest in the hollow stem at the node level (Plate 412). The male bee is generally greyer on the

Plate 411. Bamboo Carpenter Bee, *Xylocopa iridipennis* (Hymenoptera, Apidae); length 26 mm.

(Left) Plate 412. Entrance hole to the nest of Bamboo Carpenter Bee in the internode of a dead Bamboo stem.

(Below) Plate 413. Carpenter Bee, *Xylocopa collaris* (Hymenoptera Apidae); length 20 mm.

Plate 414. Entrance hole to the nest of *Xylocopa collaris* in a dead tree stump.

thorax whereas the female is all black. A second species, *Xylocopa collaris* (Plate 413), has been found living in tunnels in large old dead stumps of an unidentified tree in the Tai Po Kau Forest Reserve (Plate 414). In Malaysia and Africa the common *Xylocopa* bees typically bore tunnels in wooden house beams, particularly in those used to construct open garages and in those that over-lap the building in the eaves. Here the tunnels are considerable and measure some 12–15 mm in diameter and 10–20 cm in length. All the *Xylocopa* bees possess a large pair of biting mandibles in addition to the usual sucking proboscis (tongue) which is used for collecting nectar as food. Also in the Apidae is a number of parasitic species, many of which are parasitic on other bees, but nothing is known about these species locally.

Fig-wasps

Since the local fig-wasps are of some basic biological interest these will be considered in more detail. The relationship between the Agaonid wasps and the *Ficus* plants is truly unique in being the only recorded case of complete mutualism (mutual physiological interdependence) between an insect and a plant (see Hill, 1967). Many insects are completely dependent upon certain host plants but this is the only case where this complete dependence is reciprocated by the plant being dependent upon one species of insect. The insects can only develop inside the ovules of the ripening figs, and they in turn are the sole means of pollination for the *Ficus* plants. The life cycle of a fig-wasp is shown diagrammatically in Plate 377. The genus *Ficus* is of interest in that it is clearly a group of great antiquity which has evolved certain unusual characteristics. The fruit (fig or syconium) is an enlarged receptacle which has grown up and around the small flowers until it almost completely encloses them, except for the apical ostiole which is seen externally as the 'eye' of the fig and is effectively sealed by a series of interlocking bracts. The most primitive species of figs are the huge tropical banyan trees with small, thin-skinned figs which bear inside them scattered male flowers with the anthers and a series of short-styled gall (wasp) flowers and long styled female (seed) flowers. The only difference between gall and female flowers being the length of the stigma.

The genus *Ficus* has been studied extensively in Asia and the Pacific region by Professor E. J. H. Corner. In his taxonomic monograph (Corner, 1968), he divided the genus into four separate sub-genera, and thence into taxonomic sections, sub-sections, series and sub-series before arriving at the species level. He describes a total of 500 species, sub-species and biologically distinct varieties for the region comprising tropical Asia, Australasia, and the Pacific Islands. It is expected that including Africa and Central and South America there will be a world total of about 1,600 distinct forms (species, sub-species and varieties).

TABLE 2: **SYSTEMATIC LIST OF HONG KONG *FICUS* SPECIES**

SUBGENUS	SECTION	SUBSECTION	SERIES	SUBSERIES	SPECIES
Urostigma	Urostigma		Superbae Caulobotryae		1. *F. superba* v. *japonica* 2. *F. virens* v. *sublanceolata*
	Leucogyne				3. F. *rumphii*
	Conosycea	Conosycea	Drupaceae	Indicae	4. F. *benghalensis* 5. F. *altissima*
		Benjamina	Callophylleae		6. *F. *microcarpa*
	Stilpnophyllum				7. F. *elastica*
Pharmacosycea	Oreosycea		Vasculosae Nervosae	Vasculosae	8. *F. *vasculosa* 9. *F. *nervosa*
Ficus	Ficus	Ficus			10. F. *carica*
			Podosyceae		11. *F. *pyriformis* 12. *F. *variolosa* 13. *F. *erecta* v. *beecheyana* 14. F. *pandurata*
		Eriosycea	Eriosyceae	Eriosyceae Trichosyceae Cuneifoliae	15. F. *esquiroliana* 16. *F. *hirta* 17. F. *langkokensis*
	Rhizocladus		Plagiostigmaticae	Plagiostigmaticae	18. *F. *sarmentosa* v. *impressa* 19. *F. *pumila*
			Racementaceae Distichae	Racementaceae	20. F. *sagittata* 21. F. *hederacea*
	Sycidium	Sycidium Palaeomorphe	Scabrae Pallidae Subulatae		22. F. *aspera* 23. *F. *tinctoria* spp. *gibbosa* 24. F. *subulata*
	Neomorphe		Variegatae	Variegatae	25. *F. *variegata* v. *chlorocarpa*
	Sycocarpus		Tuberculifasciculatae	Hispidae Tuberculifasciculatae	26. *F. *hispida* 27. *F. *fistulosa*

* Species with fig-wasps.

The Hong Kong species of *Ficus* in Table 2 are a fair representation of the group. Out of the 27 species recorded by Hill (1967a), probably 16 are truly indigenous and complete with their specific fig-wasps. *Ficus* sub-genus *Urostigma* are the banyans; they are huge monoecious trees with aerial roots, often epiphytic or strangling when young, and are best exemplified by the Chinese Banyan (*Ficus microcarpa* Linn f.—not *retusa* Linn.) (Plates 415 & 416). This is an ancient group, being widely dispersed throughout tropical Africa, Asia, Central and South America. The sub-genus *Pharmacosycea* contains monoecious trees, small or large, and reach the peak of their evolution in Melanesia, though some species occur in Asia and tropical America. The sub-genus *Sycomorus* has only 12 monoecious species confined to Africa except for one species occurring in India. The sub-genus *Ficus* represents the evolutionary peak within the genus and it contains a heterogeneous assemblage of dioecious species ranging in habit from large trees of 30 metres in height to shrubs and creepers (Plate 417–422). *Ficus (Ficus)* is best developed from an evolutionary point of view on the Southeast Asia mainland. So botanically *Ficus* is quite well represented in Hong Kong and in several respects this is an ideal localty for a detailed study of wasps and the *Ficus*/wasp relationship.

As indicated in Plate 377 the female Agaonid leaves the ripe fig, bearing pollen either on her body or else in special hollows on the leg bases or in special pockets on her sternum. She then flies off to search for young figs of the same species, presumably relying largely on her sense of smell for distinguishing that species. On finding a suitable

Plate 415. Tree of Chinese Banyan, *Ficus microcarpa* (Moraceae); about 20 m high.

Plate 416. Closs-up of leaves and figs of *Ficus microcarpa*.

Plate 417.   Edible Fig, *Ficus carica*.

Plate 418.   Hairy Bush Fig, *Ficus hirta*.

young fig she penetrates the tightly inter-locking apical bracts in order to gain access to the interior of the fig.

The female wasp shows various anatomical modifications to facilitate her passage between the apical bracts. She has hooked mandibles and a spine-like process on the fourth antennal segment, both being used for lifting the outer bracts. A deep hollow in the front of the head capsule accomodates the reflexed antennal bases so as to make the already flattened and rather elongate head more streamlined. There is a ridged, file-like process attached to the mandibles under the head for

Plate 419.   Large Creeping Fig, *Ficus pumila*.

Plate 420.   Hillside Bush Fig, *Ficus variolosa*.

working between the bracts, operating rather like the barb on a fish-hook. The middle legs are reduced and very slender, and can be folded up against the sides of the thorax. The forelegs are modified to have very stout, short, heavily spined tibiae for pulling the bracts apart, and stout hind-legs with large spines on the tibiae for pushing once the wasp is inside the ostiole.

Once the female Agaonid wasp enters a young fig she will never leave. She will lay her eggs in the short-styled gall flowers and/or pollinate the long-styled female flowers and will then die. The pollen is carried into

Plate 421.   Cauliflorous figs of the Common Red Stem-fig, *Ficus variegata* var. *chlorocarpa*.

Plate 422.   Hairy-leaved Stem-fig, *Ficus hispida*.

the young figs in special pollen baskets called corbiculae, or in completely enclosed pockets on the sternum (chest) called sternal corbiculae which are opened by carefully applied pressure through a narrow slit-like aperture. During the process of forcing her way through the apical bracts the female wasp typically loses her wings and the ends of the antennae which break off during the struggle. It is quite common to find that a number of wasps fail in their efforts and they die stuck between the bracts, and for every female that reaches the interior of the fig, often there may be three or four dead in the bracts.

If the female wasp belongs to a dioecious species of *Ficus* and has entered a young female fig, then her function is just to pollinate the long-styled female flowers and she presumably does not oviposit. In a young gall fig she will sit on the flowers and oviposit quite easily down the wide stigma and short style and deposit a single egg at the side of the fig ovule. When laying an egg it is thought she injects a quantity of poison in the vicinity of the fig ovule. The poison apparently stimulates the endosperm of the single ovule to proliferate just as though fertilization had occurred, and the larva of the fig-wasp feeds upon the proliferating endosperm.

After the female wasp has laid her couple of hundred eggs, she dies and the hard parts of her skeleton can be found there even weeks later. The larvae develop until after pupation young adults are formed. At this stage the wingless, reduced male bites through the ovary wall and emerges into the fig cavity. He then wanders around looking for females in their galls. Having found a female he bites a hole through the wall by her abdomen, then he inserts the end of his abdomen and copulates with the relatively immobile female. The males go from gall to gall mating with the females, who then enlarge the hole made by the male and emerge into the fig cavity. Most of the males emerge synchronously, so after a few hours the interior of the fig is filled with fig-wasps.

In the dioecious fig species the anthers are usually grouped around the apex of the fruit where the wasps congregate; the females collect the ripe pollen from the newly dehisced anthers and store it in their pollen baskets while the males eat through the apical ostiole bracts until a clear exit is produced. In some species they bore straight through the wall of the fig instead. When the exit hole is completed the males retire inside the fig because they are repelled by bright light, while the females are attracted to light and they emerge on to the surface of the fig. After a few minutes of wing-cleaning and toiletry the females fly off in search of young figs of the same species to oviposit or to pollinate.

As can be seen in Plate 377 it is typical that in fig-wasps there is striking sexual dimorphism. In fact it is so extreme that at first sight there is no apparent relationship between the sexes of the same species. This caused great taxonomic confusion in the early days of fig-wasp research. It is still causing some problems even now, for this dimorphism also extends to a number of the 'parasite' species and in some cases the male has still not been correlated with the correct female. In addition to the pollinating Agaonidae there is also in Hong Kong a total of at least 43 other chalcid wasp species belonging to the families Torymidae,

Eurytomidae, Ormyridae, Eulophidae and Pteromalidae inhabiting the figs of the 15 common indigenous local species of *Ficus* (Table 3). At present the precise biological roles played by these chalcids are not known and for convenience they are lumped together under the title of

TABLE 3: **THE CHALCID FAU**

	Ficus SPECIES	NO. SAM-PLES	AGAONIDAE	SYCOPHAGINI	APOCRYPTINI	SYCORYCTI
Subgenus *Urostigma*	1. *F. superba* v. *japonica*	20	B. (B.) ishiiana	—	—	Sycoryctes A Sycoscapter infectorius
	2. *F. virens* v. *sublanceolata*	5	B. (B.) coronata	—	—	Sycoryctes B Sycoscapter infectorius
	3. *F. microcarpa*	20	B. (Parapristina) verticillata	—	—	Sycoryctes A Sycoryctes B Sycoscapter gajimaru
Pharmacosycea	4. *F. vasculosa*	3	Dolichoris vasculosae	—	—	—
	5. *F. nervosa*	3	B. (B.) nervosae	—	—	—
Subgenus *Ficus*	6. *F. pyriformis*	7	*B. (B.) silvestriana	—	—	Sycoscapter
	7. *F. variolosa*	22	*B. (B.) silvestriana	—	—	Sycoscapter
	8. *F. erecta* v. *beecheyana*	3	*B. (B.) silvestriana	—	—	Sycoscapter Sycoscapter
	9. *F. hirta*	14	B. (B.) javana	—	—	Sycoryctes C Sycoryctes simplex Sycoscapter
	10. *F. pumila*	4	B. (B.) pumilae	—	—	—
	11. *F. sarmentosa*	1	B. (B.) callida	—	—	—
	12. *F. tinctoria* spp. *gibbosa*	2	Liporrhopalum gibbosae	—	—	—
	13. *F. variegata* v *chlorocarpa*	27	Ceratosolen appendiculatus	Eukoebelea ? spinitarsus vc	Apocrypta A u	Sycoryctes D Sycoryctes E
	14. *F. hispida*	25	C. solmsi marchali	—	Apocrypta bakeri u	—
	15. *F. fistulosa*	10	C. c. constrictus	—	Apocrypta ? varicolor c	—
		166	15	1	3	8

† The abundance of the respective species is indicated by the following abbreviations: vc = very common;   c = common;   u = uncommon;   r = rare.

'parasites'. Some are undoubtably phytophagous, others entomophagous, and others probably hyper-parasites.

In the original studies on the fig-wasps of the Edible Fig *(Ficus carica)* in the Mediterranean, it was concluded that the single species of parasite

HONG KONG FIGS.†

...OTRY- SINI	OTITESELLINI	EPICHRYSOMAL- LINAE	EURYTOMIDAE	ORMYRIDAE	EULOPHIDAE	PTEROMA- LIDAE	NO. SPE- CIES
try- *;A* u	*Otitesella* *ako* vc	*Acophila mikii* vc *Sycobiomorpha* *bimasculinum* vc	*Sycophila* A c *Sycophila* B c *Sycophila* C u	*Ormyrus* A u *Ormyrus* B u	*Tetras-* *tich us* A u	Tridy- mine A r	14
try- *;* *~yi* r *try-* *;F* r	*Otitesella* *ako* r *Otite-* *sella* A r	*Acophila mikii* u *Sycobiomorpha* *bimasculinum* vc	*Sycophila* A r *Sycophila* B c *Sycophila* C vc	*Ormyrus* A u	—	—	13
try- *;* *~gi* c *try-* *;B* c	*Otitesella* *ako* r *Otitesella* *yashiroi* c *Otite-* *sella* A c	*Eufroggattisca* A r *Odontofrog-* *gatia* A vc *Odontofrog-* *gatia* B vc new genus (1) vc new genus (2) vc	*Sycophila* B vc *Sycophila* D vc *Sycophila* E r	*Ormyrus* C u	*Tetras-* *tichus* B r	Tridy- mine B r	20
—	—	—		—	—	—	1
—	—	—	—	—	—	—	1
—	—	—	—	—	—	—	2
otry- *'s C* c	—	—	—	—	—	—	3
—	—	—	—	—	—	—	3
otry- *'s D* vc	—	—	—	—	—	—	5
—	—	—	—	—	—	—	1
—	—	—	—	—	—	—	1
otry- *'s E* vc	—	*Neosycophila* *omeomorpha* vc	*Sycophila* F vc	—	—	—	4
—	—	—	—	—	—	—	5
otry- *'s* *losa* vc *otry-* *'s F* u	—	—	—	—	—	—	4
otry- *'s G* vc	—	—	—	—	—	—	3
8	3	4	4	3	2	2	80

Total number of
chalcid species 53

found there was in fact unable to inject poison to stimulate endosperm proliferation and so it laid its eggs in ovaries where the Agaonid female had already oviposited and injected poison. By developing more rapidly the larva of the parasite was in some way able to consume the endosperm at the expense of the Agaonid larva. A term was coined for this situation —cleptoparasitism.

All the female 'parasites' have lengthy ovipositors, in some cases a long obvious external ovipositor, but in some cases it is externally very short or non-projecting and is coiled-up inside the wasp's abdomen. It appears that there is some correlation between the length of the female parasite's ovipositor and the thickness of the wall of the fig it normally inhabits, but the details of this relationship have not been worked out yet. The studies carried out by Hill (1967) on the fig-wasps of Hong Kong is the only one made so far on a complete community, and it has been shown that locally the Agaonidae are completely host-specific. This situation is also true for most species of 'parasites', although there is some straying between closely related species of *Ficus* in a few instances.

Apart from the basic symbiotic (mutualistic) relationship between the Agaonidae and *Ficus* plants being unique, there are many other aspects of the *Ficus*/fig-wasp relationship that are biologically of great interest, particularly the mechanisms of pollination, the method of endosperm stimulation by the insect poison, the insect dimorphism in relation to their parasitic habits, and the whole parasitic complex in the fig-wasp fauna. Here is a rare opportunity for making a study of the comparative evolution of a group of plants and a series of groups of insects showing a varying degree of dependence upon the plants. *Ficus* is probably the dominant genus of plants in Hong Kong for the indigenous species are mostly widespread and very common in most parts of the Colony. Thus there is very much scope for further studies on the fig-wasps of Hong Kong.

ORDER **COLEOPTERA**
(Beetles)

This, the last group of insects, is very common throughout the whole of Hong Kong, and it is in fact the largest group of insects. There is a total of 95 families and more than 270,000 species of beetles in the world. They are generally easy to recognize in that the forewings are hardened and thickened into elytra which meet in the midline down the back. The elytra are purely protective in function and are not used for flying. The hind-wing is typically large and membraneous and is kept folded under the elytron, although a few species have only the elytra but no functional wings. The adults vary in size from minute (1 mm) to very large (70–80 mm long) with different habits, and the larvae are greatly diverse both in structure and habits. A number of species are very damaging to plants and are serious agricultural pests.

Sub-order **Adephaga**

Carabidae

The first group are all predatory and carnivorous and are regarded as being more primitive than other beetles. The Carabidae are called ground beetles, and although there are a couple of phytophagous species which feed on strawberry pips in Europe, they are characteristically predators of other arthropods in leaf-litter and soil. The larvae are elongate, agile creatures with large thin piercing jaws, to be found in leaf-litter and soil. Both adults and larvae are important ecologically as predators of soil invertebrates. Some small species parasitise Lepidoptera pupae and fly pupae. In Malaysia there are several arboreal species which tend to be more brightly coloured than the drab black ground-dwelling species. There is quite a large number of local carabid beetles, but they are mostly black and of smallish size; no one species seems to be particularly abundant, and to date only a few have been identified.

Cicindelidae

The long-legged, agile tiger beetles (Cicindelidae) are well named for they are in their small way very ferocious as they prey on other insects. The Blue Spotted Tiger Beetle *(Cicindela separata)* has elytra coloured blue-green with six creamy spots (Plate 423) and is very common along certain shaded paths, such as Lugard Road around the Peak, and many conduit and hillside paths, where it typically jumps and flies for short distances in front of a walker. It seems that a particular area is

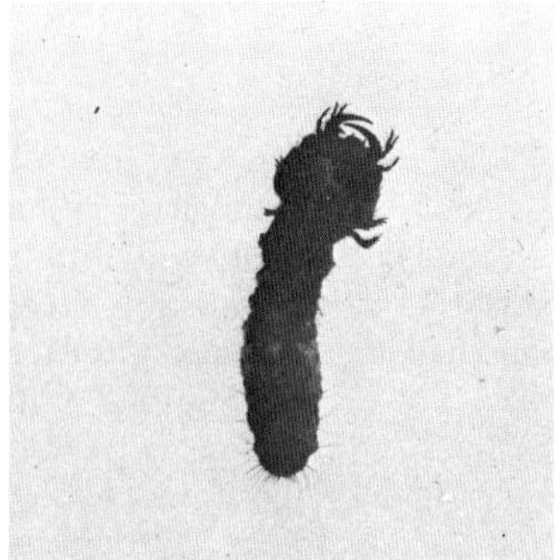

Plate 423. Blue Spotted Tiger Beetle, *Cicindela separata* (Coleoptera, Cicindelidae); length 16 mm.

Plate 424. Top view of larvae of Blue Spotted Tiger Beetle.

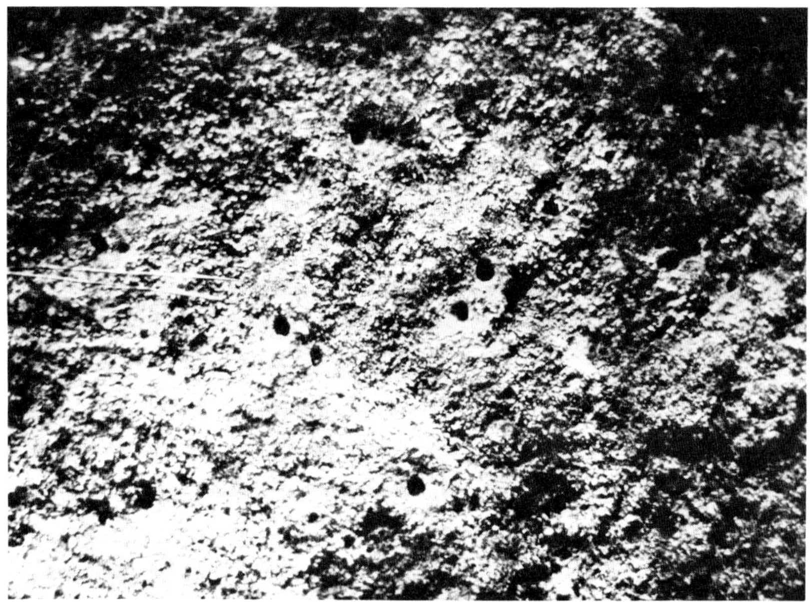

Plate 425. Larval burrows of Blue Spotted Tiger Beetle; Stonecutters Island.

patrolled by each separate individual, and if disturbed they will return to their respective areas.

The larvae (Plate 424) are soil dwellers usually, and they build a narrow vertical burrow sometimes more than 30 cm deep (Plate 425). They have a well-developed head with strong jaws, and the curiously

Plate 426. Side view of larvae of Blue Spotted Tiger Beetle.

bent S-shaped body (Plate 426) has the three pairs of thoracic legs short, medium sized, and long, in that order, and distinct forwardly-directed hook on the back of the fifth abdominal segment. They live in their tunnel, sitting at the top waiting for an unwary insect on which to prey. The prey is pounced upon, seized by the jaws, and dragged down into the burrow where it is quickly dismembered and eaten. *Cicindela anchoralis* (Plate 427) is the Beach Tiger Beetle to be found on certain sandy shores where they hunt for prey over the exposed sand-flats.

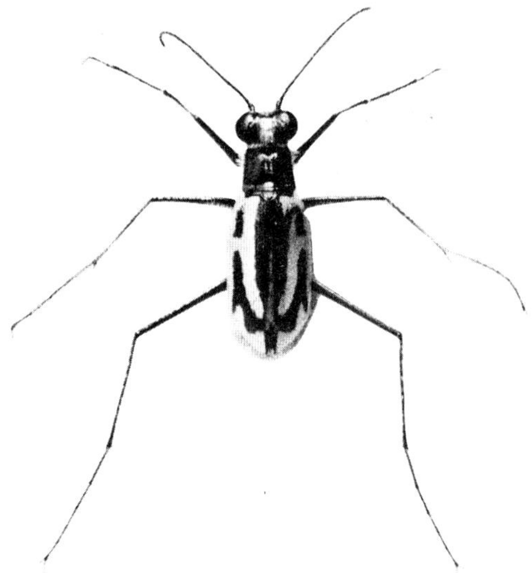

Plate 427.   Beach Tiger Beetle, *Cicindela anchoralis* (Coleoptera, Cicindelidae); length 14 mm.

In Hong Kong there are three types of tiger beetles. First, there are *Cicindela* (Plate 423) and *Heptodonta,* with their conspicously spotted,

Plate 428. Black Flightless Tiger Beetle, *Tricondyla pulchripes* (Coléoptera, Cicindelidae); length 18 mm.

metallic bodies and their ground patrolling habits. Their eggs are laid in short burrows in sandy soil and the shafts made by the larvae can be 10–25 cm deep. The second group contains *Collyrus* species which are metallic blue, without spots, and have reddish-yellow legs. They fly rapidly about small bushes and the larvae are said to burrow in living branches of these shrubs with the tunnel mouth directed downwards. A closely-related species can be found in other tropical countries boring into branches of tea, coffee, cocoa and cotton bushes. The final group are represented here by the black flightless *Tricondyla pulchripes* (Plate 428) which looks remarkably like a large ant and can often be found sitting on paths, walls, or tree trunks, but the genus is regarded as being an arboreal forest inhabitant. Although they are wingless they are able to run swiftly and they are common in all parts of Hong Kong. Little is known of their life history but it has been supposed that their larvae also make tunnels in twigs and branches.

Dytiscidae

The other two common Caraboid families are both water beetles. The True water beetles are the Dytiscidae, which are well-represented locally by several species of *Cybister* of which *Cybister tripunctatus* (The Common Large Water Beetle) (Plate 429) is very abundant. It breeds in

Plate 429.   Common Large Water Beetle, *Cybister tripunctatus* (Coleoptera, Dytiscidae); length 28 mm.

some streams, paddy fields and standing water throughout the New Territories. It is particularly abundant in the autumn when it is sold as food locally. Large numbers are captured easily because at this time of the year they disperse to seek new bodies of water and they collect at lights at night. These large, dark, smooth, streamlined insects are 20–30 mm in body length, with small forelegs but large flattened bristly hindlegs with which they swim. On land these beetles are virtually helpless, being unable to walk, but unless enticed by house or street lights they are normally completely aquatic and when dispersing they fly straight from one body of water to another.

These diving beetles are most easily spotted as they rise to the surface of the water to renew their air supply which is held among fine hairs on the upper surface of the abdomen under the elytra. Males can be distinguished from females by the presence of small adhesive suction pads on the forelegs; they are used during mating to grip the smooth body of the females. The beetles can stay submerged for ten minutes or more before surfacing for a new air supply. Eggs are laid singly in slits cut in water plants by the female's ovipositor, and the larvae live in the water.

The elongate tapering larvae are buoyant and can swim freely; they float to the surface to take in air through their terminal respiratory tube. They are as voracious as the adults and have a large pair of curved pointed mandibles, with which they kill insects, snails, worms, tadpoles and small fish. They can do considerable damage in freshwater aquaria

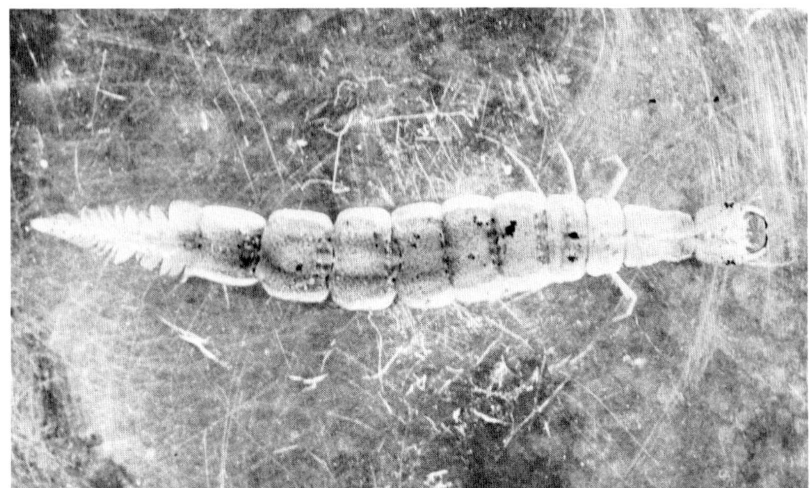

Plate 430.   Larva of *Cybister tripunctalus*; length 50 mm.

if accidentally introduced. The larvae (Plate 430) have hollow mandibles through which digestive enzymes can be passed into the body of the prey, so digestion takes place externally and the ingested body proteins are sucked back into the gut of the beetle larva. The adult feeds by biting and chewing.

Although the commonest water beetles are the large *Cybister* species there is a number of small beetles, either black or spotted dark brown in colour, to be found quite regularly. The commonest small water beetles are probably *Eretes sticticus* which has a black spotted brown body and is some 10–12 mm in length, the brown *Dineutes marginatus* and the black *Hydraticus* species. The tiny water beetles are species of *Hydrovatus* and *Canthydrus*.

Gyrinidae

The Gyrinidae are called whirligig beetles, and these are surface-dwelling water beetles, usually gregarious, and sometimes are found in large groups on the surface of still water or quiet parts of streams. They are small (4–8 mm long), smooth, shiny, black tapering beetles. If disturbed, they readily dive and hide amongst submerged vegetation. The larvae are elongate, predacious, and the abdominal segments bear long plumose, filamentous gills. These beetles are fairly common in Hong Kong and two have been definitely identified as species of

Orectochilus. Several other species have recently been collected but not yet identified.

Sub-order **Archostemata**

Micromalthidae

A separate sub-order of peculiar primitive beetles is represented in Hong Kong by the genus *Micromalthus.* Their larvae undergo a complex and unusual development and the beetles feed on wood; the local specimens were collected from tap-water (Marshall & Thornton, 1973).

Sub-order **Polyphaga**

The majority of beetles are placed in the Sub-order Polyphaga, and they do show a wide range of feeding habits; some are carnivorous and predatory, some phytophagous, and others omnivorous. Their systematic separation into families is not easy, for although the majority fall easily enough into well-defined family groups there are some beetles very difficult to place.

Super-family **Hydrophiloidea**

Hydrophilidae

The Scavenging Water Beetle is *Hydrophilus cashmirensis* (Plate 431) and is recognized by its very long maxillary palps which hang down from the mouth-parts and function as though they were the antennae;

Plate 431. Scavening Water Beetle, *Hydrophilus cashmirensis* (Coleoptera, Hydrophilidae); length 30 mm.

the hairy short antennae are in fact used to break the water surface film to allow air to pass down into a special respiratory channel made by other hairs along the side of the insect body where the respiratory spiracles are situated. These lateral tracts join up with the air reservoir under the wing cases (elytra) and an airfilm is trapped on the under-surface of the body. The hind-legs are used for swimming, but unlike *Cybister* and the other true water beetles, it paddles alternatively and so proceeds on a rather wobbly course. It also uses the middle legs to swim. This large, dark, oval-shaped beetle feeds on rotting vegetation and is not nearly such a fast swimmer as the predatory water beetles. The larvae of *Hydrophilus* are predacious, but the others are scavenging. Actually, the majority of species in this family are not really aquatic but terrestrial in damp locations. A few are pests, and *Helophorus porculus* is the Turnip Mud Beetle which is a minor pest of root crops in Europe. Several small local species have now been identified as *Laccophilus, Regimbartia,* and *Globaria* species.

Super-family **Histeroidea**

Silphidae

Some beetle groups are biologically very ill-defined which makes them difficult to appreciate by non-specialists. For example the Silphidae are largely carrion feeders and include the well-known European Burying Beetle *Necrophorus* which has distinctive orange and black colouration and buries the corpses of rats, mice, birds, rabbits, etc. by excavating beneath them. Eggs are laid in galleries leading from the buried corpse and apparently the young instars are at first fed by the female beetle. Within this group some species are predacious upon snails, others on caterpillars; *Aclypea opaca* is the Beet Carrion Beetle which is a minor pest of Sugar Beet in Europe.

Super-family **Staphylinoidea**

Staphylinidae

Staphylinidae are commonly called rove beetles and are easily recognized by being usually black in colour and having short elytra which leave most of the abdomen uncovered. Despite the small size of the elytra they do successfully cover their large, complexly-folded hind-wings, and these beetles fly quite strongly. The large black species *Ocypus olens* common in European gardens does not appear to have a local equivalent in Hong Kong. They are in general litter and soil inhabitants and sometimes are found in cattle dung in the New

Territories. One of the commonest local species is a small orange and black-banded *Paederus fuscipes* recorded as a predator of several species of aphids. Another species of *Paederus* is known in East Africa as the 'Nairobi Eye-fly' for it flies at night and comes to lights. It secretes an irritating fluid from special glands if handled, and if the fluid on the fingers comes into contact with the eyes then intense irritation ensues. Staphylinid larvae are elongate, predatory forms looking very much like carabid larvae, and are important soil predators. The larvae of the European *Aleochara* are parasites of fly pupae. The newly hatched larva bites through the fly puparium and once inside feeds on the fly pupa, emerging later as an adult beetle by rupturing the puparium wall.

Super-family **Scarabaeoidea**

The Scarabaeoidea is a very large distinctive group with many species, a large number of which are well known biologically or as agricultural pests. The larvae are of the scarabaeiform type, that is, a large, fleshy grub with well-defined head and jaws (but usually without large eyes) and thoracic legs, but a soft, white rounded C-shaped body usually slightly swollen terminally (Plate 432). They are subterranean burrowers

Plate 432.   Chafer Grub, larva of Scarab beetle; length 25 mm.

to be found in soil, rotting vegetation or decaying trees. The soil-inhabiting species tend to feed on the roots of many different plants. As a group they are renowned for the capability of sound production by stridulation in the larvae as well as the adults. The adult beetles have short clubbed antennae and toothed front tibia; in most groups the elytra finish short of the abdomen apex so that the terminal segments are uncovered.

Passalidae

The Passalidae have no common name, and are quite large, flattened black or brown beetles to be found under loose tree bark in the tropics, but not to date recorded locally.

Lucanidae

Stag beetles constitute the family Lucanidae, and like the preceeding group their elytra completely cover the abdomen. The males are very striking in their sexual dimorphism with enormous, elongate mandibles. The commonest local species is the black and brown-banded *Prosopocoilus biplagiasus* (Plate 433) which is quite abundant but the sexual dimorphism in this species is only slight. The larvae live in rotting wood, where they take at least a year to develop.

Plate 433. Male Stag Beetle, *Prosopocoilus biplagiasus* (Coleoptera, Lucanidae); length 30 mm.

Geotrupidae

The Geotrupidae are 'dor' beetles, so-called because of the loud humming noise they make in flight. This noise is also made by the scarabs, and the etymological origin of the word 'dor' is obscure. They are large (often) black beetles with coprophilous habits and are easily confused with the proper dung beetles (Coprinae). Apparently the antennae of this group are always 11-segmented whereas in the Scarabaeidae they are 8–10 segmented. Eggs are laid singly in tunnels under patches of animal dung and typically plugs of dung are used to seal off the tunnels and to provided food for the developing larva. A few species of Geotrupidae have been collected in Hong Kong but they are not as common as the Coprinae, and only a couple have been identified.

Scarabaeidae

The Scarabaeidae is generally regarded as a family by the majority of entomologists, in which case it contains a series of quite well-defined sub-families which require separate mention. The adults are called scarabs or chafers, etc., and the larvae are chafer-grubs or white-grubs.

Sub-family **Cetoniinae**

The Centoniinae are called rose or flower chafers, and the adults are typically of brilliant coloration (often green), and are diurnal in habits and often to be found feeding on flowers (nectar) or overripe fruits of bushes and trees. Their mouth-parts are very weak and they can only eat soft plant material. Locally, there are only two species of *Protaetia*, *P. orientalis* and *P. fusca*. The commoner *Protaetia orientalis* is of shiny green colour with white flecks (Plate 434), often to be seen in the summer on lawns or grass-land where they are looking for oviposition sites. The

Plate 434. Rose Chafer, *Protaetia orientalis* (Col., Scarabaeidae, Cetoninae); length 22 mm.

Plate 435. Large Green Flower Chafer, *Agestrata orichalcea* (Col., Scarabaeidae, Cetoninae); length 40 mm.

larvae live in the soil under turf as typical chafer-grubs, or in rotting wood or vegetation. These beetles are usually rather flattened, and the elytra are incurved level with the hind-legs. In flight the elytra are only raised slightly and the wings protrude through these lateral emarginations.

A spectacular species of Cetoniinae is the Large Green Flower Chafer (*Agestrata orichalcea*) (Plate 435).

Sub-family **Rutelinae**

The Rutelinae also eat flowers, leaves and fruits of a wide range of plants, and are sometimes called flower beetles or June beetles. They are very well-represented in Hong Kong by a number of species of *Anomala*; probably the commonest species are the small (12–14 mm) Brown Striped Chafer, *A. varicolor* and the Copper/Green Flower Beetle, *A. cupripes* (Plate 436) which has bright copper-coloured legs and ventral surface. The small black *A. lasiocaula* is at times found in large numbers, and like most of the other chafers it is nocturnal and flies to lights at night. There are also species of *Adoretus, Mimela* and *Pseudosinghala* recorded locally. The body is typically oval, smooth and shiny, usually with thickened tibiae on the hind-legs, and with long mobile claws of unequal length. The adults are nocturnal in habits and will fly to lights at night.

A stiking, smallish, blue-black species is *Popillia histeroidea* sometimes to be found eating the flowers of *Melastoma* (Plate 437) in a manner typical of most Rutelinae. *Popillia japonica* is a pretty little brown and

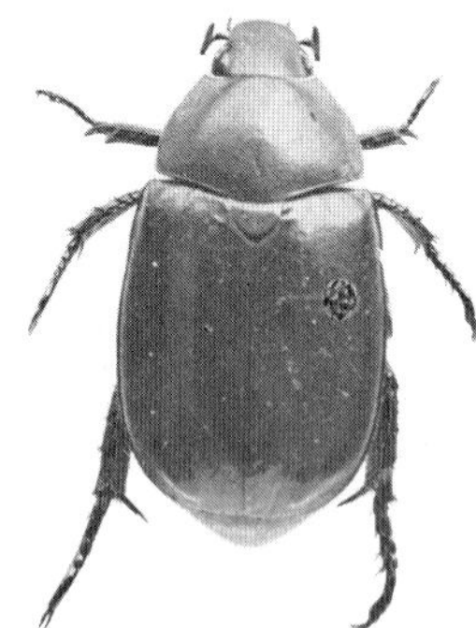

Plate 436.   Copper/Green Flower Beetle, *Anomala cupripes* (Col., Scarabaeidae, Rutelinae); length 40 mm.

Plate 437.   Flower of *Melastoma* being eaten by adult *Popillia histeroidea*.

irridescent green flower beetle known as the Japanese Beetle; it is endemic to East Asia, and is now rated as one of the most serious international pests. Great care is taken with legislative and quarantine measures to prevent its introduction or establishment into Great Britain. It is a serious defoliator of many trees. Other small species of *Popillia* are found in the New Territories, such as *P. mongolica,* but the Japanese Beetle has not yet been recorded.

Sub-family **Melolonthinae**

The Melolonthinae are the cockchafers and the common Large Brown Cockchafer is *Holotricha geilenkenseri,* some 10–14 mm in length, and the Small Brown Cockchafer, equally common, is *Phyllophaga prollaeta.* These beetles are nocturnal, and have fat, rounded bodies usually of a dull brown colour. In flight the elytra are held vertically above the back. Another small species is the Small Black Cockchafer *(Sophrops cephalotes).* The adult cockchafers bite pieces out of the foliage, flowers or fruit of many different trees, shrubs and herbaceous plants, and the chafer-grub larvae eat the roots of grasses and other plants. One or two specimens of very large chafers have been found in the New Territories some years ago, measuring some 50–60 mm long and 30 mm broad but these are undoubtably very rare.

Sub-family **Coprinae**

The Coprinae (= Scarabaeinae) are the true dung beetles, some of which live their entire lives in patches of dried animal dung, but a few are renowned for their habit of making and rolling large balls of dung. It was the ancient Egyptians who first recorded the activities of *Scarabaeus* as it rolled its dried dung balls across the Egyptian country-side, and they were reputed to regard the spherical dung ball as a symbol of the world, which led to the scarab being regarded as a sacred beetle. Sometimes the spherical dung ball is the food for the adult beetles, but in other cases the dung balls are taken into an underground chamber and an egg is laid in each ball of dung. Occasionally in parts of the New Territories the large black spectacular *Catharsius molossus* (Plate 438) can be seen, though it has not been observed rolling dung balls. A species of *Catharsius* in India makes a very large dung ball which is finally covered with a layer of clay and when first recorded it was thought that these were ancient stone cannon-balls. A number of small dung beetles can be found locally, including a couple of really quite tiny species, about 8–10 mm long, which have not been identified.

Plate 438.   Giant Dung Beetle, *Catharsius molossus* (Col., Scarabaeidae, Coprinae); length 38 mm.

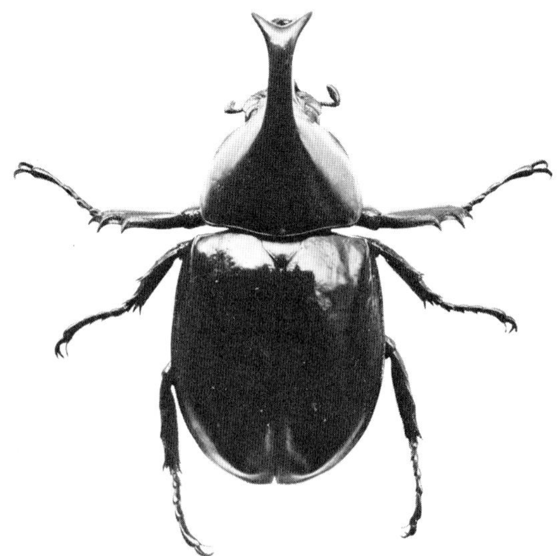

Plate 439.   Male Unicorn Beetle, *Xylotrupes gideon* (Col., Scarabaeidae, Dynastinae); length 45 mm.

Sub-family **Dynastinae**

The final scarabid sub-family is the spectacular Dynastinae, which includes some of the largest and most striking beetles: rhinoceros beetles, unicorn beetles and elephant beetles. Most are black in body colour and nocturnal in habits, but the huge Goliath Beetles in this group (body 90 mm long and 50 mm broad) have distinctive brown elytra and black and white thorax. These unfortunately are tropical rain forest canopy inhabitants and do not occur in Hong Kong. This group is almost completely tropical in distribution and is best-developed in South America. Typically the male beetle has spectacular 'horns' developed from the top edge of the prothorax and a corresponding vertical 'horn' on the head. Two large local species have been collected but only one is identified (Plate 439), and it is now known that it is *Xylotrupes gideon*.

In a few species both male and female beetle have 'horns', but the female tends to be much smaller; the horns are often toothed or bifurcated. The notorious rhinoceros beetles are species of *Oryctes*, with two species occurring in Africa, and with *Oryctes rhinoceros* in India, Southeast Asia and the Pacific Islands. This species has been recorded in South China but it is not certain that it has actually been collected in Hong Kong. It is definitely to be expected here from time to time. The adult beetles feed in the 'cabbage' (heart or growing point) of the palm, and their biting results in large triangular gaps in the leaves as they expand. This damage is characteristic and can be seen on coconut palms throughout Southeast Asia quite commonly. So far as is known *Oryctes* is quite specific to palms. As a number of palms do grow in Hong Kong so *O. rhinoceros* might be expected. The larvae live in decaying vegetable matter, showing a preference for rotting palm trunks, and so can develop in a variety of habitats.

Super-family **Dryopoidea**

Psephenidae

The next group consists of several small families of sub-aquatic beetles, usually without special respiratory structures. The larvae of the Psephenidae are quite common in the swift freshwater streams of the New Territories and Lantao Island. They are called 'water pennies' and are flattened and oval in shape, up to 12–14 mm long, and they sit on the rocks and stones in the stream (Plate 440). They use their body shape and the pressure applied by the water to maintain their position on the rocks. They have well-developed legs and move quite quickly. Respira-

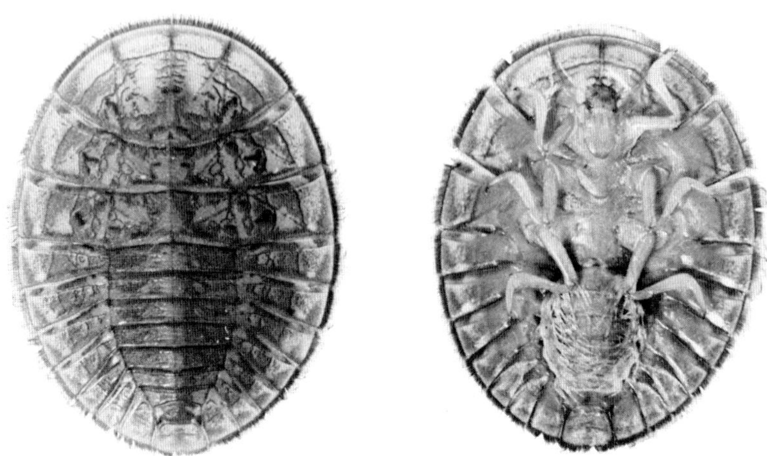

Plate 440.   Water Penny, larva of *Psephenus* sp. (Coleoptera, Psephenidae), found on rocks in freshwater streams; ventral view showing abdominal gills; length 12 mm.

Plate 441.   Aquatic larva of unidentified beetle, probably *Stenocolus* sp. (Coleoptera, Ptilodactylidae), showing the ventral abdominal gill tufts; length 35 mm.

tion is achieved by using a series of branched abdominal tracheal gills which can be seen in Plate 440. The adult beetles have not yet been recognized, but are reputed to live in the vicinity of the streams.

Ptilodactylidae

In many of the New Territories freshwater streams can be found large fat beetle larvae in the substrate and under stones. They have conspicuous ventral abdominal gill tufts and are thought to belong to the genus *Stenocolus,* but as yet have not been authoritatively identified (Plate 441).

Super-family Buprestoidea

Buprestidae

The Buprestidae are flattened timber beetles, sometimes called jewel beetles because they are often brightly irridescent, but the two locally recorded species are of a fairly dull green colour. The large local buprestid is the dark brown/green *Chalcophora japonica* some 30–35 mm in length (Plate 442), and the small species is *Chrysobothris* sp., only about 10–12 mm long. The larva has a tiny head embedded into the thorax and a greatly expanded and flattened prothorax, with a thin body posteriorly. They bore in tree branches and trunks and are very long-lived, the larger species being larvae for up to two years. Because of the shape of the larvae and the resulting shape of the emerging adult beetles, their burrows in timber and the emergence holes are oval in shape, being distinct from the more circular tunnels made by the long-horn beetles (Cerambycidae).

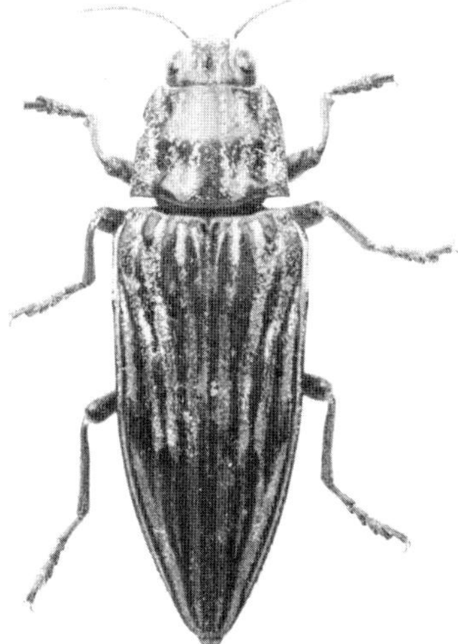

Plate 442. Large Green Jewel Beetle, *Chalcophora japonica* (Coleoptera, Buprestidae); length 35 mm.

Super-family **Elateroidea**
Elateridae

Click beetles (Elateridae) are elongate beetles, tapering slightly at each end. If disturbed in vegetation they fall to the ground on their backs feigning death, but if touched they react with an audible 'click' and spring into the air, turning over and landing on their feet. Their larvae are the notorious wireworms of cosmopolitan agricultural importance. Most agricultural countries have a wireworm problem, but in the various parts of the world the beetles are of different species. The smooth, shiny, local Green Click Beetle is *Compsosternus auratus* (Plate 443), the underneath of which is a golden green, and it is quite common in most parts of the Colony. Several small brown species, reminiscent of the European *Agriotes,* are quite common in places, but the wireworm larvae are seldom encountered. A South American genus called *Pyrophorus* is a 'fire-fly' with luminescent patches and also luminescent eggs and larvae.

Super-family **Cantharoidea**
Lampyridae

The usual 'fire-flies' and 'glow-worms' are found in the family Lampyridae, and typically both adults and larvae have bioluminescent patches on the body. Flying adults are referred to as 'fire-flies' and the terrestrial larvae to be found on soil or foliage are the 'glow-worms'.

Plate 443. Large Green Click Beetle, *Compsosternus suratus* (Coleoptera, Elateridae) length 35 mm.

The commonest local species is a small elongate brown beetle some 10–15 mm in body length, with the head more or less hidden dorsally under the edge of the thoracic dorsum called *Lynchuris analis*, and a phosphorescent patch under the distal end of the abdomen. It is sporadically common in certain localities during the summer; sometimes a dozen can be seen twinkling amongst the vegetation in the evening dusk.

Occasionally, a very large flattened black larva is found, some 30–40 mm in length, with a peculiar head and two sub-terminal luminescent organs (Plate 444). This is a species of *Luciola* and yet the adult beetle

Plate 444.   Giant Glow-worm, larva of *Luciola* sp. (Coleoptera, Lampyridae); length 40 mm.

has not been found. The larvae are predacious; their mouth-parts at rest tend to be withdrawn, and their sharp, sickle-shaped jaws are hollow. Enzymes are pumped into the snails and slugs upon which they feed and the externally digested body contents are sucked up by the mouth of the larva. The function of the bioluminescence in the larvae is unknown, but clearly in the adults it is used to bring the sexes together for mating purposes.

Cantharidae

The Lampyridae are lumped together with the Cantharidae (soldier beetles) in the Cantharoidea, but the predacious soldier beetles are few in Hong Kong, and to date none have been identified.

Super-family Dermestoidea

Dermestidae

The Dermestidae are small, rather rounded, beetles found in households and are associated with stored products; they have larvae covered with soft bristles. Some of the more important pest species are the Larder Beetle, *Dermestes lardarius* (Plate 445) which eats bacon and a wide range of dried animal and plant material, and the Hide Beetle, *D. maculatus* whose diet consist almost entirely of dried animal protein and in the wild state feeds on animal cadavers. *Attagenus piceus* is the Black Carpet Beetle to be found attacking carpets, furs and clothing. One

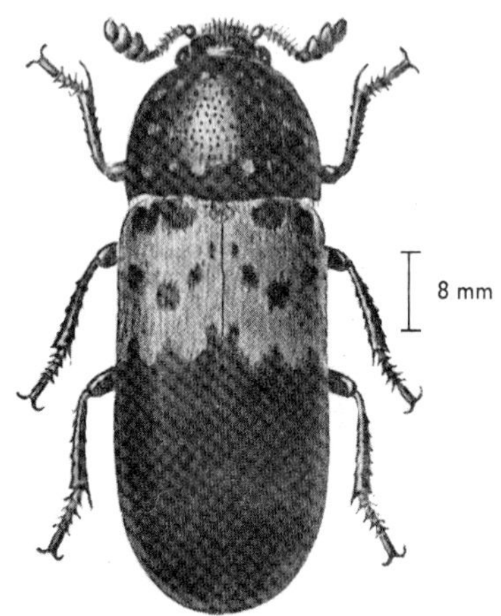

8 mm

Plate 445.   Larder Beetle, *Dermestes lardarius* (Coleoptera, Dermestidae).

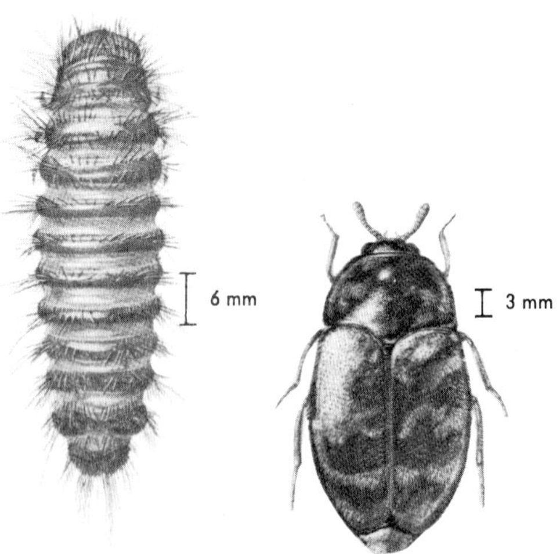

6 mm

3 mm

Plate 446.   Khapra Beetle, *Trogoderma granarium* (Coleoptera, Dermestidae).

of the very few truly herbivorous dermestid beetles is the Khapra Beetle (*Trogoderma granarium*) (Plate 446) which feeds solely on stored grains. Species of *Anthrenus* are called Museum Beetles and feed very destructively on dried natural history museum specimens. *A. scrophulariae* is the cosmopolitan Carpet Beetle (Plate 447).

Plate 447.   Carpet Beetle, *Anthrenus scrophulariae* (Coleoptera, Dermestidae).

Super-family **Bostrychoidea**

Anobiidae

Next comes a group of closely related families to be found in timber or stored products, the Bostrychoidea. The Anobiidae are wood beetles but also contains some stored products pests. Within the family are the

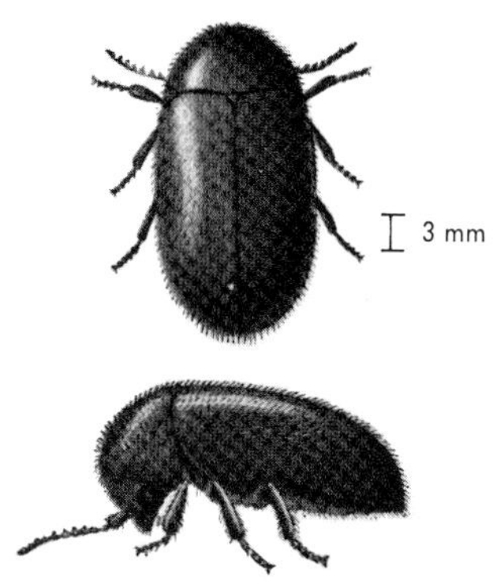

Plate 448.   Cigarette (Tobacco) Beetle, *Lasioderma serricorne* (Coleoptera, Anobiidae).

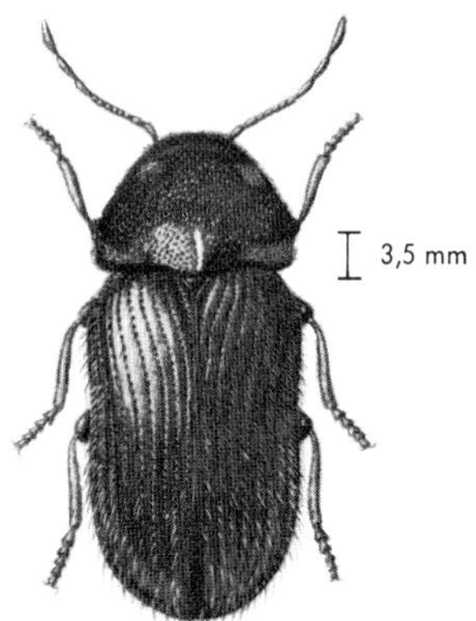

Plate 449.   Drugstore Beetle, *Stegobium paniceum* (Coleoptera, Anobiidae).

temperate timber beetles, *Anobium punctatum* (Common Furniture Beetle) and *Xestobium rufovillosum* (Death Watch Beetle); the former species occurs in Hong Kong. Two cosmopolitan pests are *Lasioderma serricorne* (Plate 448), the Tobacco or Cigarette Beetle, and *Stegobium paniceum* (Plate 449), the Drug Store Beetle; both feed on a wide range of stored foodstuffs, and the former also on dried tobacco leaves and cigarettes. *L. serricorne* is basically a tropical species and its development ceases below 18°C, but *S. paniceum* is a more temperate species and found in Europe, Japan and U.S.A.

Ptinidae

The Ptinidae are aptly named spider beetles, for with their rounded body and long legs their resemblance to spiders is quite close. Two of the commonest species to be found regularly in local godowns are *Ptinus tectus,* the Australian Spider Beetle (Plate 450) and *Niptus hololeucus,* the Golden Spider Beetle (Plate 451); both are to be found feeding on a wide range of foodstuffs, more typically in godowns rather than on domestic premises.

Bostrychidae

The Bostrychidae are wood beetles with a characteristic cylindrical body shape and an enlarged prothorax which gives them a hunch-backed

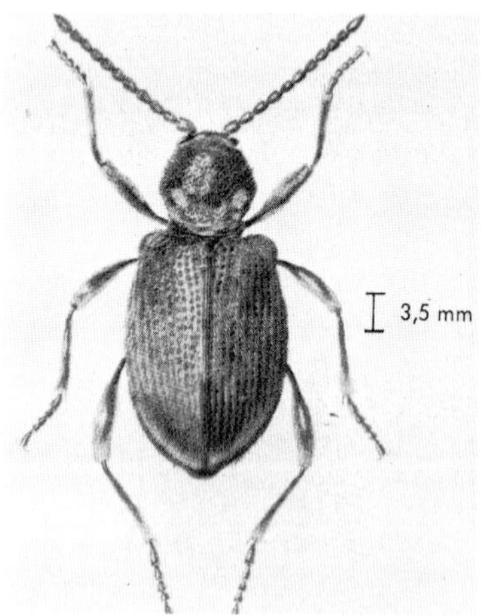

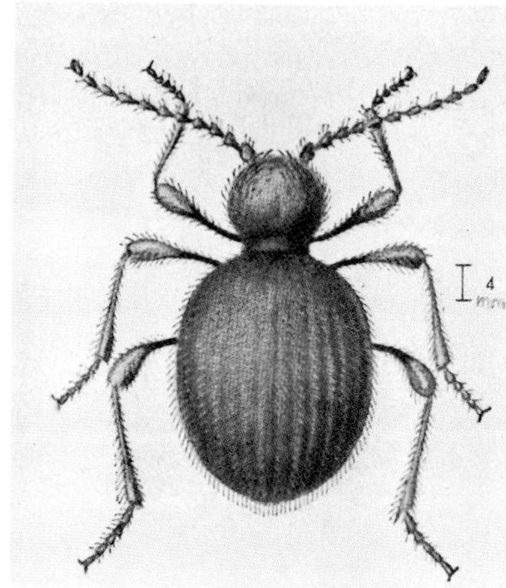

Plate 450. Australian Spider Beetle, *Ptinus tectus* (Coleoptera, Ptinidae).

Plate 451. Golden Spider Beetle, *Niptus hololeucus* (Coleoptera, Ptinidae).

appearance in profile with the head placed almost ventrally and overhung by the prothorax. The body is black and stout and because of its cylindrical shape it makes a round hole as it burrows in fallen timber or dried wood. Sometimes it attacks standing dead timber. Two important pest species belong to the genus *Apate* and they bore into coffee trunks in Africa and South America and a very similar species has been found in Hong Kong, called *Bostrychopsis parallela*. One species to be found locally in food stores in rice is the small (3 mm long) brown Lesser Grain Borer *(Rhizopertha dominica)* (Plate 452).

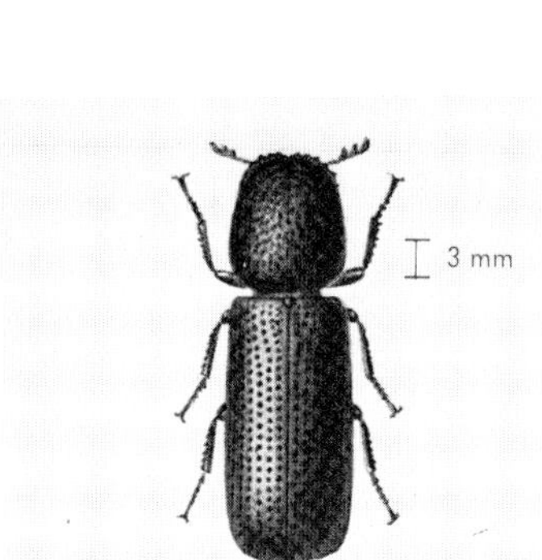

Plate 452. Lesser Grain Borer, *Rhizoper-tha dominica* (Coleoptera, Bostrychidae).

Plate 453. Copra Beetle, *Necrobia rufipes* (Coleoptera, Cleridae).

Lyctidae

Very closely related to the previous family is the Lyctidae (powder post beetles). These beetles bore into dead timber; apparently only broad-leafed trees are attacked, but to date they have not been recognized locally.

Super-family Cleroidea

Cleridae

In the Cleridae, which is a group of brightly coloured tropical beetles, is one species of particular note, *Necrobia rufipes* or the Copra Beetle (Plate 453) which is a serious pest of copra in the tropics and sometimes it is also a serious pest on bacon.

Trogositidae

Similarly in the Trogositidae there is the Cadelle *(Tenebroides mauritanicus*) (Plate 454), a large, dark, flattended beetle infesting flour, grains and other foodstuffs, and very common in local godowns.

Melyridae

A large tropical family showing superficial resemblance to Cantharidae, and it is well represented locally by several abundant species with long blue elytra, a small blue head and a small yellow prothorax. The large species is named *Idgia oculata*—nothing is known of its biology.

Super-family Lymexyloidea

Lymexylidae

An odd group of beetles with elongate soft bodies, vestigial elytra, large membraneous hind wings—a very characteristic appearance. In 1978 a single specimen was seen in Pokfulam, but it eluded capture. The larvae are said to bore cylindrical tunnels in hard wood.

Super-family Cucujoidea

Silvanidae

Following these groups in systematic sequence comes a very large, heterogeneous group called the Cucujoidea, but the actual composition of the group varies considerably from text to text. Only a few of the larger families and those in which there are particularly important pest species will be mentioned here. In the small Silvanidae are the Saw-toothed Grain Beetle *(Oryzaephilus surinamensis)* (Plate 455) and the almost indistinguishable Merchant Grain Beetle *(O. mercator)*, both of which are tropical stored foodstuff pests feeding on a wide range of foodstuffs. They are common in local rice but they tend to be secondary pests for they usually prefer to feed on rice grains which have already been damaged, so they are usually found associated with primary pests such as Rice Weevils (*Sitophilus* spp.).

Nitidulidae

In the Nitidulidae is the Dried Fruit Beetle *(Carpophilus hemipterus)* (Plate 456) which as the name indicates is a pest of dried fruits. The tiny Rust-red Grain Beetle *(Cryptolestes ferrugineus)* (Plate 457) is probably the only species of importance in the Cucujidae. The other species are to be found mostly living under loose tree bark where they prey on other arthropods, and they have the collective name Sap Beetles.

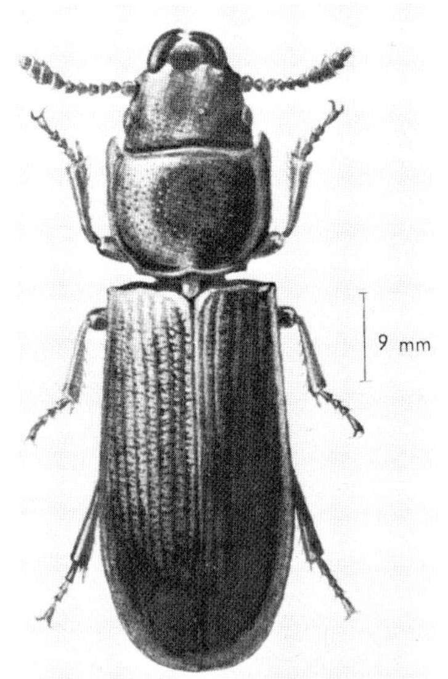

Plate 454. Cadelle, *Tenebriodes maurita-nicus* (Coleoptera, Trogositidae).

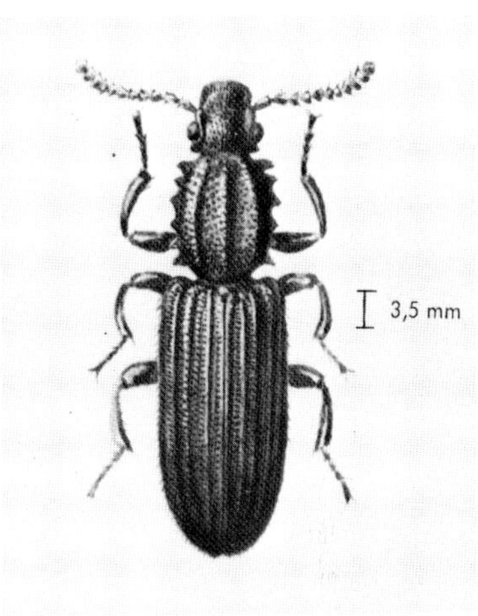

Plate 455. Saw-toothed Grain Beetle, *Oryzaephilus surinamensis* (Coleoptera, Silvanidae).

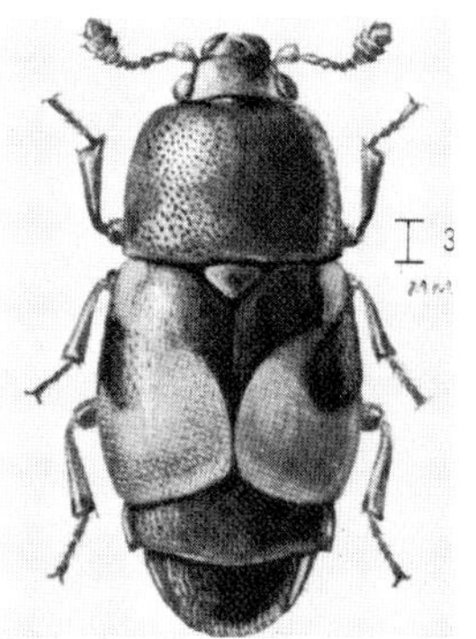

Plate 456. Dried Fruit Beetle, *Carpophi-lus hemipterus* (Coleoptera, Nitidulidae).

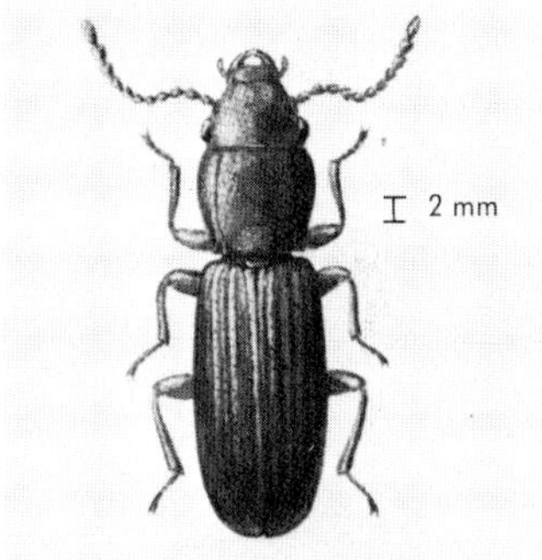

Plate 457. Rust-red Grain Beetle, *Crypto-lestes ferrugineus* (Coleoptera, Cucujidae).

Coccinellidae

One large family of importance in this group is the Coccinellidae—the ladybirds, or ladybird beetles. They are important agriculturally for both larvae and adults are active predators of aphids, scale insects, mealy-bugs and other small plant bugs; various species of Coccinellidae have been used very successfully in biological control campaigns in

Plate 458.   Adult ladybird beetles, *Rodolia* species (Coleoptera, Coccinellidae); length 6 mm.

Plate 459.   Typical Coccinellid larva on a *Hibiscus* leaf where it had been eating Cotton Aphid Nymphs.

India, Africa, Australia and the U.S.A. There are several species here in Hong Kong which were imported from India to control some local plant bug pests. There are probably several dozen different local species of ladybirds, and typical adult beetles are shown in Plate 458. Common local species are placed in the genera *Chilocorus, Coccinella, Coelophora, Rodolia,* and *Menochilus,* and Plate 459 shows a typical larva.

One genus that is completely different biologically is *Epilachna* which is phytophagous and its different species are important pests on curcurbits, legumes, and solanaceous crops in Africa, Asia and the Americas. The local species is *Epilachna sparsa,* a small brown beetle with black

(Left) Plate 460. Adult *Epilachna sparsa* (Coleoptera, Coccinellidae), on the leaf of Black Nightshade; body length about 6 mm.

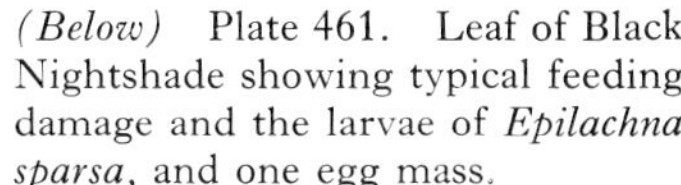

(Below) Plate 461. Leaf of Black Nightshade showing typical feeding damage and the larvae of *Epilachna sparsa,* and one egg mass.

spots (Plate 460) whose larvae bear distinctive fleshy spikes and tubercules (Plate 461). Plate 461 also shows an egg mass and the typical damage done to the leaf by both adults and larvae. It feeds locally on Egg Plant *(Solanum melongena)* and the leaves of Black Nightshade (S. nigrum).

Tenebrionidae

The Tenebrionidae contains *Tenebrio molitor,* the Yellow Mealworm Beetle, which is sometimes found in infested foodstuffs, but is more often reared in laboratory cultures as food for reptiles, amphibia and insectivorous birds. The two species of Flour Beetle, *Tribolium confusum* (Plate 462) and *T. castaneum,* are cosmopolitan stored foods pests and are very common in Hong Kong, usually in flour, biscuits and the like rather than in actual grain. The Broad-horned Flour Beetle, *Gnathocerus cornutus* (Plate 463) is also in this family. Several species of *Strongylium* are to be found on tree trunks in forested areas, and the brown *Gonocephalum pseudopubens* on several sandy beaches locally.

Meloidae

The family Meloidae are referred to as oil beetles, or blister beetles. These are long-legged, soft-bodied beetles with a deflexed head and distinct narrow neck. They are usually conspicuous in their coloration and bold in their behaviour, relying it seems upon their toxic properties for avoidance. Actually, there are two distinct sub-families, the oil

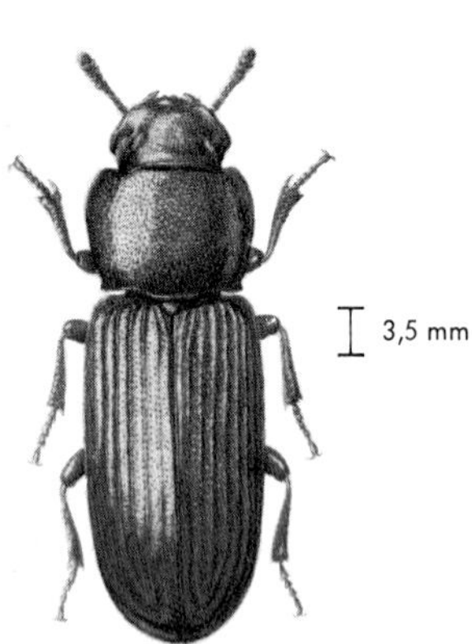

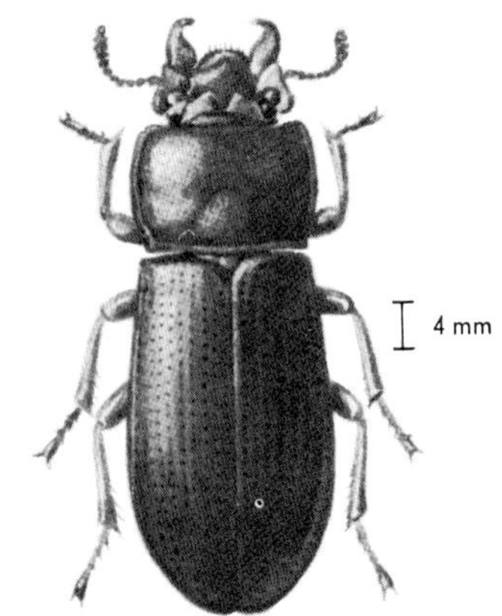

Plate 462. Red Flour Beetle, *Tribolium confusum* (Coleoptera, Tenebrionidae).

Plate 463. Broad-horned Flour Beetle, *Gnathocerus cornutus* (Coleoptera, Tenebrionidae).

beetles are wingless ground beetles with short elytra, but the blister beetles are distinctively coloured, flying insects, often to be seen eating flowers of shrubs and herbs.

The life history of these beetles is unique in that they have a very active first instar larva called a triungulin which seeks out the underground egg-pods of short-horned grasshoppers (Acrididae) or the nests of Aculeate Hymenoptera. Having found the egg-pod the triungulin moults and becomes a soft-bodied, short-legged inactive form which feeds upon the grasshopper eggs. Consequently Meloidae tend to be abundant in sandy grassy areas with a high grasshopper population density. In other countries a commonly recorded bee host is *Anthophora* which is also a common local bee in Hong Kong.

The adults can be very destructive to flowers. On Lantao Island in 1963 a fifty metre hedge of *Hibiscus* was completely deflowered by a large population of *Mylabris phalarata* beetles. As a group they are particularly destructive to legume flowers, and this makes them quite serious pests on a number of legume crops.

There are four main species to be found in Hong Kong in two quite different genera; *Mylabris phalerata* is the Large Yellow-banded Blister Beetle (Plate 464), some 25 mm in length, black with three yellowish-red bands across the elytra; whereas the Small Yellow-banded Blister Beetle, *Mylabris cichorii,* is identical in appearance, only smaller and being 12–18 mm in body length. The Striped Blister Beetle is *Epicauta gorhami* (Plate 465), a black beetle some 13 mm in length, with white longitudinal lines down the centre of each elytron and along the margin, and the apex of the abdomen usually visible dorsally. *Epicauta tibialis* is the Black Blister Beetle and is a larger version of *E. gorhami* but

Plate 464.   Large Yellow-banded Blister Beetle, *Mylabris phalerata* (Coleoptera, Meloidae); length 28 mm.

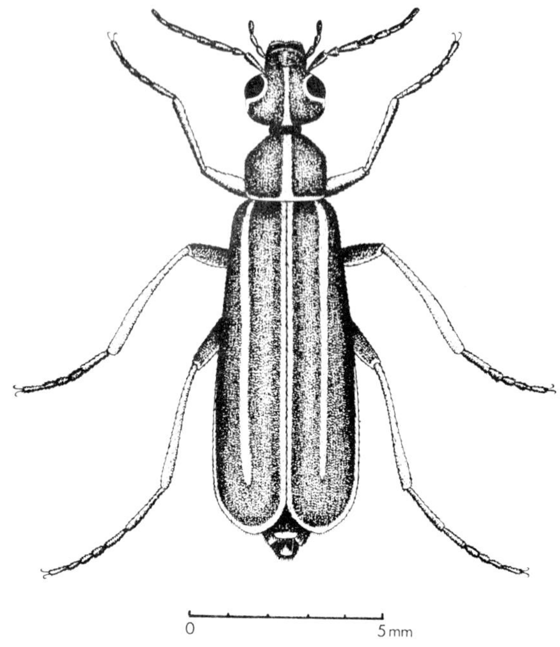

Plate 465. Striped Blister Beetle, *Epicauta gorhami* (Coleoptera, Meloidae).

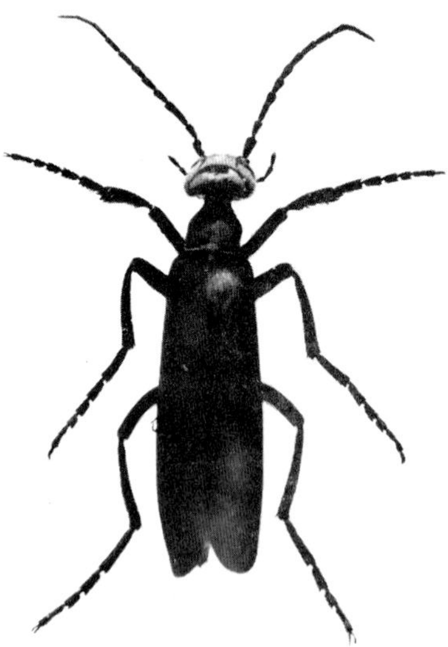

Plate 466. Black Blister Beetle, *Epicauta tibialis* (Coleoptera, Meloidae); length 23 mm.

without the white stripes and with a red coloured head capsule (Plate 466). At intervals large local populations of blister beetles can be found in certain restricted habitats, but after a few days they disperse and for the rest of the time only odd specimens are found.

Super-family **Chrysomeloidea**

The super-family Chrysomeloidea is one of the largest groups of beetles and is probably the best represented group in Hong Kong. Although they are technically grouped together satisfactorily, they represent several very distinct groups biologically.

Cerambycidae

Cerambycidae are timber borers and because of their long antennae they are commonly called long-horn beetles or longicorn beetles. They are amongst the largest insects in the world, and they undoubtably posses the greatest linear dimensions if legs and antennae are also measured. With the exception of the very rare huge chafers that were once found in the New Territories, this group certainly contains the largest local beetles.

Eggs are laid on tree trunks or branches in small holes made by the female, and the developing larvae are legless, blind and cylindrically shaped with a soft body but well-sclerotized head capsule and stout jaws. They tunnel in the heart-wood and also in the sap-wood beaneath the bark. Sometimes they have extensive tunnels in the sap-wood just under the bark but if disturbed they retreat to a tunnel that goes deep into the heart-wood where they are virtually unattainable. They spend one to three years as larvae; the time spent generally being correlated to the levels of moisture and nutriment in the wood. It has been recorded several times that wood which had been made into furniture for several years suddenly yielded adult beetles, which presumably were present there as larvae at the time of felling.

This group typically attacks living timber locally. Sometimes dead or dying trees appear to be preferred, although some small species bore the pith or roots of herbaceous plants. An inhabited tree is usually very distinctive for there are piles of fresh white frass and pieces of chewed wood under the holes, sometimes considerable mounds of frass are found if the species is a large one. Pupation takes place in the tree in a special cell where the tunnel has been extended right up to the bark level and then blocked off from the bark by a plug of wood chips and frass. Thus the emerging adult beetle has only to pull away the frass plug and then

Plate 467. Typical emergence hole of adult Longhorn Beetle (Coleoptera, Cerambycidae), in this case from the trunk of *Pinus massoniana*; diameter 14 mm.

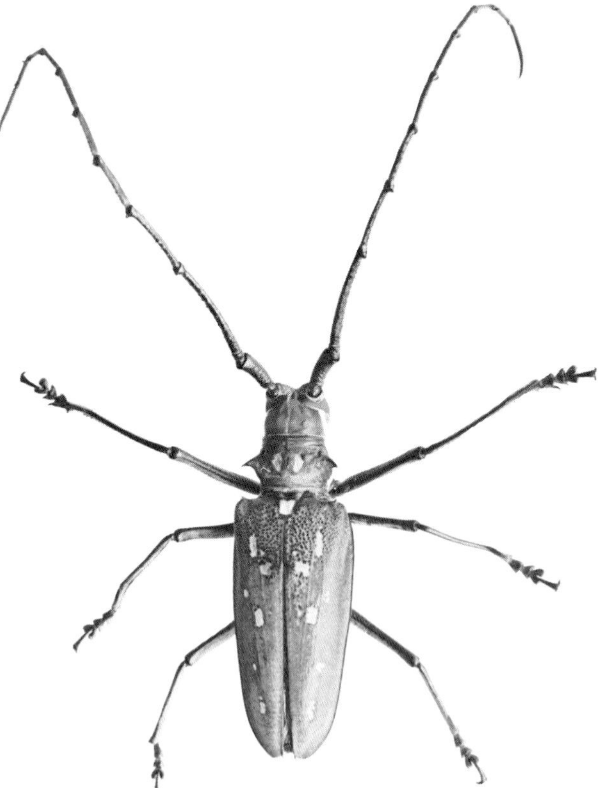

Plate 468. Giant Red-spotted Longhorn Beetle, *Batocera rufomaculata* (Coleoptera, Cerambycidae); length 70 mm.

Plate 469.   Emergence site of adult *Batocera rufomaculata* from dying tree of *Sapium discolor*.

bite through the bark to effect its emergence, and by this time the bark overlying the sapwood tunnel area often has died and cracked and started to fall off.

Since the body of the larva is cylindrical in shape the tunnel made is typically round and the final emergence hole circular (Plate 467), as distinct from the more oval-shaped tunnel of the Buprestidae. In practice sometimes hole shapes are not so precise and long-horn larva tunnels can be quite oval. This can apparently be correlated to larval habits because the species that bore under the bark often make more elliptical tunnels whereas the deep heart-wood borers are usually more cylindrical. In a few cases the adult long-horn beetle does some damage to the host tree by eating bark, but this is not common.

The most spectacular local species is the Giant Red-spotted Longhorn, *Batocera rufomaculata* (Plate 468), some 70 mm in body length and 25 mm broad with very long antennae (these are typically longer in the smaller-bodied male). This beetle is specific to the Mountain Tallow Tree *(Sapium discolor)* and in spring emergence holes and opened frass plugs can be seen on the tree trunks in several areas in the New Territories (Plate 469). In autumn and winter, piles of fresh frass can be very large and the cracked bark will indicate the location of the sapwood tunnel areas. The specific name of this beetle refers to the two orange-red spots on the thorax, but unfortunately this colour fades after death to become yellowish like the elytra spots. Another species of *Batocera* is quite common locally in the New Territories in branches and trunks of Edible Fig *(Ficus carica)* and Mango, this is the White-spotted Long-horn *(Batocera rubus)* (Plate 470). The larvae typically bore along the pith of branches (Plate 471) making holes at intervals for frass expulsion

Plate 470.　White-spotted Longhorn Beetle, *Batocera rubus* (Coleoptera, Cerambycidae); length 36 mm.

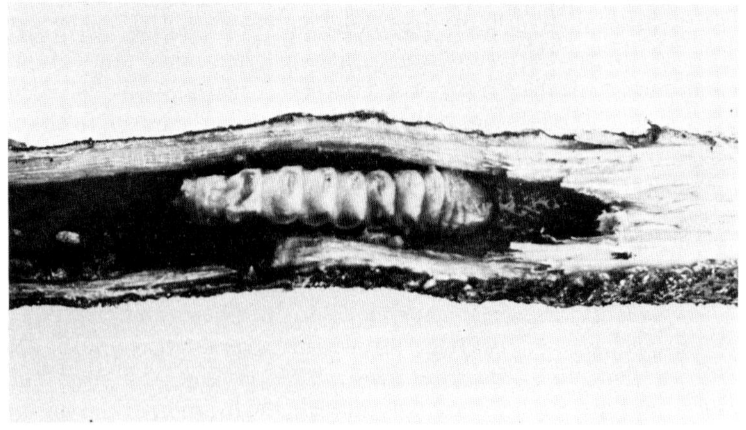

Plate 471.　Larva of *B. rubus in situ* in a branch of Edible Fig Tree.

Plate 472. Head of larva of *B. rubus* protruding from its tunnel in a cut branch.

Plate 473. Adult Jackfruit Longhorn, *Apriona germari* (Coleoptera, Cerambycidae); length 48 mm.

(Plate 471 & 472). The adult beetle does damage to the tree by sometimes eating off patches of bark.

The two commonest large species here are the grey-brown Jackfruit Longhorn *Apriona germari* (Plate 473), some 30–40 mm in length, and the Citrus Longhorn *Anoplophora chinensis* (Plate 474) which has a distinctive black body with white flecks and spots on the elytra, and long

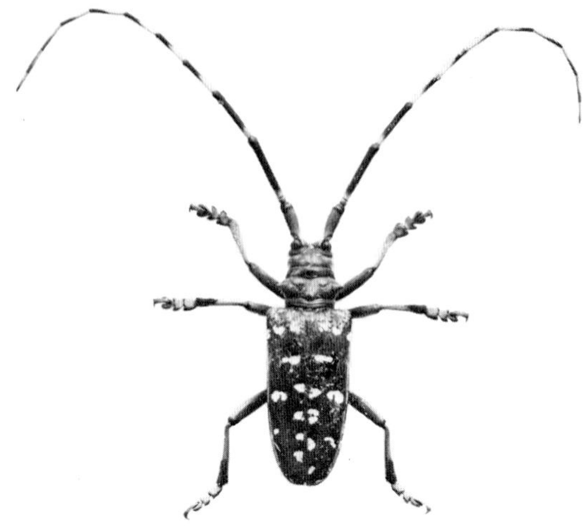

Plate 474.   Citrus Longhorn Beetle, *Anoplophora chinensis* (Coleoptera, Cerambycidae).

Plate 475.   Stem of Citrus tree damaged by the feeding of larvae of Citrus Longhorn.

banded antennae, but with a body length of about 25 mm. It sometimes completely kills small *Citrus* trees by its boring (Plate 475). A very attractive thin elongate metallic blue species is *Chelidonium sinense* (Plate 476), often to be found in large numbers in bamboo foliage, in parts of the New Territories and on Cheung Chau Island.

A small interesting species is *Chlorophorus annularis* (Plate 477) to be found boring in dead bamboo stems where the larvae eat out the nodes and emerge through small holes at intervals along the stem (Plate 478); it also bores in sugarcane stems in the New Territories. A very similar species with grey rather than yellow markings is *C. macaonensis,* commonly found on buildings; it is called the Chinaberry Longhorn and has been reared from branches of *Melia* and *Prunus* spp. A very common smallish brown and black flecked species is *Monochamus tesserula* which has a body length of 20 mm, but its host plant is unknown.

A large number of small species of Cerambycidae are quite common in Hong Kong but most have not been identified and nothing is known

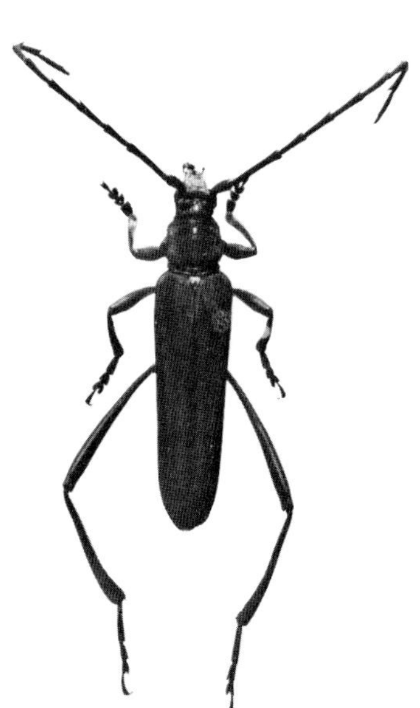

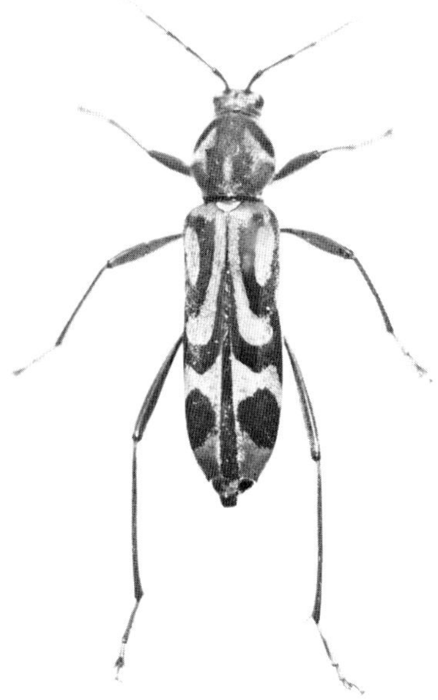

Plate 476. Blue Longhorn Beetle *Chelidonium sinense* (Coleoptera, Cerambycidae); length 25 mm.

Plate 477. Adult Bamboo Longhorn Beetle, *Chlorophorus annularis* (Coleoptera, Cerambycidae); length 14 mm.

Plate 478.   Damage to bamboo by Bamboo Longhorn Beetle, *Chlorophorus annularis* larvae.

of their biology. The smallest species collected to date is just 4 mm in length.

Bruchidae

The Bruchidae is the next family, and this does not have a vernacular name. As a group they are naturally found boring in the fruits and seeds of Umbelliferae, Convolvulaceae, and Leguminosae. They are best known as bean and pea bruchids and are pests of stored legume seeds (pulses). The Cowpea Bruchids *(Callosobruchus maculatus* and *C. chinensis)* (Plate 479) are both quite common locally on stored pulses. Occasionally the Groundnut Borer *(Caryedon serratus)* (Plate 480) is also found. These pest species, although more frequently thought of as stored products pests, do in fact infest field crops which are ripe and have the pods opened. Since they are not strong fliers, field infestations usually occur only within half a mile of an infested storehouse.

Chrysomelidae

The final family in this group is the Chrysomelidae, called leaf beetles and in most cases both adults and larvae inflict the same damage to crop plants by eating their leaves, usually making small holes over the leaf

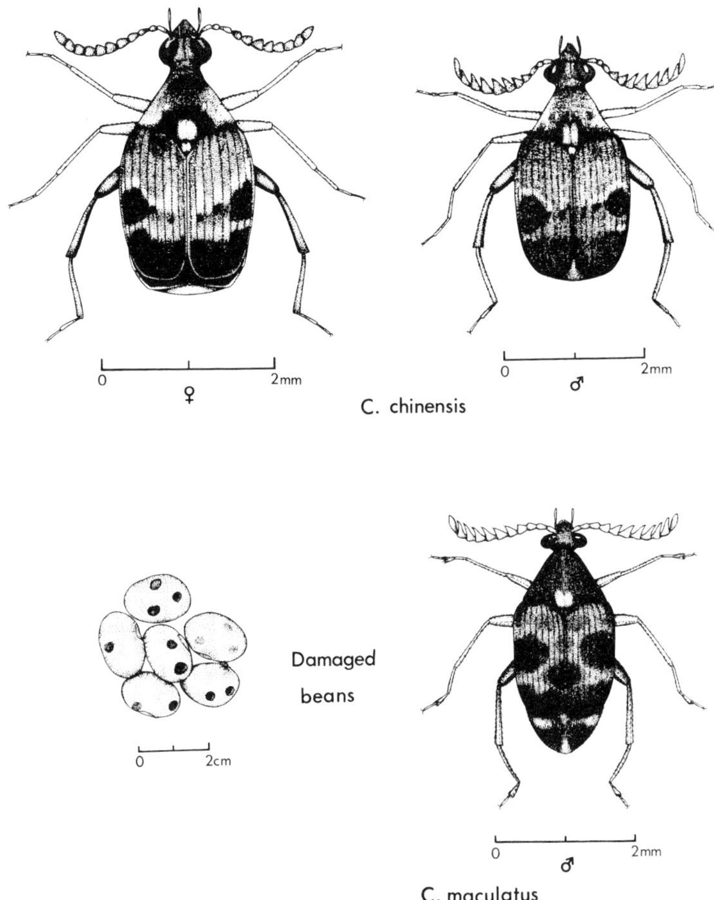

Plate 479.   Two common Cowpea Bruchids, *Callosobruchus chinensis* and *C. maculatus* (Coleoptera, Bruchidae), with damaged seeds.

surface. Within the family are at least seven very distinct groups usually referred to as sub-families, but some of these are biologically and morphologically very distinctive.

Sub-family **Hispinae**

The first group is the Hispinae, called hispid beetles, where the tiny adults are often spiny. The local species is the Paddy Hispid *(Dicladispa*

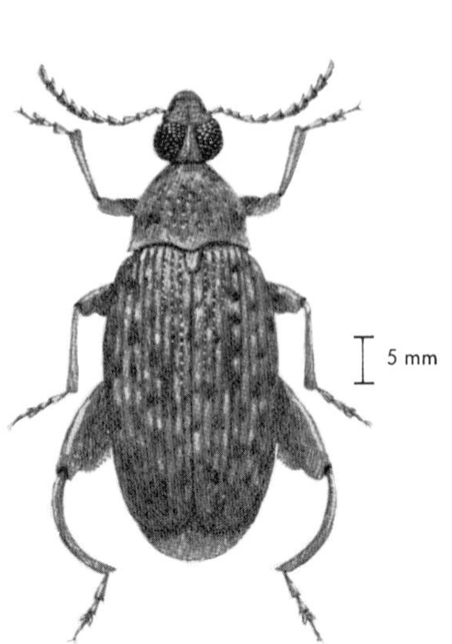

Plate 480.   Groundnut Borer, *Caryedon serratus* (Coleoptera, Bruchidae).

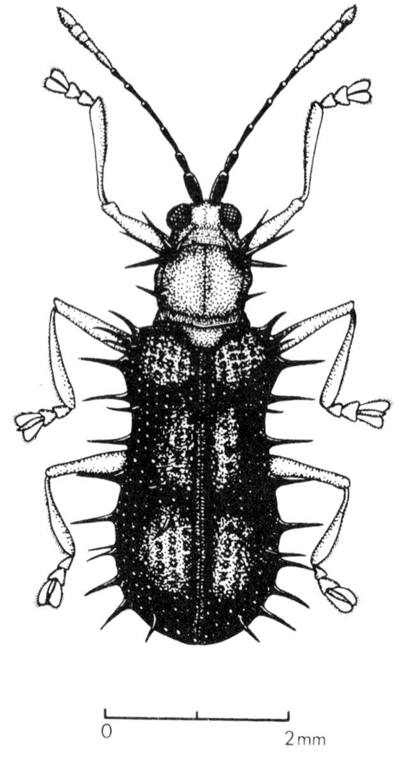

Adult

Plate 481.   Adult Paddy Hispid, *Diclaolispa armigera* (Coleoptera, Chrysomelidae, Hispinae).

armigera) (Plate 481) with its distinctive spines on the thorax and elytra. This is a pest on cultivated rice in Asia from Pakistan through to Papua and South China, where the adult feeds on the leaves making white longitudinal scarred areas, and the larvae mine in the leaves between the two epidermal surfaces making white blotches.

Sub-family **Cassidinae**

The Cassidinae are tortoise beetles, with a distinctive enlarged flattened dorsal skeleton shielding the body and head underneath. These beetles are often beautifully irridescent gold or greenish in colour, and the larva is curious in that as it grows it keeps the old exuviae (moulted skins) attached to the posterior end of the body. The damage done to the leaves of the host plant by both adults and larvae is the same, with holes eaten in the leaf lamina but usually leaving the leaf margin intact as distinct from most caterpillar or weevil damage.

In Hong Kong by far the most preferred host for tortoise beetles is the genus *Ipomoea* and at least four species are common. *Ipomea* is commonly represented here by *I. batatas* (Sweet Potato), *I. carica*, *I. digitata* (Morning Glory) and also *I. brasiliensis,* to be found on sandy sea shores above the high tide mark, and also by some less common and cultivated species. Plate 482 shows an adult and a larva of *Aspidomorpha* species which is very common throughout Africa and parts of Asia. Plates 483 & 484 show the larva and adult of the shiny green little *Metriona circumdata* on Morning Glory leaves, while Plate 485 shows the tortoise beetle damage to a Sweet Potato Leaf. The other common local species is *Aspidomorpha chinensis* which has four distinct 'arms' or 'horns' and this distinguishes it from the common Two-horned Tortoise Beetle which has been tentatively identified as *Aspidomorpha furcata* and has only two anterior 'horns' demarked in the shiny gold coloration (Plate 486).

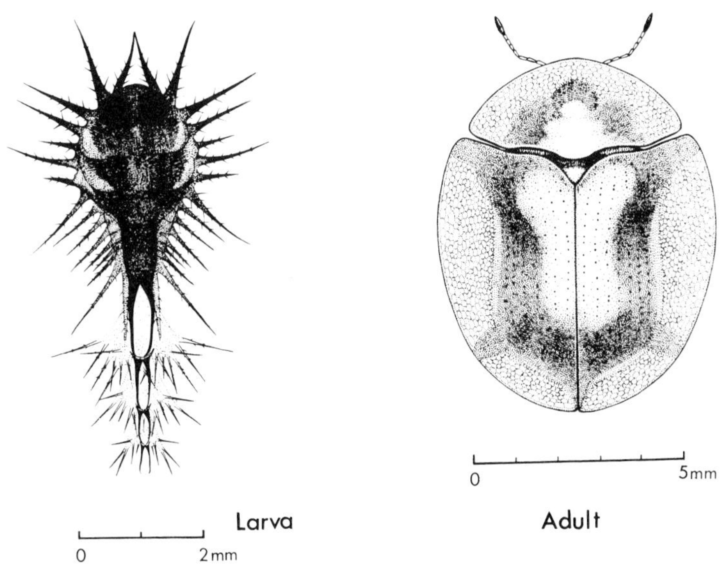

Plate 482.   Larvae and adult of tortoise beetle, *Aspidomorpha* sp.

Plate 483.   Larva of Green Tortoise Beetle, with moulted exuviae attached to posterior part of body.

(Above)   Plate 484.   Green Tortoise Beetle, *Metriona circumdata* (Col., Chrysomelidae, Cassisinae); on Morning Glory leaf; body length 5 mm.

(Left)   Plate 485.   Sweet Potato leaf eaten by tortoise beetles.

(*Above*) Plate 486. The local Two-horned Tortoise Beetle, *A. furcata* (Coleoptera, Chrysomelidae, Cassidinae).

(*Left*) Plate 487. Adult purple *Sagra purpurea* (Col., Chrysomelidae, Sagrinae); length 20 mm.

Sub-family **Sagrinae**

The Sagrinae is a little-known group of leaf beetles but occasionally the beautiful shiny purple-coloured *Sagra purpurea* is found locally. It has long, stout, curved hind-legs (Plate 487).

Sub-family **Eumolpinae**

The Eumolpinae are represented by a small rounded metallic greenish-

Plate 488.   Small Black Leaf Beetle, *Colasposma metallicum* (Col., Chryso-melidae, Eumolpinae); body length 4 mm.

black leaf beetle called *Colasposoma metallicum* (Plate 488) to be commonly found on Morning Glory leaves, and Sweet Potato, and the small black *Nodina chalcosoma* in Rose Myrtle flowers.

Sub-family **Criocerinae**

In Europe the common Asparagus Beetle is a representative of the Criocerinae, but no local species belonging to this group have so far been collected.

Sub-family **Galerucinae**

The Galerucinae are closely related to the flea beetles and they both contain a large number of important pest species. They are difficult to distinguish from some of the other sub-families. Two common local species are the small Black Cucumber Beetle *(Aulacophora lewisi)* which

Plate 489.   Ten-spotted Leaf Beetle, *Oides decempunctata* (Col., Chryso-melidae, Galerucinae).

Plate 490. Larva of *Oides decempumpunctata* feeding on leaves of *Ampelopsis* vine, Tweed Bay.

has a black body and brown head and thorax to be found on Cucurbitaceae, and the large Ten-spotted Leaf Beetle *(Oides decempunctata)* recorded feeding on grapevine and *Ampelopsis* vines (Plates 489 & 490). The tiny *Monolepta signata* can be found at times in the New Territories in large numbers, sitting on the tips of blades of grass (Plate 491).

Plate 491. *Monolepta signata* (Col., Chrysomelidae, Galerucinae) sitting atop a blade of grass in the New Territories; body length 4 mm.

Sub-family **Halticinae**

Flea beetles (sub-family Halticinae) are very characteristic with their flea-like size, long antennae, and stout hind femora with which they jump considerable distances. They are an easily recognized sub-family. The adults eat small holes in the leaves of their host plants giving them a 'shot-hole' appearance, but in heavy infestations the leaf is so damaged that holes are nearly contiguous and the leaf dies (Plate 492). Attack on seedlings can be particularly devastating and in Hong Kong during the summer young *Brassica* plants in the New Territories are at times completely destroyed by *Phyllotreta* flea beetles. *Phyllotreta* is a genus of cosmopolitan distribution; a dozen or more species are pests of Cruciferae, while the other one is on Cotton, and another on cereals.

The larvae live in the soil; they feed on the roots of the plants but do no appreciable damage, and pupation takes place in the soil. Plate 493 shows two examples of adult flea beetles, *Phyllotreta cruciferae* is shiny black with a greenish tinge, and *P. nemorùm* is indistinguishable from the local *P. striolata* with the black body and a broad yellow stripe down each elytron. Plate 494 shows two adult *F. striolata* on a damaged *Brassica* leaf in the New Territories. Apparently there are two species of

Plate 492.   Flea beetle damage to a leaf of *Mallotus spelta*, by adults of *Apthona wallacei* (Col., Chrysomelidae, Halticinae); beetles 2 mm in length.

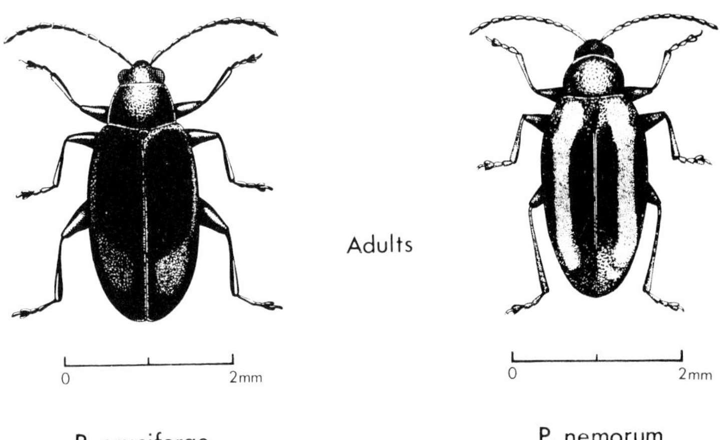

Plate 493. Adult Flea Beetles, *Phyllotreta crucifera* and *P. nemorum* which is almost identical with the local *P. striolata*.

Plate 494. Adult Flea Beetles *F. striolata* feeding on a leaf of a *Brassica* plant (Col., Chrysomelidae, Halticinae).

Plate 494A.   Citrus Flea Beetle, *Prodagricomela nigricollis,* a flea beetle whose adult eats holes in the leaves and the larva mines the leaves; adult beetle eating a leaf of pummelo at Tai Po Kau.

Plate 494B.   Citrus Flea Beetle larva *in situ* at the end of it tunnel.

Argopistes occurring locally on the leaves of *Citrus* spp. and in 1973 an interesting infestation was recorded at Pokfulam Reservoir where a large and a small tree of *Mallotus apelta* had very serious and extensive leaf damage inflicted by a tiny black flea beetle identified as *Apthona wallacei.* It is unusual for a tree (6 m high) to be attacked by flea beetles as they normally infest herbaceous plants or shrubs. A recent record was of *Prodagricomela nigricollis* (Citrus Flea Beetle; Citrus Leaf Miner) attacking *Citrus* trees at Tai Po Kau, the small dark adults were eating holes in the leaves (Plate 494A) and the orange coloured larvae were making wide tunnel mines (Plate 494B). A similar species was found at Wu Kwai Sha in 1980 attacking Chinese Privet (Plate 494C).

Plate 494C.   Privet Flea Beetle, larval leaf tunnels, Wu Kwai Sha, 1980.

Super-family **Curculionoidea**

The ultimate group of beetles is the Curculionoidea (weevils *sensu lato*), which on the basis of number of species can be regarded as one of the largest, and in several respects it is considered as the most highly evolved group of beetles. The antennae are always clubbed apically and usually elbowed, and a large number of species have a long thin snout bearing the mouth-parts terminally. The larvae are all completely legless (apodous) but with a well sclerotized head capsule and large powerful jaws. The first two groups are not well studied, and in Hong Kong only odd unidentified specimens have been encountered.

Anthribidae

The Anthribidae is a tropical group found associated with dead timber and they have a short, broad snout and long thin antennae.

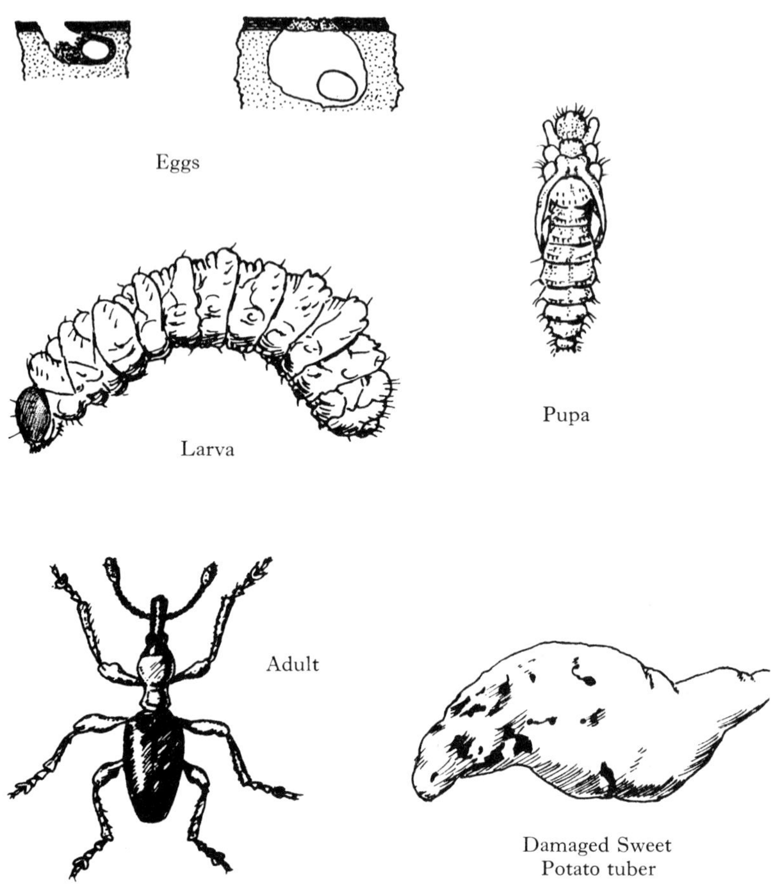

Plate 495.   Sweet Potato Weevil, *Cylas formicarius* (Coleoptera, Apionidae); immature stages, adult beetle, and damaged Sweet Potato tuber.

Brenthidae

The Brenthidae are also confined to wooded tropical regions and in body form they are elongate and thin, with quite long, thin antennae.

Apionidae

The Apionidae are sometimes regarded as separate weevils but often they are placed within the Curculionidae as just a sub-family. They are easily recognized by their non-elbowed (non-geniculate) antennae, and often they have a long pointed snout tapering down from a stout body. *Cylas formicarius* (Plate 495) is the local Sweet Potato Weevil; both adults and larvae spoil sweet potato tubers with their tunneling and burrows. Damage is done in the field but will continue after the harvest, and with more than a very light infestation the tuber is ruined for commercial purposes. Very tiny black, long-snouted *Apion* weevils are found throughout the world on a large range of plants, but the flowers of Leguminosae are preferred. There are 80 species of *Apion* in Britain alone.

Curculionidae

The Curculionidae are the weevils proper *(sensu stricta)* and as adults they all have clubbed, elbowed (geniculate) antennae, with the face drawn out into either a long narrow snout (long-nosed weevils) or a short broad one (broad-nosed weevils). The important palm pest, the Asiatic Palm Weevil *(Rhynchophorus ferrugineus)* has not yet been recorded, but might be expected to occur in Hong Kong. The adults feed on the palm crown and oviposit there, and the larvae bore in the crown and later in the stem of the various species of palm tree. Each region of the tropics has its own species of *Rhynchophorus* as a major palm pest. *R. ferrugineus* is particularly important on oil palm trees in Malaysia where it is a serious pest.

The other large local weevil is the fairly common Bamboo Shoot Weevil *(Cyrtotrachelus longimanus)*. This is a spectacular insect about 35–45 mm in body length, orange-brown in colour with a black head and various patches on the body. The snout is long and thin and flattened distally where the mandibles are borne; the forelegs are particularly long and fringed with long bristles, and the tibiae end in a long stout curved spine (Plate 496). The purpose of these anatomical modifications is revealed by observation of the female during her feeding/oviposition activities (Plate 497) when it is seen that she sits head-downwards on a thick young shoot of the bamboo some 20–30 mm in diameter and firmly grasps the stem with her forelegs. She then uses her mandibles to bite a hole through the stem of the plant, presumably feeding on the plant sap while doing so, and her snout is just long enough to pierce through to the pith of the stem where subsequently an egg is laid. The adults can be seen sitting on the young bamboo shoots in many different parts of the Colony during August each year.

The egg develops rapidly into a white, legless grub which when fully

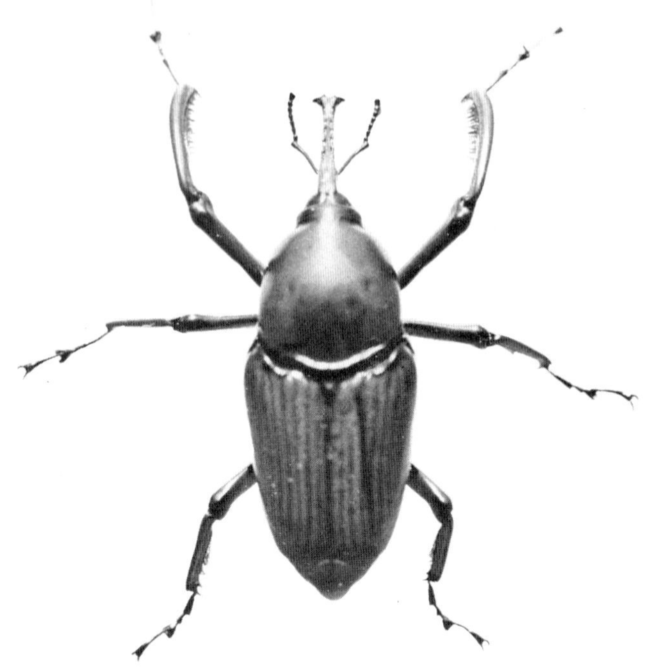

Plate 496.   Adult Bamboo Shoot Weevil, *Cyrtotrachelus longimanus* (Coleoptera, Curculionidae); body length 32 mm.

Plate 497.   Bamboo Shoot Weevil, female, feeding and making an oviposition site in the shoot of a young bamboo stem, probably *Sinocalamus oldhami*.

Plate 498. Larva of Bamboo Shoot Weevil inside tip of bamboo stem showing the eaten out growing tip.

grown completely fills the hollowed out growing point of the bamboo shoot. In fact all that remains are the apical leaf sheaths enclosing the large fat larva (Plate 498). It is thought that pupation takes place in the soil.

After the growing point is destroyed the plants make secondary growth by the development of lateral buds into small branches (Plate 499). In many parts of Hong Kong, whole stands of bamboo show this distinctive secondary branching, sometimes only a few metres from the

Plate 499. Secondary branching of bamboo stems following apical shoot destruction by Bamboo Shoot Weevil.

ground but other times the branching may be 10–15 m from the ground. Only certain species of bamboo appear to be attacked. The several very large local species and the several small species appear to be exempt from infestation whereas the most frequent host is the common *Sinocalamus oldhami* with its 20–30 mm thick shoots. Fortunately most stands of bamboo that are attacked in August usually show a second period of growth in November when there are no weevils about to damage the shoots.

A fairly common pest species of weevil is the Banana Stem Weevil *(Odoiporus longicollis)*, most easily distinguished from the Banana Weevil *(Cosmopolites sordidus)* by the fact that the elytra do not quite cover the abdomen leaving the terminal fifth or sixth segments uncovered dorsally (Plate 500). *Cosmopolites* occurs throughout Africa, India, Australasia, and Malaysia and bore in the rhizome of the banana plant. *Odoiporus* is the Southeast and East Asian counterpart but restricts its boring activities to the 'stem' of the banana, that is to the leaf bases. In Malaya both weevils occur and their joint effect on a banana plant is devastating. The larvae bore in the pseudostem making extensive tunnels which often become infected by fungi and bacteria (Plates 501 & 502) and pupation takes place in a fibrous coccoon within a large wide tunnel. Lightly infested plants do not appear to suffer any ill-effects, but more heavily damaged plants may suffer a pseudostem break as the fruit stem developes and the weight of fruit increases; very heavily damaged plants break when quite small as a result of the physical weakening of the

Plate 500.   Adult male *(left)* and female *(right)* Banana Stem Weevil, *Odoiporus longicollis* (Coleoptera, Curculionidae); body length 13 mm.

Plate 501.   Banana Stem Weevil, adult, larva, and pupa inside pupal coccoon.

Plate 502.   Stem of Banana plant with outside leaf base removed, showing extensive Stem Weevil damage.

Plate 503.   Young Banana plant, with stem broken resulting from heavy Banana Stem Weevil infestation.

pseudostem (Plate 503). Actual 'stem' breakage is uncommon in Hong Kong, but is far more frequent in Malaysia where the *Odoiporus* infestations are heavier.

Among the local broad-nosed species is a common small dusty-yellow grey species, *Hypomeces squamosus* (Plates 504 & 505) found commonly on leaves of *Citrus*, Sweet Potato and Morning Glory. The typical damage done by this type of leaf-weevil is that they eat the leaf margin in a series

Plate 504.   The Gold Dust Weevil *Hypomeus squamosus* (Coleoptera, Curculionidae); body length 12 mm.

Plate 505.   *Hypomeus squamosus* feeding on the leaf of a Grapefruit bush.

of notches, and different species do almost identical damage to different crops in all parts of the world. In Plate 506 can be seen such characteristic damage done by this type of weevil to the leaves of an unidentified plant. Another species commonly found on *Ipomoea* leaves is *Blosyrus herthus*. It is grey coloured with a squat, rounded, lumpy body and broader snout.

Two important pests of stored grains are Maize and Rice Weevils, *Sitophilus zeamais* and *S. oryzae* (Plate 507), almost undistinguishable in size and shape. They are small (2–4 mm long), dark brown or blackish

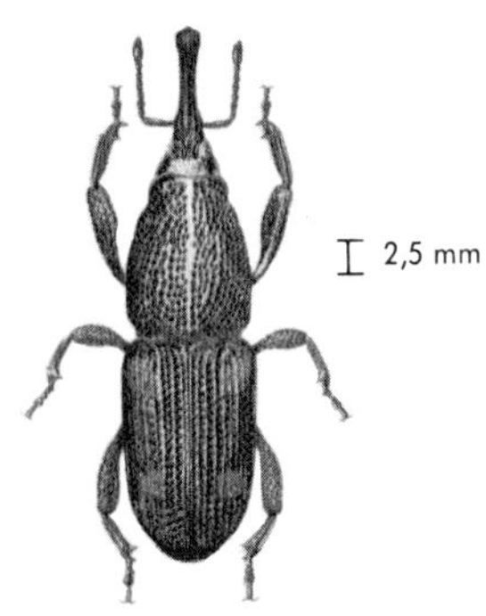

Plate 506.   Typical weevil leaf-notching damage done by feeding adults, usually broad-nosed species.

Plate 507.   Rice Weevil, *Sitophilus oryzae* (Coleoptera, Curculionidae).

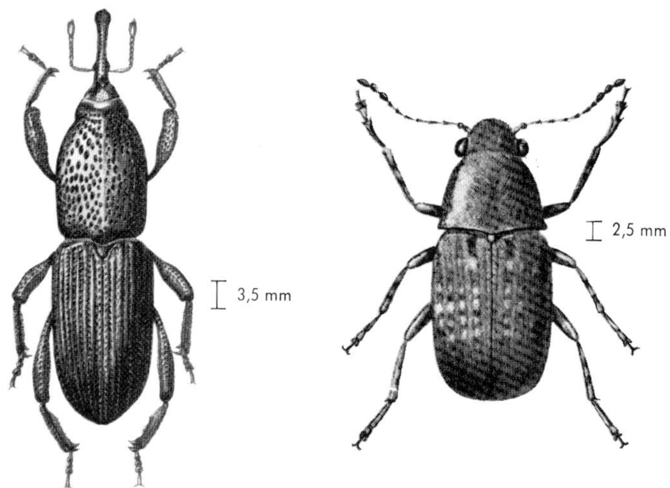

Plate 508. Grain Weevil, *Sitophilus granarius*.

Plate 509. Coffee Bean Weevil, *Araecerus fasiculatus* (Coleoptera, Anthribidae).

and are primary grain pests in that they attack undamaged grains, and they do extensive damage in locally stored rice. The temperate Grain Weevil *Sitophilus granarius* (Plate 508) occurs occasionally on grains shipped from more northern regions.

One broad-nosed weevil pest of stored products is the Coffee Bean Weevil *(Araecetus fasciculatus)* (Plate 509) which is well distributed throughout the warmer parts of the world and is common locally. It flies strongly and is to be found in many flats; it is also found attacking the drying seeds of Nasturtiums grown as pot plants and in flower beds. A few species of weevils have larvae that bore tree trunks, including the local *Sipalinus hypocrita.*

Scolytidae

The final group of beetles, very closely related to weevils and some-times included within the family is the Scolytidae which are mostly timber pests boring under the bark of living trees. One species is at present responsible for serious damage in Europe. On its body carries spores of the fungus which causes Dutch Elm Disease, and thousands of mature elm trees have been destroyed in Southern England in the last few years by this disease. Some scolytid beetles do occur here and the typical larval and breeding galleries can be seen under the bark of several local trees, particularly on *Pinus massoniana.*

CHAPTER SIX

COMMON ARACHNIDS

CLASS **ARACHNIDA**

ORDER **ACARINA**
(Mites and Ticks)

The Acarina are not insects, as can be seen easily by their four pairs of legs (instead of three), lack of wings, and lack of definite head, and having the body divided into two segments, rather than the head, thorax and abdomen of the insects. So they are classified in the Class Arachnida together with the spiders and scorpions. However, if one deals with insects on plants, which in the professional sense becomes agricultural entomology, or insects parasitic upon animals or man, then it is customary for the entomologist to include the Acarina in his sphere of activity. Similarly, for the person who cultivates plants in his garden or on his verandah, then from time to time the pest that will attack the flowers and ornamental trees will be plant-parasitic mites rather than insects; sometimes mixed infestations of both mites and insects will be present. The study of Acarology (i.e. Acarina) is particularly difficult because of the minute size of the mites and their specialized nature, and little work on mites and ticks has been done in Hong Kong. It is hoped that this small chapter will serve as an introduction.

Unlike the fairly stable classification for the insects, Acarina classification tends to differ markedly according to the textbook used. Adopting the scheme whereby the Acarina is regarded as an order there are seven distinct groups regarded as sub-orders, which together contain more than 200 different families and many thousand species. A large number of these families and species will most certainly be represented locally but to date they have not been studied.

Onychopalpidae

The Onychopalpidae is a small group of primitive, predatory, cryptozoic mites.

Sub-order **Mesostigmata**

The Mesostigmata is by far the largest group and is sometimes lumped together with the sub-order Ixodides and called the Parasitiformes because of the parasitic nature of most of the species contained therein. The Mesostigmata are small parasitic or predatory species, well armoured and with the typical four pairs of legs. It should be mentioned that in the life-history of many Acarina the stage which hatches from the egg is called the larva and has only six legs. After the first moult it acquires the typical eight legs and is then referred to as a nymph.

Parasitidae

The Parasitidae are small predatory mites found in litter and decaying vegetation. They feed on small insects and other arthropods.

Dermanyssidae

The Dermanyssidae are known as red mites, and are medium-sized parasitic mites, the best known of which is the cosmopolitan blood-sucking Red Mite of Poultry *(Dermanyssus gallinae)*. This is a serious fowl pest for a heavy infestation may kill the birds, and in addition it is naturally infected with several viruses. Given the opportunity it will feed on man. There are also to be found *Ornithonyssus bursa,* the Tropical Fowl Mite, and *O. bacoti,* the Tropical Rate Mite, which is in fact cosmopolitan on rats and man. Several other species are to be found on wild birds and rodents.

Phytoseiidae

The family Phytoseiidae is a group of mites both of interest agriculturally and academically. They are predatory species to be found usually on plant foliage. *Phytoseilus reigeli* is a large, active, red-coloured, long legged predator upon red spider mites (Tetranychidae) which are serious agricultural crop pests on many crops throughout the world, and it has been used successfully for biological control programmes in Europe, Africa and North America. In fact this mite is now being bred specially for use in biological control and several companies in U.S.A. and Europe offer it for sale. Similar species belonging to this family can be found on local vegetation. Most of the other 63 families in this group are small and not of particular importance.

Sub-order Ixodides

The Ixodides are the largest Acarina and are called in the vernacular—ticks. They are all parasitic, mostly on vertebrates, but many on mammals and birds, including man if given the opportunity. They are large in size with a well-developed, piercing proboscis-like mouth structure called a hypostome, and armed with successive rows of recurved teeth. Once the hypostome is fully inserted into the host body it is fixed in position; any human pierced by a large tick is advised to induce the tick to withdraw its hypostome by, for example, touching the back of the tick with a lighted cigarette end or putting kerosene on its body. If the tick is just pulled off, then the hypostome invariably breaks off at the base and remains fixed *in situ* where it often causes a localized infection. Ticks are divided clearly into two separate families, the Argasidae (soft ticks) and the Ixodidae (hard ticks). Fortunately in Hong Kong the tick fauna is very sparse owing to the lack of natural reservoir hosts (wild mammals or even domestic cattle), but in Africa, India, and many parts of North and South America the local tick problem can be very serious, particularly since they are vectors for a number of protozoan and viral diseases.

Argasidae

The Argasidae are soft-bodied with the mouth-parts hidden ventrally; they are cosmopolitan, feeding on many hosts (for example, mammals, birds, snakes, turtles), and most frequently living in the nests or burrows of the host. They frequently feed intermittently and may drop off the host between feeds. *Argas persicus* is the Fowl Tick which is common in warm and temperate countries on a range of birds. This is a serious pest for the tick feeding disturbs and upsets the birds and also causes anaemia in addition to spreading several diseases. Several species of *Ornithodorus* occur on wild and domestic animals and birds, and this genus is a vector of several medical and veterinary diseases in parts of tropical Asia and the U.S.A.

Ixodidae

The hard ticks as their name suggests have a large dorsal skeletal plate and hence are regarded as being armoured. They are large, oval bodied ticks with the mouth-parts clearly visible at the anterior end of the body. The dorsal skeletal plate (scutum) extends over the whole body in the male tick but only over the anterior part in the female. Some are very ornate and have coloured, enamel-like areas on the body. Eggs are laid together in batches of up to 18,000 under stones or in crevices, and the female tick then dies. The newly hatched larvae (with only six legs) are called 'seed ticks' and they climb up into the foliage to wait for a passing host. In grassland they form a little brown clusters on the terminal leaf tips and they wait for a passing host to brush against the plant whereupon they firmly hold on to the fur, feathers or clothing of the host. Once on the host the larval ticks take a blood meal, and when engorged they usually drop off to moult on the ground.

Actually, hard ticks are biologically classified as one-host, two-host, or three-host ticks, according to the number of hosts required during the life-cycle. The one-host ticks remain on the same host all the time, with moulting taking place on the body of the host. The two-host ticks have the larvae feeding and moulting on the host but the nymph drops off to moult on the ground, and the adult tick has to find a new host. The most common type is the three-host tick where each stage in the life-history (larva, nymph, and adult) requires a separate host. Typically at each feed the tick engorges itself and physically swells up enormously, often ending up finally as large as a pea seed when fully engorged.

The common European Castor Bean Tick *(Ixodes ricinus)* does not apparently occur here. In Hong Kong the common tick on domestic dogs is *Rhipicephalus sanguineus* (Brown Dog Tick) (Plate 510) which is originally from Africa but now is almost quite cosmopolitan. This is a three-host tick which is a vector of several diseases, and is very abundant on some local dogs. In some houses and flats these ticks can be seen at times literally in hundreds, crawling up the walls as they search for new hosts.

Plate 510.   Brown Dog Tick, *Rhipicephalus sanguineus* (Acarina, Sxodidae).

Sub-order **Trombidiformes**

The Trombidiformes contains both plant and animal parasites. They are small to tiny, and a few are predators. Out of the 35 families about six are of importance. In some groups there is considerable morphological reduction and the number of legs can be reduced to only two pairs.

Eriophyidae

The Eriophyidae are minute plant gall mites, only visible under a microscope and of worm-like appearance with only two pairs of anterior small legs. Although microscopic in size they produce very visible

Plate 511.   Ventral surface of leaf of *Schefflera octophylla* showing erinia produced by a species of *Eriophyes* (Acarina, Eriophyidae).

outgrowths on plant leaves, usually on the undersurface, and the warty or hairy-looking outgrowths are called erinia. Very small erinia are often sunken, but as they become larger they swell and protrude and between the hair-like outgrowths the minute worm-like mites can be seen with a microscope. A local species of *Eriophyes* causes large erinia on the leaves of *Schefflera octophylla* (Ivy Tree) (Plate 511) and on some small trees the leaves are so heavily attacked that they become completely twisted up and distorted.

Typical eriophyid erinia can be found locally on a wide range of plants. Several species of Eriophyidae are serious agricultural pests. One local species, Citrus Rust Mite *(Phyllocoptruta oleivora)* (Plate 512) produces

Plate 512.   Citrus Rust Mite, *Phyllocoptruta oleivora* (Acarina, Eriophyidae) and damaged fruit.

disfiguring rusty-coloured patches on young citrus fruit.

Tarsonemidae

The Tarsonemidae are also plant mites but without a vernacular name. They are small and not so reduced anatomically but are still rather elongate and the legs rather short. Some are free-living, some scavenging, some parasitic on insects and several are important crop pests, especially on strawberries, but to date none has been identified locally.

Tetranychidae

The Tetranychidae are the red spider mites so well-known to horticulturalists and agriculturalists. They are small, just visible to the unaided eye, red or green in colour, and found on plant foliage, usually the leaves. They lay large globular red eggs on the leaves. Sexual dimorphism is evident (Plate 513), and there are several very important

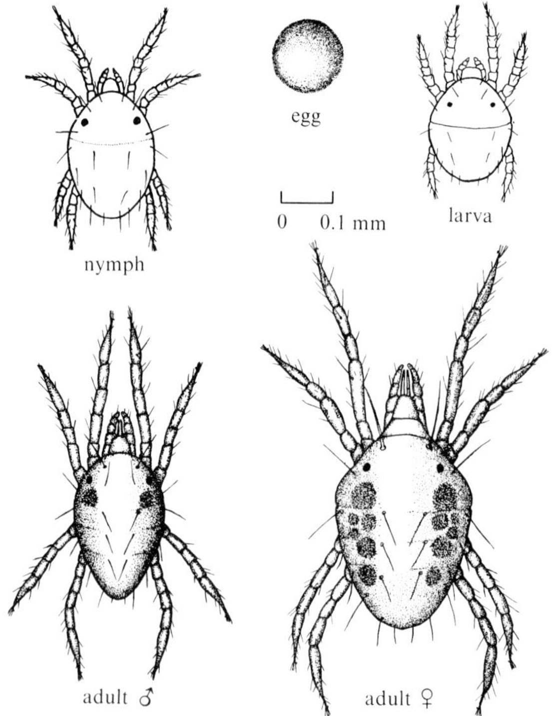

Plate 513. Tropical Red Spider (Carmine) Mite, *Tetranychus cinnabarinus* (Acarina, Tetranychidae), showing immature stages and adults.

species. The mites scarify heavily the leaves as they rupture the epidermal cells to suck out the sap. In heavy infestations there is sometimes a fine silk webbing produced over the foliage, and the leaves if seriously damaged curl up, become brown and fall prematurely. They are active mites with no limb reduction. The Tropical Red Spider Mite (Red Cotton Mite) *Tetranychus cinnabarinus* is pan-tropical in distribution and polyphagous in feeding habits (Plate 513); it is scarcely distinguishable from its temperate counterpart *T. urticae*. Very heavy local infestations can be seen on the leaves of *Oxalis corymbosa* (Plate 514), papaya, orchids and other plants. Two other important species occur in Hong Kong, they are the Citrus Red Spider Mite *(Panonychus citri)* (Plate 515) found on local *Citrus* and peach trees, and the Oriental Mite *(Eutetranychus orientalis)* (Plate 516), common here on the leaves of frangipani where it produces a white scarification (Plate 517).

Tenuipalpidae

The Tenuipalpidae (= Phytoptipalpidae) contains tiny, reddish, plant mites of which a few are crop pests (but not very serious).

Demodicidae

Demodicidae are tiny, worm-like mites called hair follicle mites because they inhabit the skin and hair follicles of various mammals; in

Plate 514. *Oxalis corymbosa* leaf, ventral surface, with a heavy infestation of Tropical Red Spider Mite.

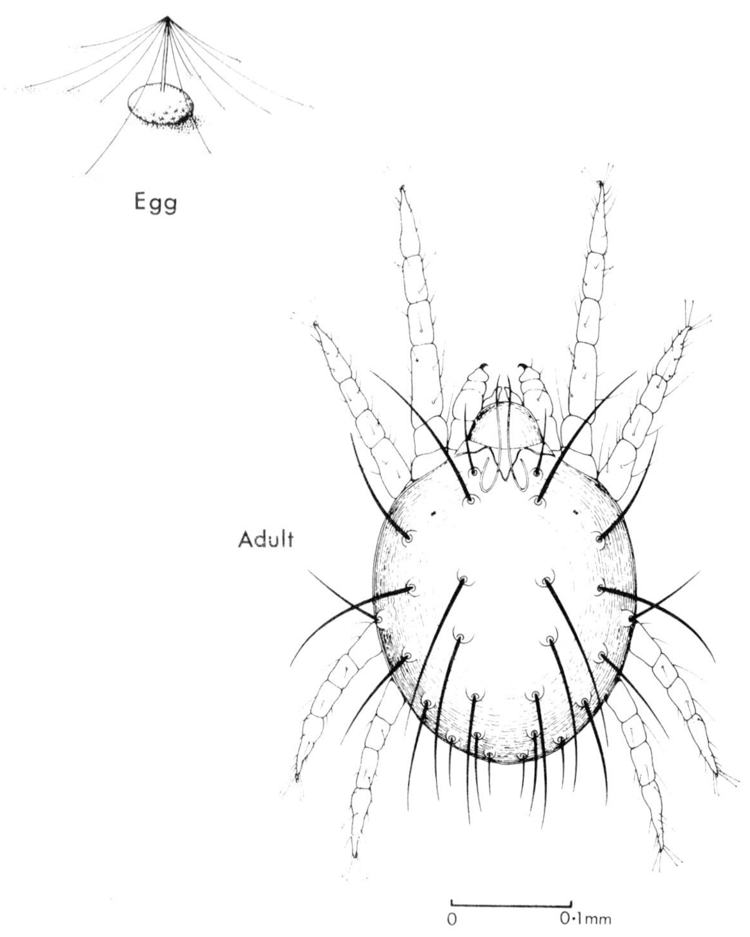

Plate 515.   Citrus Red Spider Mite, *Panonychus citri* (Acarina, Tetranychidae).

man they inhabit particularly the hair on the face. They usually do not appear to cause any harm. *Demodex canis* is the species on dogs, and *D. folliculorum* is the Human Hair Follicle Mite, which is reputed to be present in almost all human facial hair follicles.

Trombiculidae

The Trombiculidae are a curious group in that the nymphs and adults

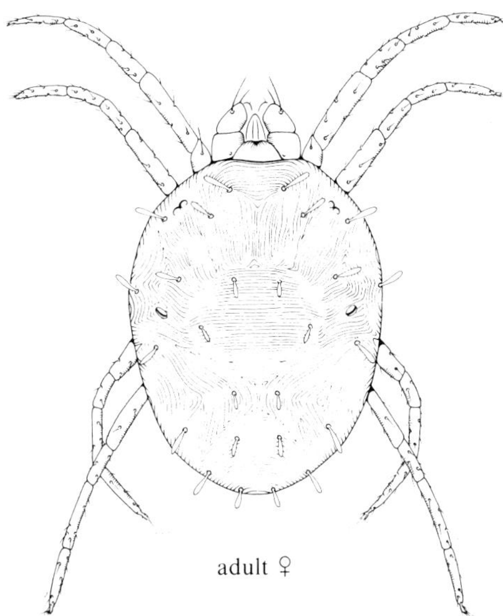

Plate 516. Oriental Mite, *Eutetranychus orientalis* (Acarina, Tetranychidae).

Plate 517. Frangipani leaf; dorsal surface with heavy infestation of Oriental Mite.

are free-living in the tropics and southern U.S.A., but the tiny larvae of *Trombicula* spp. are skin parasites of man and animals and they are reputed to be the most irritating (itching) ectoparasite known to man. Rodents and birds are the natural hosts, and these mites are common in S.E. Asia and their distribution does extend up to Hong Kong where they have been collected from local birds. The ectoparasitic larvae arc known as chiggers or redbugs (not to be confused with the Jigger Flea that burrows into the skin of the feet of man in Africa).

Sub-order **Hydrachnelle**

Hydrachnidae

The sub-order Hydrachnelle contains only one important family and that is the Hydrachnidae which are water mites. The many species of *Hydrachna* are large, plump, red-bodied, long-legged, active, predatory mites to be found in the local freshwater streams and reservoirs. They have two pairs of eyes which are easily visible under a microscope. The mites themselves are often up to 2 mm in body length. Several tiny dark grey or blackish species of water mites are abundant occasionally.

Sub-order **Sarcoptiformes**

The Sarcoptiformes is another group with rather diverse habits biologically. They are generally either free-living or parasitic.

Acaridae

The Acaridae are free-living and some cause damage as food pests. *Acarus siro* (Plate 518) is the cosmopolitan Flour Mite, very common in most parts of Hong Kong on a range of flour and milled grain products. This is basically a scavenger and it is to be found naturally in birds' nests and animal lairs. Also in this group are species of the cosmopolitan Cheese Mites (species of *Rhizoglyphus*) to be found on many different foodstuffs.

Sarcoptidae

In the Sarcoptidae are the itch and mange mites. *Sarcoptes scabei* is the minute, reduced mite with only two pairs of legs which burrows into human skin to cause the itching which earns it the name of the Human Itch Mite. Another variety of the same species is the Mange Mite (Scabies) found on animals. Mange or scabies is one of the most prevalent disorders affecting local cats and dogs, as any casual observation will immediately reveal (Plate 518A). *Cnemidocoptes* is a very common skin mite on birds and is common on chickens locally.

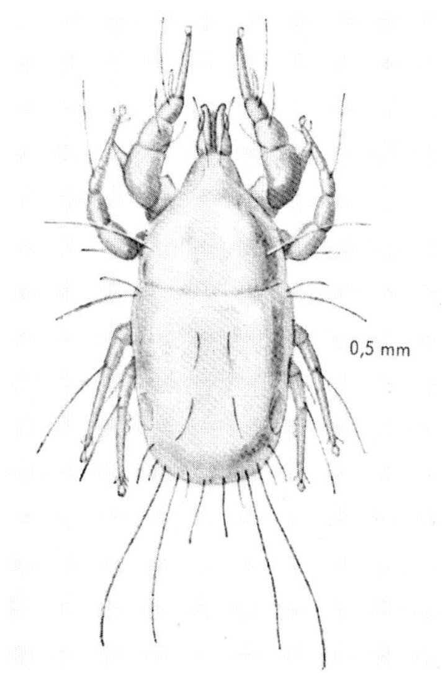

Plate 518. The cosomopolitan Flour Mite, *Acarus siro* (Acarina, Acaridae).

Plate 518A. Mange Mite *(Sarcoptes scabei)* on local dog, Wu Kwai Sha, 1980.

Psoroptidae

The Psoroptidae are Scab Mites on mammals; the genus *Psoroptes* is a widespread pest of importance on cattle, and *Otodectes* is the mite causing ear mange in the Carnivora and to be found in the ears of local cats and dogs. Several other families in this group are elsewhere found in the skin and feather follicles of wild birds, but they have not yet been collected in Hong Kong.

Sub-order **Oribatei**

Oribatidae

The final group of mites in this classification is the Oribatei, containing some 35 families, most of which are soil mites and of ecological interest because of their abundance. They feed on organic debris generally, and seven families are now known to act as vectors for various tapeworms in other countries, for example, the Sheep Tapeworm *Moniezia*. They are sometimes called beetle mites or armadillo mites because of their rounded appearance and well sclerotized skeleton, and they are very abundant in local soils and leaf litter.

ORDER **ARANEIDA**
(Spiders)

The final major order of Arachnida is the Araneida, the spiders, a group known to everyone. They all have abdominal silk glands which are visible externally as spinnerettes, and all have the first major pair of appendages, the chelicerae, modified into large poison fangs connected to internal poison glands. The mouth is small, and the pharynx suctorial, and they imbibe the body fluids of their prey by sucking. The male is invariably smaller than the female, and often is eaten by the female either during or immediately after the mating ritual. The family classification of spiders seems to be rather in a state of flux according to the different sources consulted.

Epeiridae

Most of the spiders of Hong Kong belong to the family Epeiridae (= Aranidae). They are the orb-web spiders which build a web of fine sticky silken strands suspended in vegetation in which they trap flying insects. Typically they have a silken tunnel or retreat in which they sit and wait for struggling prey to vibrate the web. The eggs are generally laid in a silken coccoon in the female retreat. The sub-family Argiopinae are the worldwide common webspinners, being characterized by their large vertical web and large female spiders.

Three common local genera are *Leucauge* spp., *Araneus* spp., and *Argiope* spp. (Plate 519), to be found in wooded areas with large webs up to nearly a metre in breadth. The very common black and yellow woodland spider of Hong Kong is *Nephila maculata* (Plate 520) which occupies a separate sub-family of its own—Nephilinae. The big female has a body length of up to 40 mm and a leg-span of 10–14 cm; she mostly sits in the centre of her large web which is up to 1 m in diameter and strung between adjacent trees, bushes, or across paths (Plate 521). Almost indistinguishable spiders are found in large numbers in Malaysia, the Seychelles, and along the coast of Kenya and Tanzania where they are especially abundant in the telephone wires there. The female has huge chelicerae and is presumed to be quite poisonous. In 1963 a soldier from Sek Kong camp was hospitalized for several days after being bitten on the arm by such a spider. In Tai Po Kau forest a large female *Nephila* was found eating the remains of a small bat *(Pipistrellus abramus)* that had been trapped in her web. Sometimes smaller webs are found with a small black delicate spider sitting in the centre and these are usually the males.

Plate 519.   Large Woodland Web Spider, *Argiope* sp. (Araneida, Epeiridae, Argiopinae); leg-span 6 cm.

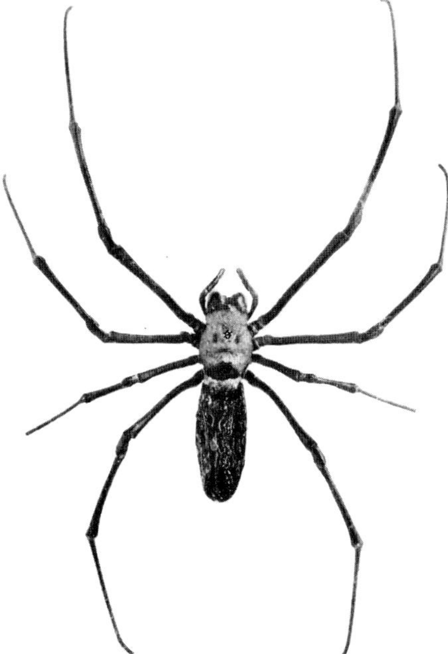

Plate 520.   Large Woodland Spider, *Nephila maculata* female (Araneida, Epeiridae, Nephilinae); leg-span longitudinally 14 cm.

Plate 521.   Female Large Woodland Spider, *Nephila maculata,* sitting in her web.

Generally, the orb-web spiders are rather sluggish and do not move fast, but because of the size of the poison-fangs (chelicerae) it is strongly recommended that they are left alone, unless they are present in a garden where they are a definite hazard to small children, then they should be destroyed (most household aerosol pesticides will kill spiders).

Gasteracantha in the separate sub-family Gasteracanthinae is a very small rounded black spider. It has a body diameter of 5–8 mm with several distinctive 'spines' projecting posteriorly from the opisthosoma and is to be found sitting in a small web.

Thomisidae

The Thomisidae are called crab-spiders because they hold their walking legs sideways and they move transversely in the manner of a crab. Some species are brightly coloured and they often sit inside opened flowers waiting for visiting insects.

Lycosidae

The Lycosidae are wolf spiders which hunt by speed and surprise and

Plate 522.   Large Brown House Spider, *Heteropoda venatoria* (Araneida, Lycosidae); leg-span 10 cm.

build no webs. They do make silk in which they keep their egg mass, and large females can often be seen carrying the silken coccoon under their body. The most spectacular local species is the large Brown House Spider, *Heteropoda venatoria* (Plate 522), with a body length of about 25 mm and leg-span of 10 cm. It is very abundant locally, and is most common in older houses and buildings but is now becoming well-established in the new blocks of flats. It is active at night when it preys on cockroaches, but can be seen sitting in corners during the daytime. The poison-fangs are very large and it gives a very painful 'bite'; they move with extreme rapidity. It is very much a domestic species and is not usually found outside human dwelling places. Smaller species of wolf spider are common in woodland and in litter but have not been studied at all.

Attidae

The final family of spiders is the Attidae (= Salticidae) which as the name suggests are jumping spiders. They are small, often distinctively patterned, webless, and they progress and hunt by means of a series of jumps. Again they have not been studied locally, although they are common here as a group.

OTHER ARACHNIDS
(Kingcrabs, Scorpions, Chelifers, etc.)

The Arachnida are a class of Arthropoda (taxonomically equivalent to the Insecta) characterized by their four pairs of jointed legs and a lack of a definite head, instead they have the body divided into two major parts, the anterior prosoma and the posterior opisthosoma. They lack the compound eyes of insects and have instead a group of ocelli distributed over the anterior part of the prosoma. Most species do not have a proper mouth and jaws for biting their food but have a small and strong sucking pharynx with which they imbibe liquid food or small chopped-up pieces of prey. They are mostly predacious and carnivorous and some have poison fangs with which they kill their prey.

These other Arthropoda are included here because they are closely related to insects and also because many people think of them as virtually being insects.

Order **Xiphosura**

The great majority of arachnids are terrestrial. In Hong Kong there are two species of king-crabs belonging to the order Xiphosura; the large species, *Tachypleus gigas,* which grows up to a breadth of 30 cm or more, is usually found in deeper water and in Deep Bay, and the smaller *T. tridentatus* can be found burrowing shallowly on some of the local sandy beaches in the sub-littoral zone of the shore. They both burrow in the sand or mud looking for worms and small clams which they chew up with their rather weak mouth-parts; the 'mouth-parts' are actually the spiny bases of the legs, called gnathobases.

Order **Scorpionida**

In the order Scorpionida are the scorpions, characterized by having a poison-sting at the end of their 'tail'. The sting is used to kill the prey which is then held by the large chelate pedipalps and chopped into small pieces by the small chelate chelicerae which work rather like scissors. The small chopped pieces of prey are then sucked up into the mouth by the suctorial pharynx. In some tropical countries scorpions grow to a length of 15 cm or more, but the only species recorded from Hong Kong is only about 25 mm long when adult. Since the publicity given to the first captures in 1971 and 1974/1975, many specimens have now been found—all from the Tai Po Kau Forestry Reserve in the New Territories where they were found under stones or large clods of soil. It has been identified as *Hormurus australasiae* and presumably feeds on tiny soil

Plate 523.   Hong Kong Scorpion, *Hormurus australasiae* (Arachnida, Scorpionida); body length 25 mm. including 'tail'.

and litter arthropods and worms. Since it is so small it can be assumed that the sting is too tiny to cause any pain to man even if the sting can pierce human skin (Plate 523).

Order **Pseudoscorpionida**

The Pseudoscorpionida are sometimes referred to as chelifers, and in fact most of the species belong to the genus *Chelifer*. They are small, mostly from 5–10 mm in length, and as the name indicates they resemble tiny scorpions, but have no 'tail' and of course no sting. They live gregariously in small colonies under loose tree bark or in leaf litter. They are presumably carnivorous and feed upon smaller weaker arthropods, and are very abundant in most parts of Hong Kong.

Order **Opiliones**

The Opiliones, or sometimes called Phalangida, are another order of Arachnida which contains the harvestmen. These are curious little creatures having a small rounded body some 3–5 mm in diameter, with four pairs of long delicate legs 10–20 mm in length. They are generally found in litter or walking over soil, or on tree trunks or in foliage, and are said to be predacious on small insects. There are several different species to be found locally but so far only *Gagrella splenders* (Plate 524) with its distinctive blue body has been identified. A similar species with a reddish body and very long legs is quite common in some wooded areas on tree trunks and in the foliage.

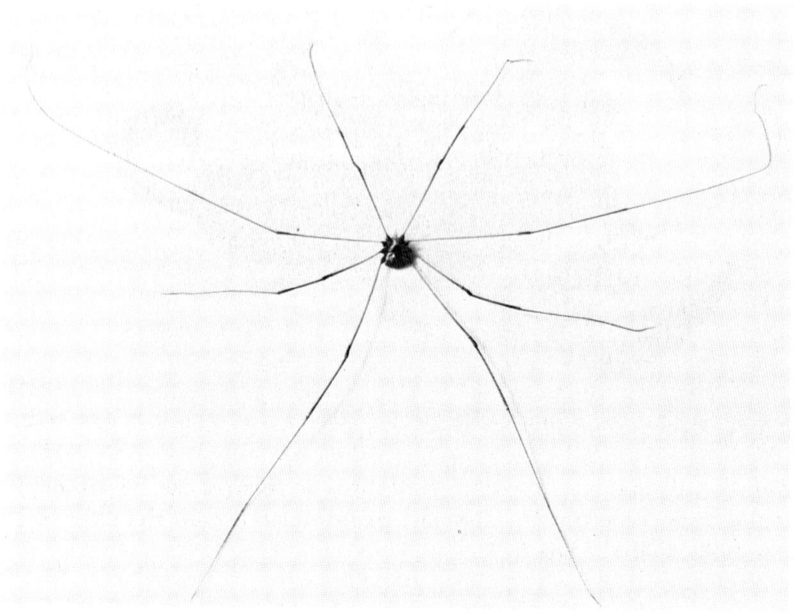

Plate 524.   Harvestman, *Gagrella splendens* (Opiliones); legspan 6 cm.

CHAPTER SEVEN

COMMON MYRIAPODS

CLASS **MYRIAPODA**
(Millipedes and Centipedes)

The last group of terrestrial Arthropoda common in Hong Kong is the class Myriapoda. They are characterized by having a large number of relatively uniform body segments with each segment bearing a pair of appendages (usually walking legs).

Order **Diplododa**

The order Diplopoda are the millipedes. They do not really have a thousand legs but do have a large number of several dozen pairs. As a group they are elongate, mostly black or dark brown in colour, rounded in section with short antennae, and with apparently two pairs of legs per body segment. In reality each apparent body segment consists of two fused body somites. Each segment bears a pair of glands which secrete hydrocyanic acid, and this accounts for the unpleasant smell most millipedes generate. When alarmed, most species are capable of rolling themselves up in a tight coil and lie completely immobile.

Millipedes have weak biting mouth-parts and feed either on detritus or soft vegetable matter. Within the order are three widespread families and although to date no local millipede has been authoritatively identified it is thought that there are local representatives of each.

Millipedes are generally found in cool damp places, such as leaf litter, holes in trees, holes in rocks, caves, under rocks, fallen trees and in rotting vegetation. They are only in the surface layers of soil generally—rather on the soil than in it.

Julidae

In the family Julidae (round millipedes), *Julus* is the most common genus and in many countries is the most abundant (Plate 525). Several local species are common; some are brown or black in colour but a few are bright reddish.

Polydesmidae

In the Family Polydesmidae (flat millipedes), *Polydesmus* and allied genera are not round in their body section but distinctly flattened, with an enlarged flat dorsal plate on each body segment. In Borneo and Malaysia there are short, broad and flat millipedes that look very much like large wood-lice; some have now been found locally.

Plate 525. Round Millipede, possibly *Julus* sp. (Diplopoda, Julidae); body length 34 mm.

Blaniulidae

The family Blaniulidae are spotted millipedes. *Blaniulus* is an elongate and thin (rounded) millipede with a series of conspicuous red spots along the body.

Two other orders of Myriapoda which are associated with millipedes are the Pauropoda and Symphyla. These are generally small animals, slender, with pale delicate bodies and they are often found actually burrowing in the soil. It appears that they need the added protection provided by a subterranean life. So far no specimens have been definitely collected from local soils, but they can be expected to occur here.

Order **Chilopoda**

The order Chilopoda contains the centipedes, a very vigorous, active group with their usually bright reddish-brown bodies, long jointed legs, longish antennae, and large black-tipped poison-fangs (chelicerae). As their behaviour suggests they are fierce predators, and if handled some will give a very painful 'bite'. Three common families can be easily distinguished.

Geophilidae

The family Geophilidae are thin centipedes. *Geophilus* is a thin, very elongate and rather delicate animal, usually secretive by nature and seldom seen.

Scolopendridae

The family Scolopendridae contains the centipedes proper. *Lithobius* is the common and widespread European genus and it is thought that the very common local smaller centipedes (body lengths 4–8 cm) may well be species of *Lithobius*. They are very common and widespread throughout the Colony, and are often found under flotsam and jettsam on the beaches. In life they are often rather bluish in colour, and they bite painfully if handled. *Scolopendra* is the local Giant Centipede (Plate 526) attaining a length of 16–18 cm or more—quite an alarming

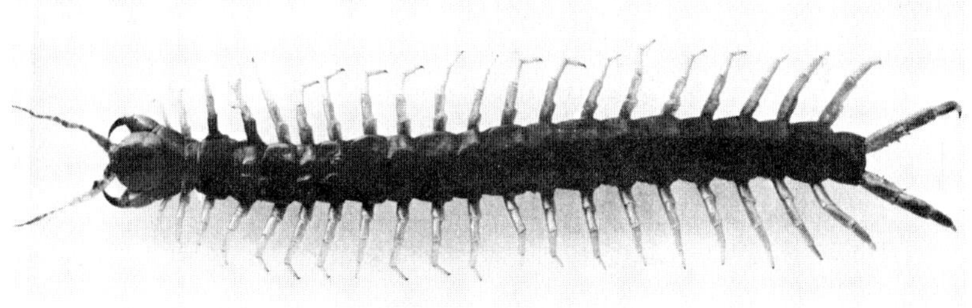

Plate 526. Giant Centipede, *Scolopendra* sp. (Chilopoda, Scolopendridae); body length 18 cm.

creature to encounter: *Scolopendra multidens* is the Urban Giant Centipede, to be found still in fair numbers in the older dwelling places in Hong Kong, such as the old houses in Wanchai and parts of Kowloon. They live, generally on the ground floor, in warm, damp and dark places which will of course harbour the cockroaches upon which they often feed; the poisoned prey is chewed up by the biting mandibles. *Scolopendra dehaani* is the almost identical Forest Giant Centipede to be found typically in woodland and forest areas. *Scolopandra morsitans* is a small Forest Centipede seldom more than 8–10 cm in body length. It is recorded that the bites of these giant centipedes are very painful. *Scolopendra* has been recorded in other countries killing and eating mice and lizards as well as insects and smaller animals.

Scutigeridae

The family Scutigeridae is the final common group of myriapods to be found locally. In some textbooks they are referred to as 'Long-legged

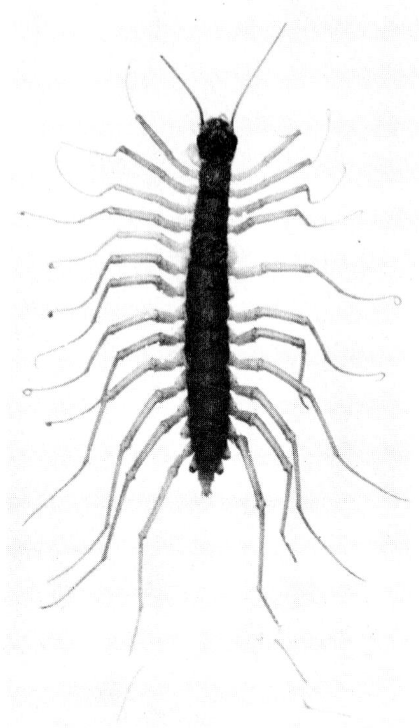

Plate 527.   Long-legged Centipede, *Theuropoda clunifera* (Chilopoda, Scutigeridae); body length 30 cm.

House Centipedes' but they are usually seen in forested areas or in leaf litter or under fallen tree trunks in Hong Kong. They have a short body (30–50 mm) but very long, delicate and fragile legs, and if captured they will invariably break off their legs at the slightest touch. Some specimens have been identified as *Thereupoda clunifera* (Plate 527) *(= Scutigera clunifera)*, but in view of the size range observed in local specimens it is possible that there may be several similar species differing mostly in size. They are alleged to feed on small insects, and large specimens have obvious 5-segmented poison fangs (chelicerae) held at the side and under the mouth. The mouth bears a distinct pair of mandibles for chewing up their prey. It is not known whether they can be harmful to man.

NOTES ON COLLECTING AND STORING INSECT SPECIMENS

Collecting

Insects occur almost everywhere in large numbers, except in the sea, and so can be collected anywhere. But certain types and species will only be found in certain habitats and at certain times of the year, for example, adult Bamboo Shoot Weevil will be found on some of the larger woodland and cultivated bamboos only during the period from the end of July through August. As already indicated, many species are restricted in their choice of host plant (or animal) so often the most effective way to locate these insects is first to find the host plant. Finding young insects on their food plant is a most effective method of collecting, for then they can be reared in captivity until metamorphosis or pupation takes place and perfect newly-emerged adults are obtained.

Rearing cages can be made from a number of different types of container, but plastic sandwich boxes are as effective as anything else for smaller insects. Cages should have tight-fitting lids but should have air holes for adequate ventilation and to prevent moisture condensation.

Professional entomological equipment is generally unavailable locally and has to be purchased from England or U.S.A. Recommended firms are listed below:

Griffin International, Griffin & George Ltd., P.O. Box 14, Wembley, Middx., HAO 1HJ, England.

Watkins & Doncaster, The Naturalists, 110, Park View Road, Welling, Kent, England.

BioQuip Products, P.O. Box 61, Santa Monica, California, 90406, U.S.A.

Nets Most collecting is done using an aerial insect net. This is generally a large, lightweight net with a mouth of 20–40 cm in diameter and an elongate light handle; the net bag is of nylon or fine mesh netting and at least twice as long as wide. Expensive types are usually made with the frame and handle of aluminium, but this type of net is not difficult to make for oneself. A sweep net is designed to brush through grass and light vegetation, and hence has a short stout handle and a much stronger frame and bag. The nets used for collecting freshwater insects are either long-handled with a small round mouth for scooping, or the bottom net has a D-shaped frame with the handle on the curved part so that the flat edge can be pushed along the bottom. A plankton net is generally of fine nylon mesh and with a plastic tube at the end of the bag for the collection of the small insects and crustacea in water. Collections of aquatic insects and larvae are best sorted out in a white enamel tray or pan where they are more easily seen.

For someone with a specific interest in insects it may be necessary to construct a special type of net, for example, the children in the Cameron Highlands of Malaya who collect the local birdwing butterflies for sale use a light net of half a metre diameter on a long bamboo pole about 2–3 m in length.

Beating   Certain small insects found in vegetation can be collected by beating the branch or stem of the plant over a tray made of a folding wooden frame with a short handle and a sheet of white cloth; or an inverted umbrella can be used but this is not so effective because of the dark colour of the material. The base of a yellow or white plastic bowl, preferably square or rectangular, is very effective; about 2–3 cm of the side walls should be left intact.

Baits   Some insects can be attracted to baits of different types. Wasps, flies and bees will come to a sugary solution or to fermenting fruit, and some butterflies will come to rotting fruit or decaying animal faeces. Either the collector lurks in the vicinity of the bait with a net or some other suitable collecting device, or the bait can be incorporated in a trap in which the insects will catch themselves.

Traps   Traps are generally constructed so that the insect can get in easily enough, but very rarely can find its way out. The nature of the trap and the bait or attractant depends upon the type of insect being sought. Pitfall traps can consist of just a jar buried to the rim in soil, and will catch ground beetles, ants, and other insects that walk on open ground. Flies, wasps and some butterflies are attracted to decaying fruit, and animal faeces. Light traps are most effective if they use ultra-violet light from a mercury-vapour lamp, and large numbers of moths, and some Hymenoptera, grasshoppers, mantids, Neuroptera, beetles, Hemiptera, etc. can be caught. It is necessary to fill the receiving container of the trap with devices such as paper egg-boxes or egg-trays so that the captured insects will sit quietly in the dark under these trays and will not fly round inside until they beat themselves to pieces. With an ordinary tungsten lamp the catch will generally be small and scarcely worth the effort.

Soil Extraction   For insects living in soil and leaf-litter it is necessary to extract them from the soil bulk for collection. Large insects can actually be sifted out using a sieve of appropriate-sized mesh. For small insects the more usual method is to take the soil sample back into a laboratory and use a Berlesé or Tullgren funnel apparatus. This extraction technique consists of placing the soil on a perforated tray at the top of a large funnel, then fixing a lighted electric bulb over the soil, and a flask containing water or alcohol under the spout of the funnel. The bright light and the drying effect of the bulb combine to drive the small soil insects down away from the surface until they pass through the perforations and fall into the flask below.

Killing Insects

A killing jar is best made from a wide-mouthed glass jar with an easily removed but close-fitting lid. Ideally, different sized jars should be used for different insects, with a mouth diameter of about 3, 6, or 10 cm respectively. The traditional killing agent has been potassium cyanide, but after a number of tragic accidents over the years, this has generally been replaced by ethyl acetate. The killing jar is made by first placing an absorbent layer of about 1 cm in thickness of cotton wool, sawdust, or filter-paper at the bottom, then a layer of wet plaster-of-Paris is poured in on its top, again about 1 cm thick. As the plaster dries small holes are pierced through to the absorbent material. After drying, the jar is charged by pouring in about 4 ml of ethyl acetate which will pass through the holes and quickly disappear, leaving the inside of the jar dry but lethal. Moisture in the jar must be avoided as small insects will become stuck in the film, and subsequently damaged. The advantage of ethyl acetate over other volatile liquids such as chloroform, ether, carbon tetrachloride, petrol, etc., is that it leaves the dead insects relaxed and suitable for mounting, whereas ether, etc., leaves the insect body rigid. This type of jar is afe to handle, but because of the volatility of the ethyl acetate it will frequently require recharging.

Large beetles and wasps which take a long time to kill and are for a while active in the jar should not be included together with smaller and more delicate insects. Really large beetles are best killed by injecting them with ethyl acetate in a glass syringe or with 5–10% formalin in a small plastic syringe, using a fine needle. Alternatively they can be easily killed in very hot water. Similarly, the very large moths (Sphingidae, Saturniidae, etc.) are best injected with a syringe otherwise their wings may be severly damaged. Butterflies can be killed by pinching the thorax laterally between the finger and thumb, but this is not generally successful with the Danainae which are very robust butterflies. For the vast majority of insect species the use of a killing jar is the best method of killing them, unless they are larvae or aquatic species which are going to be stored in a liquid, then they might be dropped directly into the fixative which will quickly kill them.

Handling

Large insects can be handled provided care is taken not to damage the more delicate organs, but many will bite if the opportunity presents itself. Many insects will remain alive for a day or two at least if they are kept in a suitable box or plastic tube with air holes; they can be sustained by including a piece of cotton-wool soaked in water or sugar solution. Butterflies have such delicate wings that they are best carried in small paper triangles and they can be carried alive for a while in this manner. Once collected the insects should be kept out of the sun which will kill them quickly by overheating and will also cause condensation inside the container. Medium sized insects are probably best handled using fine forceps. The smallest insects when alive can be picked up in a pooter (or aspirator) by suction, and when dead a fine camel-hair paint brush will pick them up easily.

Labelling

As soon as possible after collecting and killing, the specimens should be labelled. Specimens without adequate data labels are of very limited value scientifically, although they still make interesting or even beautiful decorative specimens. Data should include place, host, date, and the name of the collector, all clearly stated. In fact if one is making an extensive local collection it is worthwhile having special labels printed. For pinned specimens the label needs to be small and inconspicuous, such as (A) below.

(A)

HONG KONG:
.   . 19
ex
leg.   D.S. Hill
Zoology Dept., H.K.U.

(B)

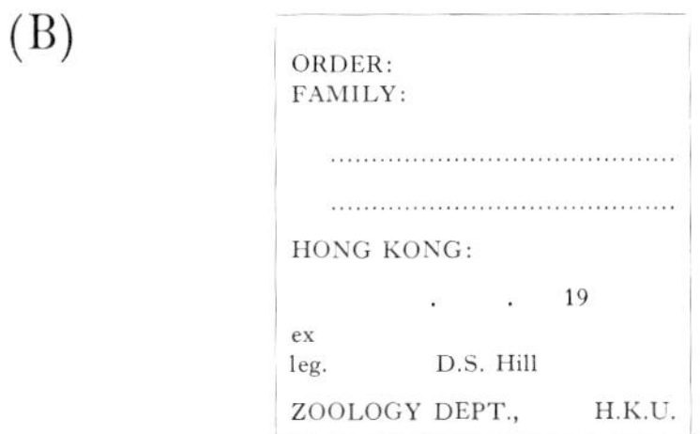

For specimens stored in alcohol, in Macartney Bottles or Wheaton Snap-Cap Bottles, a larger label (B) some 5 cm square on which the additional data are typed can be used. 'Ex' refers to the host plant, or wherever collected, and 'leg.' is a latin abbreviation meaning 'collected by'. Labels should preferably be typed (if large), or written in Indian Ink or soft dark pencil.

Mounting and Preservation

Dried specimens Most larger specimens are best preserved by pinning, and then drying in an oven for a few days. The pins must be special stainless steel (or brass) entomological pins which will not rust in that humid tropical climate. Pins of different sizes can be obtained for the different sizes of insects. Most larger insects should be pinned in the centre, or slightly to the right, of the middle thoracic segment (mesonotum), except the Coleoptera because most beetles have the thorax too hard to penetrate; beetles are traditionally pinned through the front part of the right elytron.

Smaller insects are often stuck on to a small card with 'Gloy' or a water-soluble glue, which is in turn mounted on the board bearing data labels. Certain very small insects have to be specially treated and mounted on microscope slides so that their diagnostic features can be seen and the insect can be identified; this applies to Aphididae, smaller Coccoidea (scale insects), some other small bugs, and some minute parasitic Hymenoptera.

Setting When insects are pinned they should be set in such a manner that their important diagnostic features are easily visible and they look as attractive as possible. Usually a combination of utility and beauty can be achieved with a little care. Thus legs should be spread, antennae extended and winged insects should have the wings spread out as though in flight. Butterflies and moths should have the wings held out so that the hind margin of the forewing is at right angles to the body, and the hind-wing brought up just underneath so as to make a continuous surface area. Setting is done on special setting boards which can be purchased from a biological supply house, or on a cork sheet or a sheet of polystyrene foam. If the insect body is large then a groove can be cut in the polystyrene to accommodate the body depth so that the outspread wings are dried at right angles to the body. If setting boards are purchased then it is desirable to buy a set of three or four of different sizes for use with different sized insects. To hold the wings in the outspread position small strips of stiff translucent paper (e.g. architects' drawing paper) are placed over them and held in position with two pins.

Once pinned the specimens have to be dried out and hardened; in this condition they will stay intact for many years provided that they are kept in a dry atmosphere and protected from insects and fungal attack. In museums and laboratories it is usual to have special incubator ovens for drying the mounted insects. For example at the University of Hong Kong, the pinned insects are kept in an oven at 40°C for a week or two for complete drying out. This process dries the insect body, and if a specimen is badly mounted, afterwards it can be placed in a relaxation box where it will absorb water from the saturated atmosphere and will become limp and flexible again. The specimen can then be remounted and dried again. Commercial relaxing fluid has fungicides incorporated so that the insect is protected from fungus attack during the day or two spent in the relaxing box. If the mounted dried insects are kept in display boxes somewhat exposed, they will often gradually absorb moisture from the atmosphere, causing the legs to fall and the wings to drop, and eventually the specimens may have to be reset, and dried out again.

Liquid Preservation Many small insects, the larvae of larger species, and most of the soft-bodied aquatic insects are best preserved in liquid, thus protecting their delicate body and organs such as antennae from damage. Alcohol at 70% or 80% by volume is as good as anything, and some people prefer to add 5% glycerine. In general formalin is not too successful as it hardens the specimen unduly, and FAA (formol acetic acid and alcohol) is preferred by some workers. For a study of internal organs alcohol generally does not fix the tissues adequately, and it is necessary to use histological fixatives such as Bouin's Fluid, or Calcium Formol.

Storage

Some insects are collected for decorative purposes and these are generally stored in glass-fronted boxes that hang upon a wall. The lid should be as close-fitting as possible and ideally should be air-tight. A small block of camphor or some crystals of PDB (para-dichlorobenzene) should be included to protect the specimens from mould and insect attack. Such display boxes are best hung on walls sheltered from direct sunlight because sunlight quickly fades the bright colours of most insects. For a serious insect collection, the specimens should be kept either in storage boxes or in a special cabinet with glass-topped drawers made specifically for insect specimens. The best store boxes and drawers in cabinets are virtually air-tight, and as such are difficult to manufacture. The best ones are made in U.K. and U.S.A. and have to be imported (from for example, the companies previously named). But store boxes can also be made satisfactorily locally. Essentially they are flat, rather shallow boxes with both lid and base lined with sheet cork and covered with thick white paper. The lid and base are of equal dimensions, and are closed by two small brass hooks. There should be a small container or compartment which can be filled with para-dichlorobenzene or camphor for preservation. The obvious damaging insects are museum and carpet beetles, mites and clothes moths. The minute colourless psocids (booklice) are also a main danger, for despite their tiny size they can destroy completely the dried insects; also because they are small and colourless, they are easily overlooked.

Insect store boxes can be bought in Hong Kong from Pao Sun Company, 15, Queen's Road East, Hong Kong. The boxes were made of teak, about $19'' \times 13'' \times 4\frac{1}{2}''$, with an additional tray internally, and lined with cork (but not paper) and cost about HK\$120.00 each (in 1973); they are not so air-tight as those from U.K. and U.S.A. Possibly the best insect storage cabinets are those formerly made by Hill & Son, of London, now obtainable from Grange & Griffiths Ltd., Joinery Specialists Woodstock Works, 239, Hanworth Road, Hounslow, TW3 3UA, England. Alternatively the American company "BioQuip" offers an attractive range of more utilitarian insect cabinets.

Fluid specimens should be stored away from strong light which fades them quite rapidly, and should be kept as cool as possible at an even temperature. In Hong Kong this usually means keeping the specimens in an air-conditioned room during the warmer half of the year. Ideal containers for the storage of specimens in alcohol are small glass tubes which contain the insects and are sealed with a plug of cotton-wool and placed upside down in wide-mouthed glass jars with ground-glass stoppers; in this way both the tube and the jar is filled with alcohol and offers maximum protection. Dried-out specimens are often permanently damaged and ruined as specimens. Metal lids are undesirable because of eventual corrosion, except the $3'' \times 1''$ Macartney Bottles (used for mycological cultures and obtainable locally) which to date have proved very successful for larger insects.

An excellent publication from the Centre for Overseas Pest Research (Ministry of Overseas Development), London on this subject is *Insect collecting in the tropics* (1976) by D. N. McNutt.

GLOSSARY OF ENTOMOGICAL TERMS

Abdomen	The posterior part of the three main body divisions.
Acarina	Order of Arachnida containing the mites and ticks.
Aedeagus	The male intromittent organ, or penis.
Aestivation	Dormancy during a hot or dry season.
Agamic	Parthenogentic reproduction; without mating.
Algae	Freshwater and marine aquatic non-flowering plants, usually either green, brown or red; some species are unicellular, and some are terrestrial.
Ametabolous	Without metamorphosis; insects that have nymphs which closely resemble the adults.
Anal area	The posterio-basal part of the wing, including the anal veins.
Antenna(e)	The paired sensory appendages on the head near or between the eyes; primitively long and tapering.
Antibiosis	The resistance of a plant to insect attack by having a thick cuticle, hairy leaves, toxic sap, etc.
Anus	The posterior opening of the alimentary tract.
Apical	At the end, or tip, or the outermost part.
Appendage	A stucture, or organ, usually paired, articulating from a body (trunk) segment; limb, antennae, mouth-parts, etc.
Apterous	Wingless; without wings.
Aquatic	Living in water for part or all of the life cycle.
Arachnida	Class of Arthropoda, characterized by having four pairs of walking legs, with body divided into prosoma and opisthosoma, chelicerae as poison fangs, and terminal spinnerettes.
Arboreal	Living in the foliage of trees or bushes; climbing.
Arista	Large bristle, located on the dorsal edge of the apical antennal segment in the Diptera.
Arolium	Adhesive pad-like structure at the apex of the last tarsal segment between the claws.
Atrophied	Rudimentary; reduced in size; vestigial.

Bacterium	Microscopic organisms, either free-living or parasitic, found in soil, air, and water, important for causing decay of organic material.
Beak	Proboscis; stylet; protruding composite mouth-parts of a sucking insect, usually used for piercing prior to sucking.
Bisexual	Having both males and females.
Brachypterous	Having short wings that do not cover the abdomen.
Bristle	Long stiff hair-like process; large chaeta (seta).
Bromatia	Swellings on a fungal mycelium in a fungus garden used as food by termites and ants.
Brood	Individuals hatching from the same clutch of eggs all at more or less the same stage of development at the same time.
Budworm	Common name in the USA for various tortricid larvae.
Bug	Common name for a member of the order Hemiptera.
Cannibalistic	Feeding on other individuals of the same species.
Carina	Keeled, or ridge-shaped.
Carnivore	Animal feeding on the flesh of other animals.
Carrier	Animal containing pathogenic organisms and capable of transmitting them to other animals or plants.
Caste	A form or type of adult social insect, as in termites and ants.
Caterpillar	Eruciform larva; larva of a moth, butterfly or sawfly.
Cell	A space on the wing membrane partly (open cell) or completely (closed cell) enclosed by veins.
Cercus	One of a pair of non-sexual appendages at the end of the abdomen.
Chaetotaxy	The arrangement and nomenclature of the bristles (chaetae) on the insect exoskeleton of both adults and larvae.
Chilopoda	Order containing the centipedes of the Class Myriapoda.
Chromosome(s)	Paired structures within the nucleus of a cell which carry the genes responsible for genetic inheritance.
Chrysalis	The pupa of a butterfly.
Clavate	Club-like, or enlarged at the end.
Clubbed	Having the distal portion abruptly enlarged.
Coarctate pupa	A pupa enclosed inside a hardened shell formed by the previous larval skin.
Cocoon	Silken case inside which a pupa is formed.

Commensal	Two organisms living together, neither of which is harmed and at least one benefits by the association.
Compound Eye	Eye composed of many individual hexagonal facets or ommatidia, typical of most adult insects.
Corium	The elongate thickened basal portion of the fore-wing of Heteroptera.
Cornicle	Siphunculus; one of a pair of dorsal tubular processes on the posterior abdomen of aphids, used for wax secretion.
Cosmopolitan	A species occuring very widely throughout the major regions of the world.
Costa	A longitudinal wing vein, usually forming the anterior margin (leading edge).
Coxa	The basal (first) segment of the insect leg joined to the thoracic sternum.
Crawler	The active first instar (dispersive stage) or a scale insect (Coccoidea).
Crepuscular	Animals that are active in the twilight, pre-dawn and at dusk in the evenings.
Crochets	Small hooked spines at the tip of the prolegs of lepidopterous larvae.
Crossvein	A transverse vein connecting adjacent longitudinal veins.
Cuticle	The hard chitinized outer part of the arthropod integument, which is periodically shed at ecdysis.
Deflexed	Bent downward.
Dentate	Toothed, or with tooth-like projections.
Depressed	Flattened dorso-ventrally.
Desmid	Microscopic green alga—plankton in freshwater only.
Diapause	Period of physiologically arrested development, often during the winter period.
Diatom	Unicellular algae plant, either freshwater or marine, with cell wall impregnated with silica.
Digestion	Breakdown of food material by enzyme action.
Diplopoda	Order of the Class Myriapoda, containing Millipedes.
Dispersal	Movement of individuals out of a population (emigration), or into a population (immigration).
Diurnal	Active during the daytime.
Dormant	Alive but not growing or moving; quiescent; inactive.
Dorsal	The top, back, or uppermost part of the body.

Ecdysis	The moulting (shedding of the cuticle, or skin) of larval arthropods from one stage of development to another, the final moult leading to the formation of the puparium or chrysalis.
Eclosion	The hatching of the first instar larva from the egg.
Ecology	The study of all the living organisms in an area, their interrelationships, and their physical environment.
Ecosystem	The interacting system of the living organisms of an area and their physical environment.
Elateriform	A larva resembling a wireworm with a slender body, heavily scolerotized, with short thin thoracic legs and few body bristles.
Elytron	The thickened forewing of the Coleoptera (Beetles).
Emergence	The adult insect leaving the last nymphal skin or pupal case; the appearance of a population in the field after a period of adverse conditions.
Emigration	The movements of individuals out of a population.
Endocarp	The part of a seed, not including the embryo, containing the store of food for the developing embryo.
Endoparasite	A parasite that lives inside its host.
Endopterygote	Having the wings developed internally; with complete metamorphosis (=holometabolous).
Entomophagous	An animal (or plant) that feeds upon insects.
Entomophilous	A plant pollinated by insects.
Environment	The sum-total of external influences acting on an organism.
Erinium	A growth of distorted hair-like epidermal structures on a plant leaf resulting from the attack of certain Eriophyidae (Acarina).
Eruciform larva	Caterpillar; a larva with a cylindrical body and well developed head with both thoracic legs and some abdominal prolegs.
Exarate pupa	A pupa in which the appendages are free and not glued to the insect body.
Excretion	The removal of the waste products from the body of the insect.
Exopterygote	Having the wings developed on the outside of the body; incomplete metamorphosis (=hemimetabolous).
Exuvium	The cast skin of Arthropoda after moulting.

Femur	The third leg segment between the trochanter and the tibia.
Fertilization	The fusion of a spermatozoan with an ovum; sometimes also used for the act of impregnation.
File	The file-like ridge on the ventral side of the fore-wing used in sound production by some Orthoptera.
Flagellum	The distal part of the insect antenna.
Fly	Entomologically restricted to the Order Diptera, insects having one pair of wings and the hind pair modified into halteres.
Fossorial	In the habit of digging or burrowing; modified for digging.
Frass	Wood fragments made by a wood-boring insect, usually mixed with the faeces; can refer to other boring insects also.
Frons	The area of the insect face including the median ocellus.
Furcula	The forked 'tail' or springing organ of the Collembola.
Gall	An abnormal growth of plant tissues caused by the stimulus of another animal or plant.
Gaster	The rounded part of the abdomen in Hymenoptera posterior to the petiole or 'wasp-waist'.
Gena	The area on the insect head below the eyes; the cheek.
Gene	The tiny unit of inheritance to be found on the chromosomes.
Generation	The period from any given stage in the insect life cycle (usually adult) to the same stage in the offspring.
Gill	Respiratory evaginations of the body wall, or hind gut in aquatic insects; may be tracheal or blood gills.
Gland	Organ in the body of an insect which secretes a particular useful substance.
Grub	Larva of a beetle (Coleoptera), or Hymenoptera.
Grub (White)	A scarabaeiform larva; thick-bodied with a well-developed head and thoracic legs, without abdominal prolegs, and usually sluggish in behaviour.
Haltere	A small knobbed structure on either side of the metathorax in place of the hing-wing in the Diptera with a function as a balancing organ.
Head	The anterior body region composed of six fused body somites bearing the eyes, antennae, and mouth-parts.
Hemelytron	The partly thickened forewing of Heteroptera.

Hemimetabolous	Insects having a simple metamorphosis without a pupal stage as found in the Orthoptera, Hemiptera, etc.
Herbivorous	Phytophagous; feeding on plants.
Hibernation	Dormancy during the winter.
Holometabolous	Insects having a complete metamorphosis with a pupal stage as found in Lepidoptera, Diptera, Hymenoptera and Coleoptera.
Honey-dew	Liquid with very high sugar content discharged from the anus of some Homoptera.
Hornworm	Term used in USA for larva of Sphingidae which typically possess a fleshy dorsal spine or horn on the last abdominal segment.
Host	The organism on which a parasite lives; and the plant on which an insect feeds.
Hyaline	Clear; transparent; often refers to insect wings.
Hypermetamorphosis	A type of complete metamorphosis in which the different larval instars represent two or more different types of larvae.
Hyperparasite	A parasite whose host is another parasite.
Imago	The adult or reproductive phase of an insect.
Immigration	The movement of individuals into a population or area.
Infect	To enter and establish a pathogenic relationship with a plant (host); to enter and persist in a carrier.
Infest	To occupy and cause injury to either a plant, animal, or to soil or to stored products.
Inquiline	An insect that lives in the nest of another species.
Insect	An arthropod belonging to the Class Insecta, characterized by having three pairs of legs, one pair of antennae, and a body divided into head, thorax and abdomen.
Insecticide	A toxin (poison) effective against insects.
Instar	The form of an insect between successive moults.
Internode	The part of a plant stem between adjacent nodes, or between the bases of two successive leaves.
Juvenile	Immature stage of an insect; young insect, either nymph or larva.
Labium	One of the insect mouth-parts—the median 'lower lip'.
Labrum	The 'upper lip', lying just below the clypeus.

Larva	The immature stages of an insect having complete metamorphosis between the egg and pupa; the six-legged first instar of the Acarina.
Larvicide	Toxicant (poison) effective against insect larvae.
Larviparous	An insect that gives birth to larvae rather than laying eggs.
Leaf-miner	An insect larva which lives in and feed upon the cells between the upper and lower epidermis of a leaf, being in the Diptera, Lepidoptera or Coleoptera.
Lichen	Composite plant body produced by mutualistic relationship between an alga and a fungus.
Life cycle	The various phases through which an individual species passes to maturity.
Life table	The separation of an insect population into its different age-group components (e.g. egg, larvae, pupa, adult).
Looper	A caterpillar of the family Geometridae, with only one pair of abdominal prolegs in addition to the terminal claspers, and which locomotes by looping its body.
Macropterous	Large or long-winged; having normal wings.
Maculate	Spotted; having areas of dark pigmentation.
Maggot	A vermiform insect larva, legless and without a distinct head, characteristic of the higher Diptera.
Mandible	One of the anterior pair of mouth-parts of jaws, typically bearing the cutting edges for biting.
Maxilla	One of the paired mouth-parts immediately posterior and lateral to the mandibles.
Membrane	A thin, flexible layer of tissues; that part of the wing surface between the veins.
Mesonotum	The dorsal sclerite of the mesothorax.
Metamorphosis	The change in body form during development.
Metanotum	The dorsal sclerite of the metathorax.
Migration	A change of habitat, according to the season; strictly refers to the double movement both away and returning to the original habitat.
Mimic	To assume, usually for protection, the habits, colour or structure of another organism.
Model	The organism being mimicked.
Monophagous	An insect restricted to a single host (plant) species (or genus) for feeding purposes.

Mortality	Death-rate of insects in a population; dying.
Moulting	The process of shedding the skin; ecdysis.
Mouthparts	The paired appendages and median structures, located around the insect mouth and used in feeding.
Mutualism	A form of symbiosis in which both parties derive advantage without either sustaining injury; it can be complete physiological interdependence.
Myiasis	The invasion of living tissues by larvae of certain Diptera, either cutaneous or intestinal, etc.
Myriapoda	The arthropod class characterized by the animals having a large number of walking legs.
Nasute	A termite caste in which the head narrows anteriorly into a snout-like projection, often for squirting formic acid.
Necrosis	Death of part of a plant or animal body.
Nematode	A phylum of small worm-like animals, often parasitic on or in plants and animals, but some freeliving (predaceous) in soil, freshwater, or the sea.
Nocturnal	Active at night.
Notum	The dorsal surface of an insect body segment, usually referring to the thoracic segments.
Nymph	The immature stage of an insect that does not have a distinct pupal stage (Hemimetabola); also the immature stages of Acarina that have eight legs.
Obtect pupa	A pupa in which the appendages are more or less glued to the body (Lepidoptera).
Ocellus	A simple eye, more characteristic of insect larvae than adults.
Oligophagous	An insect feeding upon only a few, closely related, host plant species (or animals).
Onisciform larva	A flattened platyform larva looking like a woodlouse in appearance.
Oötheca	Eggcase; or covering of an egg mass, as in the Dictyoptera.
Osmeterium	A fleshy, tubular, eversible Y-shaped scent gland at the anterior end of certain caterpillars (Papilionidae).
Oviparous	Reproduction by laying eggs.
Oviposition	The process of laying or deposition of eggs by the female insect, usually involving complex site selection.
Ovipositor	The egg-laying external genitalia of some female insects.

Ovo-viviparous	Insects which retain the developing egg within the maternal body so that eventually living larvae are deposited.
Ovum	Mature egg cell; egg; female germ cell.
Paedogenesis	The production of eggs or young by an immature or larval insect.
Palp	The segmented paired process borne on the maxilla or labium, and used for sensory and feeding purposes.
Pantropical	A species occurring widely throughout the tropical and subtropical parts of the world.
Parasite	An organism living in intimate association with another living organism (plant or animal), called the host, from which it derives material essential for its existence while conferring no benefit in return.
Parthenogenesis	Reproduction without fertilization.
Pathogenic	Disease-causing.
Pectinate	With processes like the teeth of a comb; comb-like.
Pest	Animal or plant causing damage to man, his crops, animals or possessions.
Pest density	The population level at which a pest species causes economic damage.
Pest load	The total number of pests attacking a particular plant or host.
Pest management	The careful manipulation of a population, after extensive consideration of all aspects of the life system as well as ecological and economic factors.
Pest spectrum	The complete range of pests attacking a particular plant or crop.
Pesticide	A chemical that by virtue of its toxicity (poisonous properties) is used to kill pest organisms.
Petiole	A constricted waist ('wasp-waist') formed in most Hymenoptera by the attenuation of the second and third abdominal segments.
Pheromone	Ectohormone; a substance secreted by an insect to the exterior causing a specific reaction in the receiving insects; usually either sexual or social.
Phytophagous	Herbivorous; plant-eating.
Phloem	The soft bast of vascular bundles consisting of sieve-tube tissue for the conduction of food materials in plant stems.

Planidium larva   A type of first instar larva in certain Diptera and Hymenoptera which undergo hypermetamorphosis.

Plumose   Feather-like; usually refers to plumose antennae.

Polyembryony   An egg developing into multiple embryos.

Polymorphism   Occurrence of different forms of individuals in the same species.

Polyphagous   An animal feeding upon a wide range of hosts, or food types.

Predator   An animal that attacks and feeds on other animals, usually smaller or less powerful than itself.

Predaceous   Feeding as a predator.

Preference   The factor by which certain plants are more or less attractive to insects by virtue of their colour, texture, aroma or taste.

Pre-oviposition   The period of time between the emergence of an adult female insect and the start of its egg laying.

Prepupa   Quiescent stage preceeding the pupal stage in some insects, particularly Diptera.

Proboscis   The extended beak-like mouth-parts of some insects.

Proleg   A fleshy, abdominal leg found in caterpillars, bearing a characteristic arrangement of terminal crochets.

Pronotum   A dorsal sclerite of the prothorax.

Propodeum   The hind part of the thorax in Hymenoptera-Apocrita, which is really the first abdominal segment fused with the thorax.

Protonymph   The second instar of mites (Acarina).

Prothorax   The anterior of the three thoracic segments.

Protozoa   Microscopic, single-celled animals; Phylum Protozoa.

Pterostigma   A thickened opaque or dark spot along the costal margin of the insect wing near the tip, as in Odonata, Hymenoptera, etc.

Pupa   The stage between larva and adult in insects with complete metamorphosis; a non-feeding and usually inactive stage.

Puparium   The case formed by the hardened last larval skin in which the pupa of higher Diptera is formed.

Pupiparous   Female insects giving birth to mature larvae that are ready to pupate.

Quarantine   All operations associated with the prevention of importation of insect pests into a territory, or their exportation from it.

Quiescence	A period of dormancy, usually imposed by inclement environmental conditions.
Raptorial	Predaceous; preying; modified for seizing prey.
Resistance	The natural or induced capacity of plants to avoid or repel insect attack; also the ability of insects to withstand the toxic effects of a pesticide, often by metabolic detoxification.
Rostrum	The beak or proboscis of Hemiptera.
Scale	Minute, flattened projection from the cuticle.
Scale Insect	Member of the Homoptera, Coccidae (Soft Scale) or Diaspididae (Hard Scales) with the body under an expanded dorsal shield (Scale).
Scarabaeiform larva	Grub-like beetle larva, characteristic of the Scaraboidea, with thickened cylindrical C-shaped body, well-developed head and thoracic legs, without abdominal legs, and sluggish in behaviour.
Scavenger	Animal that feeds on dead animals and plants, on decaying matter, or animal faeces.
Sclerite	A hardened body skeletal plate bounded by sutures or membranous areas.
Sculpturing	Ornamentation on the surface of the cuticle of certain insects, especially in the Hymenoptera and Coleoptera.
Scutellum	Sclerite of thoracic notum, the meso-scutellum appears as a large triangular sclerite behind the pronotum, characteristic of some Hemiptera and Coleoptera.
Segment	A sub-division of the body or appendage between joints or articulations.
Semi-looper	Caterpillar of the sub-family Plusiinae (Noctuidae) with two or three pairs of prolegs, and locomotes in a somewhat looping manner.
Sessile	Attached or fastened; incapable of movement from place to place.
Seta	Small bristle, or hair-like structure arising from the cuticle, sometimes extensively covering the insect body.
Siphunculi	Cornicles; the paired dorsal abdominal protruding organs of the Aphididae, through which a waxy secretion is extruded.

Species   A group of naturally interbreeding animal popula-
                   tions, which does not breed with other groups and
                   possessing certain characteristics in structure and
                   physiology.
Spine   A thorn-like out-growth of the insect cuticle.
Spiracle   Breathing aperture of insects; external office connect-
                   ing the tracheal system to the exterior.
Spur   Articulated spine, often on a leg segment (usually
                   tibia), typical of certain insects.
Sting   Modified ovipositor in some female insects used for
                   injecting poison.
Stadium   The period between moults in a developing insect.
Stridulation   Production of sound ('song') by various insects.
Sub-species   A sub-division of a species, usually a geographical
                   race; different sub-species usually intergrade with
                   each other and are capable of interbreeding.
Stylet   Modified mouth-part (usually mandible or maxilla)
                   elongated and used for piercing; usually part of
                   proboscis.
Symbiosis   The association of two different living organisms in
                   a more or less intimate association.
Synanthropy   Organisms living in association with man and human
                   dwellings.
Tarsus   The distal part of the leg beyond the tibia, consisting
                   of several small segments ending in a pair of claws.
Tegmen   The thickened leathery forewing of Orthoptera and
                   Dictyoptera.
Tergum   The dorsal surface of any body segment.
Thorax   The body region behind the head, and before the
                   abdomen.
Tibia   The fourth segment of the leg between the femur and
                   the tarsus.
Tolerance   The ability of a plant to endure infestation by a
                   particular insect pest without showing severe
                   symptoms of distress.
Trachea   A tube of the respiratory system lined with cuticular
                   ctenidia, ending externally at a spiracle and internally
                   in tracheoles.
Triungulin larva   The active first instar larva of Meloidae (Coleoptera)
                   and Strepsiptera.

Trochanter	The second segment of the leg between coxa and femur.
Tymbal	The sound-producing organ found in cicadas.
Tympanum	A vibrating membrane used as an 'ear-drum' for auditory reception in certain insects.
Valves	The two paired structures comprising the ovipositor in some female insects.
Vector	Organism able to transmit viruses, parasites or other pathogens, either directly or indirectly into new hosts.
Vein	A thickened line in an insect wing containing tracheae and blood; main types are longitudinal veins and crossveins.
Venation	The arrangement of veins in an insect wing.
Ventral	On the underside of an insect body; lower; underneath.
Vermiform larva	A legless (apodous), headless (acephalic), worm-like larva typical of some of the higher Diptera.
Vestigial	Non-functional; degenerate; poorly developed; small.
Virus	Minute organism of crystalline structure, which only develops within a certain host organism; most are pathogenic.
Viviparous	Giving birth to living young, rather than laying eggs.
Voltinism	The number of generations of an insect occuring in a year or season (e.g. univoltine, bivoltine, or multivoltine).
Waist	The constriction between the apparent thorax and abdomen, typical of the Hymenoptera Apocrita.
Warning Colour	Distinctive, bright, contrasting colour pattern used to make the insect conspicuous and to advertise its venomous nature or unpleasant taste; most common is yellow/black banding.
Wireworm	Elateriform larva; larva of Elateridae (Coleoptera) with long slender, well-sclerotized body, thoracic legs but no prolegs, and a few setae.
Xylem	The vascular tissue of plants responsible for transport of water and mineral salts throughout the plant body, and providing some mechanical support.

REFERENCES

Asahina, S. 1965. The Odonata of Hong Kong. *Konytû* 33: 493–506.

——. 1965. Taxonomic notes on Japanese Blattaria, III. On the species of the genus *Onychostylus* Boliver. *Jap. J. Sanit. Zool.* 16: 6–15.

——. 1967. The cockroaches of the genus *Rhaboloblatta* of Japan, the Ryukyus and Taiwan. *Jap. J. Med. Sci. & Biol.* 20: 425–42.

Chan, T. D., and L. B. Trott. 1972. The collembolan, *Oudemansia esakii*, an intertidal marine insect in Hong Kong, with emphasis on the floatation method of collection. *Hydrobiologia* 40: 335–43.

Cheng, L., and D. S. Hill. 1982. Marine insects of Hong Kong. In Morton, B. S. (ed.). 1982. *The marine flora and fauna of Hong Kong and southern China.* Hong Kong, Hong Kong University Press.

Cheung, W. W. K., and A. T. Marshall. 1973. Water and ion regulation in Cicadas in relation to xylem feeding. *J. Insect Physiol.* 19: 1801–16.

Clarke, C. A., Sheppard, P. M., & I. W. B. Thornton. 1968. The genetics of the mimetic butterfly *Papilio memnon* L. *Phil. Trans. Roy. Soc. Lond. (B)* 254: 37–89.

Corner, E. J. H. 1965. Checklist of *Ficus* in Asia and Australasia with keys to identification. *Gdns. Bull., Singapore* 21: 1–186.

Dudgeon, A. D. M. 1981. Studies on the biology and ecology of freshwater benthos in Hong Kong, with particular reference to a forest stream. (Ph.D. thesis, University of Hong Kong).

Eliot, J. N. 1953. New records and a checklist of butterflies from Hong Kong. *Mem. H.K. Biol. Circle* 2: 1–14.

Hanson, H. 1963. *Diseases and pests of economic plants of Central and South China, Hong Kong and Taiwan (Formosa).* Washington, Amer. Inst. Crop Ecology.

Harris, W. V. 1963. The termites of Hong Kong. *Mem. H.K. nat. Hist. Soc.* 6: 1–9.

Hill, D. S. 1967. *Figs (Ficus spp.) of Hong Kong.* Hong Kong, Hong Kong University Press.

——. 1967. Fig-wasps (Chalcidoidea) of Hong Kong. I Agaonidae. *Zool. Verh., Leiden* 89: 1–55.

——. 1967. Figs (*Ficus* spp.) and fig-wasps (Chalcidoidea). *J. nat. Hist.* 1: 413–434.

——. 1975. *Agricultural insect pests of the Tropics and their control.* Cambridge, Cambridge Univ. Press. (2nd edn., 1982.)

——, 1982. *Hong Kong insects,* vol. II. Hong Kong, Govt. Printer.

——, B. Gott, B. Morton and I. J. Hodgkiss. 1978. *Hong Kong Ecological habitats: flora and fauna.* 2nd ed. Hong Kong, H.K.U., Dept. of Zoology (Occas. Pub., No. 2).

——, and W. W. K. Cheung. 1978. *Hong Kong insects.* Hong Kong, Govt. Printer.

——, G. Johnston and M. J. Bascomb. 1978. Annotated checklist of Hong Kong butterflies. *Mem. Hong Kong nat. Hist. Soc.* No. 11: 1–62.

——, and K. Phillipps. 1982. *A colour guide to Hong Kong animals.* Hong Kong, Govt. Printer.

Hodgkiss, I. J. 1976. *A students guide to the study of aquatic ecology in Hong Kong.* Hong Kong, Dept. Education.

Hong Kong Govt. 1968. *Land utilization in Hong Kong.* Hong Kong, Govt. Printer.

Hong Kong Herbarium. 1975. *Checklist of Hong Kong plants.* 2nd ed. Hong Kong, Govt. Printer.

Hsu, Yin-Chi. 1936. Mayflies of Hong Kong with descriptions of two new species (Ephemeroptera). *H.K. Naturalist* 7: 233–38.

Imamura, T. 1976. Water mites from Hong Kong. *Proc. Jap. Soc. Syst. Zool.* 12: 21–3.

Imms, A. D. 1960. *A general textbook of entomology.* 9th ed. London, Methuen (see under Richards & Davies for the 10th edition, 1977).

Johnston, G. and B. Johnston. 1980. *This is Hong Kong: butterflies.* Hong Kong, Govt. Information Service.

Kano, R., K. Kaneko, and S. Shinonaga. 1968. Synanthropic flies in Hong Kong. *Kontyû* 36: 75–87.

Kershaw, J. C. 1907. *Butterflies of Hong Kong.* Hong Kong, Kelly & Walsh.

Kimoto, S. 1967. A list of the chrysomelid species from Hong Kong, with descriptions of three new species. *Esakia* 6: 55–63.

Lai, Y. L. 1972. An introduction to the Odonata of Hong Kong. *New Asia Coll. Bull., H.K.* 13: 109–56.

Lee, H. Y. L. and R. Winney. 1979. *A checklist of agricultural insects of Hong Kong.* Hong Kong, Government Printer.

Marsh, J. C. S. 1968. *Hong Kong butterflies.* 2nd ed. Hong Kong, Shell Co.

Marshall, A. T. 1964. Spittle production and tube-building by Cercopoid nymphs (Homoptera). 1. The cytology of the Malpighian tubules of spittlebug nymphs. *Quart. J. micr. Sci.* 105: 257–62.

——. 1970. External parasitism and blood-feeding by the lepidopterous larva *Epipyrops anomala* Westwood. *Proc. R. ent. Soc. Lond. (A)*. 45: 137–40.

——, and P. M. Marshall. 1966. The life history of a tube-dwelling Cercopoid: *Machaerota coronata* Maa (Homoptera: Machaerotidae). *Proc. R. ent. Soc. Lond. (A)* 41: 17–20.

——, and I. W. B. Thornton. 1963. *Micromalthus* (Coleoptera: Micromalthidae) in Hong Kong. *Pacific Insects* 5: 715–20.

McClure, E. and N. Ratanawarabhan. 1972. *Some ectoparasites of the birds of Asia.* Bangkok, M.A.P.S.

McNutt, D. N. 1976. *Insect collecting in the Tropics.* London, H.M.S.O.

Peterken, G. F. 1967. *Guide to the check sheet for I.B.P. areas.* Oxford, Blackwell.

Peters, W. L. 1963. A new species of *Habrophleboides* from China (Ephemeroptera: Leptophebiidae). *Proc. R. ent. Soc. Lond. (B)* 32: 41–43.

Richards, O. W. and R. G. Davies. 1977. *Imm's general textbook of entomology.* 10th ed. London, Chapman & Hall. 2 vols.

Romer, J. D. 1955. Domestic fleas known to occur in Hong Kong. *Mem. H.K. Biol. Circle* 3: 7–8.

Ross, E. S. 1978. The Embiidina of China. *Mem. H.K. nat. Hist. Soc.* 13: 1–8.

Silvestri, F. 1946. Prima nota su alcuni Termitofili dell' Indocina. *Boll. Lab. Ent. Agr., Portici* 6: 313–30.

So, Pui-Yip. 1967. A preliminary list of the insects of agricultural importance in Hong Kong. *Agric. Bull.. H.K.* (for revised edition see under Lee & Winney, 1982) 1: 1–39.

Spencer, K. A. and D. S. Hill. 1977. A new species of *Ophiomyia* (Diptera: Agromyzidae) causing leaf-galls on *Ficus microcarpa* in Hong Kong. *Agric. H.K.* 2: 419–23.

Thornton, I. W. B. 1959. A new genus of Philotarsidae (Corrodentia) and new species of this and related families from Hong Kong. *Trans. R. ent. Soc. Lond.* 111: 331–49.

——. 1961. The *Trichadenotecnum* group (Psocoptera: Psocidae) in Hong Kong with description of new species. *Trans. R. ent. Soc. Lond.* 113: 1–24.

——. 1962. The Peripsocidae (Psocoptera) of Hong Kong. *Trans. R. ent. Soc. Lond.* 114: 285–315.

——, A. T. Marshall, W. H. Kwan, and Q. Ma. 1975. Studies on lepidopterous pests of rice crops in Hong Kong, with particular reference to the Yellow Stem-borer, *Tryporyza incertulas* (Wlk.). *P.A.N.S.* 21: 239–52.

——, and S. K. Wong. 1968. The Peripsocid fauna (Psocoptera) of the Oriental Region and the Pacific. *Pacific Ins., Mono.* 19: 1–158.

Tang, Kam-tong. 1970. On a collection of Hong Kong butterflies. *New Asia College Acad. Annual* 12: 37–142.

Tsui, P. T. P., & W. L. Peters. 1970. The nymph of *Habrophlebiodes gilliesi* Peters (Ephemeroptera: Leptophlebiidae). *Proc. R. ent. Soc. Lond.* (A) 45: 89–90.

Wong, M. H. and T. D. Chan. 1977. The ecology of the marine rove beetle, *Bryothinusa* spp. (Coleoptera: Staphylinidae) in Hong Kong. *Hydrobiologia* 53: 253–56.

INDEX OF SPECIES

Reproduced with the permission of the Crown Lands and Survey